Migratory Small Ruminant Farming System in the Himalayas-Peculiarities, Problems, Prospects, Climate Effects, Nutrition and Health

S. P. Singh • J. K. Malik

Migratory Small Ruminant Farming System in the Himalayas-Peculiarities, Problems, Prospects, Climate Effects, Nutrition and Health

Springer

S. P. Singh
Former Director, ICAR-Central Institute for Research on Cattle (ICAR-CIRC)
Meerut, Uttar Pradesh, India

J. K. Malik
Former Joint Director (Research)
ICAR-Indian Veterinary Research Institute (ICAR-IVRI)
Izatnagar, Uttar Pradesh, India

ISBN 978-981-96-6690-4   ISBN 978-981-96-6691-1 (eBook)
https://doi.org/10.1007/978-981-96-6691-1

This Springer imprint is published by the registered company Springer Nature Singapore Pte Ltd.
The registered company address is: 152 Beach Road, #21-01/04 Gateway East, Singapore 189721, Singapore

Preface

Mountain regions cover over 27% of earth's land surface and are home to nearly 13% of the world population. But there is something special about the Himalayas not possessed by any other mountain range, something unimaginable in terms of unique ecosystem, biodiversity, charm, and beauty. In India and Nepal, Himalayas are also considered a seat of eternal peace and divinity. The Himalayas begin in Afghanistan and stretch over 3500 km across eight countries, viz. Afghanistan, Pakistan, India, Nepal, Bhutan, China, Bangladesh, and Myanmar—the area famously known as Hindu Kush Himalayan (HKH) region. The vast mountain rangelands are the source of livelihood for an estimated population of around 25–30 million livestock-dependent people/pastoralists who rear and raise sheep, goats, cattle, and other animals on these precious ranges. The transhumant pastoralists have transformed the Himalayan ecosystem into economically productive asset where conventional agricultural farming is not possible.

For millions of sheep and goats that graze on natural alpine grasslands, the Himalayas have become lifeline for survival during warm summer months when there is serious scarcity of feed and fodder in the plains of most countries of HKH region. For instance, in north-western states of India, the pastoral shepherds belonging to the traditional mountain society make their living by rearing large flocks of sheep and goats and other animals that follow seasonal movement in the annual migratory cycle. As the summer approaches, shepherds and animals start ascending and travel long distances to the alpine pastures, stay there for around 3–5 months for grazing, and head back to the base camps located in the foothills/Shivaliks with the onset of winter. It is estimated that in the northwestern Indian states around 70% of sheep and goat population is reared under the migratory system and substantial number of farming families are involved in this occupation. This kind of annual seasonal movement to alpine grasslands is prevalent in most countries of HKH region but in some countries the pattern and location of migration to alpine pastures vary.

The scenario of mass migration of animals in such a large number back and forth between alpine meadows and base camps every year is certainly a novel and unique animal rearing system in the Himalayas. Plenty of literature is available on the cultural dimension, historical accounts, socioeconomic profile of nomads/tribes, and animal husbandry practices but hardly any on real-time problems and constraints faced by migrating animals and herders, principal minerals limiting their production, nutritional composition of alpine pastures, adaptational and behavioral responses of animals to altitude, parasitic infestations and other diseases and disorders of migratory animals, impact of changing global climate and perceptions of shepherds on this century-old system, and the current status and future of the pastoral migratory system.

The badly scattered literature in this field has hardly served the purpose in improving this profession. The purpose of this book, therefore, is to provide answers to many of the related questions and make the information available at one place. The book has been compiled into 5 parts containing 11 chapters. The first two parts comprising five chapters focus on uniqueness and basic features of hill agriculture and migratory small ruminant farming system and history, status, and problems/constraints of transhumant pastoralism including adaptational responses to altitude. Part III comprises two chapters and highlights the impact of changing climate on small ruminant migratory system and pastoralists' perceptions of global warming on their lives and livelihoods in major Hindu Kush Himalayan countries. The last two parts comprising four chapters deal with minerals limiting productivity and nutritional deficiencies and status of common Himalayan pastures and tree foliage. In Chap. 8 we report an interesting phenomenon of intense salt hunger in the sheep flock after their prolonged stay at the high altitude. Heavy parasitism in the migratory sheep and goats is of serious concern and, keeping this in view, a separate chapter (Chap. 10) outlines in detail the extent of parasitic infestations including other disorders of migratory sheep and goats including parasitic or verminous bronchitis, pneumonia, persistent abortions, and Peste des Petits Ruminants (PPR). The last chapter (Chap. 11) discusses use of vaccines against gastrointestinal parasitic infections available for migratory small ruminants.

Due emphasis has been given for easy comprehension, and to facilitate reading, each chapter has a short abstract in the beginning which gives a quick gist of the material covered in the chapter.

We express our gratitude to many colleagues for valuable suggestions, and source material to undertake this formidable task. Scientific information concerning the topic has been retrieved from various sources such as research papers, monographs, handouts, annual reports, reviews, websites, personal communications, and our own Research Project "Studies on adaptational aspects of migratory sheep and goats of Himachal Pradesh" of which the first author (SPS) was the principal investigator for major duration of the project. Substantial part of the research work, part of which is included in Chaps. 2 and 4, was conducted at Himachal Pradesh Sheep Breeding Farms, Tal and Jeory, and funded by the Indian Council of Agricultural Research (ICAR) for which we are grateful to the Council. Sincere thanks are also due to my former fellow teaching and laboratory staff who worked with me (SPS) in the Project at H.P. Agriculture University, Palampur. I owe special thanks to my faculty colleagues, Dr. Naresh Kumar, Dr. Ravinder Kumar, and Dr. K.B. Sharma. Director Animal Husbandry, Himachal Pradesh, arranged all the working facilities for the project for which we are grateful to him. I (SPS) am also grateful to my former student, Dr. Surinder Singh Chauhan, currently Senior Lecturer at the Faculty of Veterinary and Agricultural Sciences, University of Melbourne, Australia, and

Dr. Dhanwanter Singh, former Professor and Head, Veterinary Microbiology, H.P. Agriculture University, Palampur, for their valuable feedback and encouragement to write on the migratory small ruminants of hill regions.

We are thankful to many investigators and sheep lovers for permission and provision of source material including illustrations, figures, experimental data, and photographs. The permission to reproduce certain parameters has been given by individual authors for which we express deep sense of gratitude to Professor Gustava Zubieta-calleja of Bolivia, Dr. R. Johnston of USA, and Dr. V. Sejian of Bangalore, India. We are greatly indebted to Dr. O.P. Yadav (Director) and Dr. S. Chauhan (Senior Scientist), ICAR-Central Arid zone Research Institute (ICAR-CAZRI), Jodhpur, India, and Dr. Arun Kumar (Director) and Dr. O.H. Chaturvedi (Principal Scientist), Garsa Animal Farm, ICAR-Central Sheep and Wool Research Institute (ICAR-CSWRI), Avikanagar, India, for permission to reproduce photographs of grazing sheep and goats on the hills of Jammu and Kashmir and Himachal Pradesh, respectively. Three social enthusiasts-cum-sheep lovers permitted us to use their hobby photographs of grazing sheep and goats in the meadows of Jammu and Kashmir (Sankar Sridhar), Himachal Pradesh (Bharat Bhushan Negi), and Uttarakhand (Vipin Kumar). We are individually grateful to them and others who helped us. Nandhini Viswanathan, Production officer (Books), Springer Nature, has been of constant help to us in advising on technicalities of manuscript preparation for which we are grateful to her.

We sincerely hope that this book will cater to the needs of students, teachers, researchers, environmentalists, administrative planners, and practicing veterinarians as well as to the officers of the Central and State Animal husbandry Departments and those managing pastures and grasslands. Keeping abreast of recent trends, current status and constraints is of prime importance so that the concerned professionals and planners can respond appropriately and expeditiously and thereby continue to serve the best interest of the migratory small ruminants and hardy graziers.

Finally, we salute the brave and hardy shepherds of Himalayan countries because of whom this novel work of transhumance is continuing in the interest of animals and the Himalayan ecosystem in spite of all hardships. We also express our heartfelt gratitude to our family members for giving constant moral support and bearing the hardships and our seclusion when we were busy in writing the book for the past many months.

Meerut, Uttar Pradesh, India — S. P. Singh
Izatnagar, Uttar Pradesh, India — J. K. Malik

Contents

Part I Seasonal Animal Migration- a Unique Animal Farming System in the Himalayas, Features, Nomadic Pastoral Communities, Hardy Shepherds, Migratory Patterns and Migratory Routes, Problems and Constraints, Fate and Future

1 Animal Migration/Nomadic Pastoralism: A Unique Animal Farming System in the Himalayas. 3

2 Salient Features of Small Ruminant Migratory Farming System . . . 27

3 Migratory Patterns and Migratory Routes of Pastoral Small Ruminants in the Himalayas. 45

4 Problems/Constraints Faced by the Migratory Animals and Pastoralists, Strategies for Their Amelioration, and Current Status of Migratory Pastoralism. 67

Part II High Altitude Acclimatization and Adaptation of Small Ruminants

5 Adaptive Responses to Altitude and Walking Stress in Small Ruminants 93

Part III Impact of Climate Change on Transhumance and Herders' Perceptions to Changing Climate on the Migratory System

6 Effects of Changing Global Climate on Small Ruminant Production System 115

7 Himalayan Pastoralists' Perceptions to Changing Climate and How It Is Affecting the System and Their Livelihoods 145

Part IV Minerals and Nutritional Status of Alpine Grasses and Tree Leaves of Indian Himalayas

8 Principal Minerals Limiting Production of Migratory Small Ruminants: Salt Hunger in Grazing Sheep 165

9 Nutritional Status of Common Himalayan Pastures and Tree Foliages 185

Part V Parasitic Infestations and Disorders of Small Ruminants on Indian Himalayas

10 Helminths and Other Diseases Affecting Migratory Small Ruminants 209

11 Vaccines Against Gastrointestinal Parasitic Infections for Sustained Health and Production of Small Ruminants 247

About the Authors

S. P. Singh is the former Director of the Central Institute for Research on Cattle (ICAR) and led the national effort to develop the high-yielding Frieswal crossbred cattle. He has served as Visiting Adjunct Professor at SVP University of Agriculture and Technology, Meerut, Founder Professor and Head of Veterinary Physiology at HP Agricultural University, Palampur, and Associate Professor at Pantnagar University.

Dr. Singh made pioneering contributions to thermal acclimation in calves, draft capacity of working bullocks, and a diagnostic test for sodium status in large ruminants. His work is cited in key international books on livestock nutrition and veterinary hematology. He has received several research and teaching awards and is a Fellow of the National Academy of Veterinary Sciences (NAVS).

With extensive experience in research proposal formulation, he has been the principal investigator of multiple national projects and has evaluated numerous research initiatives. He has served on university, national, and international committees and was Vice-President of the Society of Animal Physiologists of India. Dr. Singh has over 35 years of experience in teaching, research, and administration, with more than 90 publications in reputed journals.

J. K. Malik served as Joint Director (Research) at the Indian Veterinary Research Institute (IVRI) and as Head of the Division of Pharmacology and Toxicology. He was also a Professor at Gujarat Agricultural University. He conducted pioneering research in toxicology and pharmacology in Germany and contributed significantly to pharmacokinetics, chemotherapy, and biochemical pharmacology.

Dr. Malik has received numerous awards, including the Alarsin Research Award, IVRI Best Teacher Award, and prestigious fellowships. He has held leadership roles in various scientific societies and served on editorial boards of leading journals. He has worked as an FAO consultant and an expert for the United Nations Industrial Development Organization.

With over 150 research articles and 130 additional publications, he has also authored and contributed to numerous book chapters in internationally recognized scientific publications.

Part I

Seasonal Animal Migration- a Unique Animal Farming System in the Himalayas, Features, Nomadic Pastoral Communities, Hardy Shepherds, Migratory Patterns and Migratory Routes, Problems and Constraints, Fate and Future

Animal Migration/Nomadic Pastoralism: A Unique Animal Farming System in the Himalayas

1

Abstract

Home to the highest peak on the planet, the Himalayas are nature's gift to the mankind. The snow-clad mountain ranges, variable climates, rich vegetation, flowing rivulets, and gurgling waterfalls mesmerize the visitors with charm and tranquility. An amazing feature of high Himalayas is the presence of alpine grasslands with highly nutritious grasses that serve as a boon to the hilly domestic small ruminants during summer months. For centuries, migratory pastoralism has been going on in the Himalayas in search of animal feed, fodder, and other purposes. In the annual migratory cycle, the animals move to the alpine pastures during summer, stay there for 4–5 months, and return to the plains with the onset of winter to spend about 5–6 months in the foothills or plains. Millions of sheep and goats participate in this annual movement especially from the north-western states of India including union territories (UTs) of Jammu and Kashmir (J&K) and Ladakh and the states of Himachal Pradesh (HP) and Uttarakhand (UK). In fact, the seasonal animal migration has become a lifeline for the poor and marginal tribal farmers of these areas. Which communities constitute the pastoralists and which small ruminants do they rear? What is the purpose of animal migration? Is it a necessity or just an age-old social habit?

This chapter presents the unique features of the prevailing animal farming system in the hills, place of small ruminants in hill farming, small ruminants genetic resources, breeding policies and a brief overview of the history of pastoralism, hardy shepherds and their life, the scarcity of animal feed, the grazing rights and the Forest acts, and lastly, the future prospects of this farming system. Is it going to stay or would it dwindle with time?

S. P. Singh, J. K. Malik, *Migratory Small Ruminant Farming System in the Himalayas-Peculiarities, Problems, Prospects, Climate Effects, Nutrition and Health*,
https://doi.org/10.1007/978-981-96-6691-1_1

Keywords

Pastoralism · Himalayas · Shepherds · Transhumance · Grazing rights

Abbreviations

CB	Crossbred
DAH	Department of Animal Husbandry
DAHD	Department of Animal Husbandry and Dairying
FAO	Food and Agriculture Organization
GA	Geographical area
GOI	Government of India
HP	Himachal Pradesh
ICAR	Indian Council of Agricultural Research
IFAD	International Fund for Agriculture Development
IGFRI	Indian Grassland and Fodder Research Institute
IHR	Indian Himalayan Region
J and K	Jammu and Kashmir
NCF	National Commission on Farmers
SWDBU	Sheep and Wool Development Board Uttarakhand
UK	Uttarakhand
UTI	Union territories of India
UTs	Union territory

1.1 The Himalayas

For centuries, Himalayas have remained the marvel of mountaineers, climbers, trekkers, geographers, historians and poets. For them, the mountain may be exciting wonders and object of pleasure, but for the inhabitants of tiny mountain villages and their animals, they are a source of bread, butter, and a lifeline for their very existence. Most countrymen consider mountains as God Shiva's abode and the famous shrines such as Mt Kailash Amarnath and Mani Mahesh lie in the lap of high Himalayas.

The extensive mountain chain stretches from Afghanistan to eastern part of India covering nearly 3000 km long and 350 km wide area between Indus river in the west to the Brahamputra river and beyond in the east. Himalayas lie between 72° and 97° E longitude and 23° to 36° N latitude with elevations as high as 8848 m. The Indian Himalayan Region (IHR) spans 12 Indian states starting from union territories (UTs) of Jammu and Kashmir (J&K) and Ladakh and ending with Arunachal Pradesh covering an area of about 523,000 km^2 (ENVIS Bulletin 2010). This chapter provides an overview of available information and summarizes the current knowledge on the subject largely in three north-western states: Himachal Pradesh (HP), Uttarakhand (UK), and UTs of Jammu and Kashmir and Ladakh, which cover nearly 68% of the geographical area of the Indian Himalayas.

Himalayas are considered the loftiest mountain complex on earth with 31 peaks, several of which are prominent (Gurung 1990). Everest with a height of 8848 m is the most daunting and fascinating peak. Himalayas have extreme variation in climate and topography creating unique situations for adaptability and survival of domestic animals. They are gifted with almost whole range of world's climate, flora, and fauna within four or five vertical miles. The northern most areas have temperate climate with frozen slopes, whereas the areas bordering plains (Shivaliks) have subtropical sub-humid climate. The mid Himalayas are characterized with mixed climate. The forests in lower hills are alive with insects and other animal life, but the upper ones are characterized by near silence. With rise in elevation, the Sal trees give place to Pine, Pine to Oak, and Oak to Juniper and *Rhododendron* and then come the end of tree line. An amazing feature of high Himalayas is the presence of natural alpine grasslands with nutritious grasses. This nature's gift is a boon to the migratory small ruminants every summer season when they are taken there for grazing.

1.2 Geographical Divisions of Indian Himalayas

The most acceptable geographical division of Himalayas includes three main ranges, viz., western Himalayas, central Himalayas, and eastern Himalayas (ENVIS Bulletin 2010). The western Himalayas comprise of UTs of Jammu and Kashmir and Ladakh and states of Himachal Pradesh and Uttarakhand and occupy an area of 331,392 km^2 which is about 62.1% of the Indian Himalayan Region (IHR). The eastern Himalayan states of Sikkim and Meghalaya occupy an area of 29,525 km^2. Table 1.1 presents the geographical and other salient features of the two north-western states and two union territories and two north-eastern hill states.

The main focus of the present chapter is north-western Himalayan states of Himachal Pradesh (HP) and Uttarakhand (UK) and UTs of Jammu and Kashmir and Ladakh. Some aspects of animal keeping in the eastern Himalayan states of Sikkim and Meghalaya relating to heavy parasitism have also been discussed in Chap. 10. There is a special emphasis on Himachal Pradesh where the first author was the Principal Investigator of an Indian Council of Agricultural Research (ICAR) sponsored Research Project on Migratory Small Ruminants. Several thousands of these animals are taken to alpine pastures for grazing every summer and are brought back to the lower plains (base camps) with the onset of winter. Topics that will be emphasized in this chapter include geographical division of Himalayas, features of hill agriculture and animal husbandry system, place of small ruminants in hill agriculture, small ruminant genetic resources in different hill states, and their contributions.

Table 1.1 Total geographical area, land use pattern, and other features of UTs of Jammu and Kashmir and Ladakh and some north-western and north-eastern Himalayan states (ENVIS Bulletin 2010; Rawat 2004)

	North-western states/union territories			North-eastern states	
Parameters	J&K and Ladakh	HP	UK	Sikkim	Meghalaya
Geographical area (GA) (km^2)	222,236[a]	55,673	53,483	7096	22,429
Area of total IHR (%)	41.65	10.43	10.02	1.33	4.20
Total forest area (% of GA)	10.14	66.5	71.05	47.8	69.5
Area under permanent pasture and other grazing land (km^2)	1190	14,960	1990	–	–
Alpine vegetation (%)	5.93	31.06	15.9	–	–
Major pastoral tribes	14	9	5	1	3

[a]Area including under Pakistan

1.3 Features of Hill Agriculture and Animal Husbandry System

Agriculture is the main source of livelihood for majority of the hill inhabitants, and next to agriculture, livestock are most productive assets for the poor farmers. Of the total area of Himalayas (53 million hectares), around 15.8% is used for marginal farming and rest occupied by rangelands, forests, pastures, waste lands, and permanent snow (Partap 2011). As per the report presented in the brainstorming session by the ICAR-IGFRI technical group (Sandhu et al. 2016), the country is divided into 15 regions on the basis of agro-climatic attributes and the north-western Himalayan region includes states of Himachal Pradesh, Uttarakhand, and union territories of Jammu and Kashmir and Ladakh. Depending primarily upon elevation from the sea level, each of the state/union territory has further identified itself into three to four agro-climatic zones (Table 1.2).

According to one estimate, Indian Himalayas support about 50 million domestic animals including 15.8% goats and 10.4% sheep (Partap 2011). Owing to limited feed and other resources in higher mountains areas, the animal rearing is pastoral and the herds are migratory in nature. This novel annual feature of animal migration is an age old practice in the Himalayan states.

Agriculture and livestock production systems on the hills have some distinctive features vis-à-vis the plains some of which are as follows:

(a) The land holdings are meager and inadequate which force the farmers to seek alternate means of earning. Majority of farmers in the Himalayan states have land holdings less than 0.5 hectare with average holdings in Himachal Pradesh is about 1.2 hectares (Partap 2011). The land is in terraces and has diverse fertil-

Table 1.2 Agro-climatic zones of the north-western Himalayan states

Agro-climatic zones of north-western Himalayan states/union territories					
Jammu and Kashmir including Ladakh		Himachal Pradesh		Uttarakhand	
Zone	Feature (Elevation meters)	Zone	Feature (Elevation meters)	Zone	Feature (Elevation meters)
I	Sub-tropical (Jammu area)	I	Low hill subtropical (<650 m)	A	Tarai Up to 1000 m
II	Sub-temperate (mid zone)	II	Mid hill sub humid (650–1800 m)	B	Mid hill south 1000–1500 m
III	Temperate (Entire Kashmir valley)	III	High hill wet temperate (1800–2200 m)	C	High hills 1500–2400 m
IV	Arid temperate (Leh and Kargil)	IV	High hill dry temperate or alpine (>2200 m)	D	Very high hills >2400 m
	Bhat and Salam (2016)		Gupta et al. (2017)		ICAR-IGFRI (2021)

ity and comparatively less productive and, therefore, unsuitable for conventional agriculture.

(b) Agriculture on the hills faces two significant concerns, viz., crop land scarcity and water scarcity.

(c) Pastoral livestock system utilizes 25% of the terrestrial land using mostly dry land or hills where conventional farming is not possible.

(d) Most hill areas are rain fed and there are hardly any sources of irrigation. Extreme weather conditions serve as deterrent to profitable agriculture.

(e) Owing to limited opportunities in the farming system and lack of industrialization in mid and high hills, the younger population is forced to seek employment in Government jobs and services in the plains. A substantial section of population work in the defense services, hotel, automobile, and other industries in metropolitan cities.

(f) A distinctive feature of hill agriculture is the dominant role and participation by women in farming operations, livestock rearing, and feed and fodder transport.

(g) Farmers face lots of problems in proper marketing of the agricultural produce and livestock products.

(h) Feed resources are limited due to constraints of land and cropping area and other inputs which force the livestock owners to obtain feed and fodder at a higher price from the plains.

(i) Livestock rearing in the hills is of sedentary, semi-migratory, as well as migratory nature. Animal migration especially of sheep and goats is a means to avoid harsh summer climate and feed scarcity conditions. Upper regions of Himalayas are endowed with lands having grasses (Themeda-Arundinella type) that are highly nutritious and serve as an excellent feed source for migratory small ruminant species such as sheep and goat in favorable times.

(j) Remoteness and inaccessibility of resources is also a distinct feature.

(k) Besides natural forests and grasses, agro-forestry constitutes tremendously to fodder supply. Unlike plains, farmers obtain green leafy fodder from tree leaves, shrubs, grasses, and a limited supply from crop residues. A large variety of trees and shrubs are found in agro-ecosystems of Himalayas, which serve as a source of fodder and wood. The forest cover in the states of Himachal Pradesh and Uttarakhand is significantly higher than in Jammu and Kashmir. Also the permanent pastures in the state of Himachal Pradesh are very abundant (Fig. 1.1).

1.4 Transhumance/Animal Migration—A Unique Animal Farming System in the Himalayas

Since migration of livestock, especially sheep and goats, has been continuing as an essential activity in the Himalayas, it is desirable to provide insights on various facets of this farming system and examine its current status and constraints with a view to improve the productive potential of these animals. The annual pattern of animal migration shows that the animals migrate to high altitude pastures in summer and start backward journey for the base camps located in the plains before the

Fig. 1.1 The State of Uttarakhand is rich in quality meadows. (Photograph contributed by Mr Ashutosh)

onset of winter. Often, questions arise if it is a conventional practice to maintain age-old cultures or a need-based activity without which the sustainability of livestock would be jeopardized. The basic reason of maintaining annual migratory practice is economical to overcome the scarcity of feed and fodder for the animals and also escape from hot summer months. In this process, the vast natural vegetation wealth of the hills is converted into human-usable items such as milk, meat, wool, and hide. It is obvious that most north-western states and union territories have very limited feed resources. The union territories of Jammu and Kashmir and Ladakh and the states of Himachal Pradesh and Uttarakhand have shortage in dry feed to the tune of 27.3%, 55%, and 16.3%, respectively. The shortage of greens in union territories of Jammu and Kashmir and Ladakh and the states of Himachal Pradesh and Uttarakhand was estimated at 67%, 13%, and 92.3%, respectively. All these regions suffered badly from shortage of concentrate (Mir et al. 2016; Gupta et al. 1998).

Several studies have shown that pastoralists make significant contribution to the local, regional, and global economies and world food security and in various other ways. Pastoral livestock system produces approximately 10% of world meat consumed supporting 200 million households (Uddin and Kebreab 2020; Arjjumend 2018).

1.5 Place of Small Ruminants in Hill Agriculture

Sheep and goats occupy a special place in the hill economy for many reasons. In general, number of livestock declines with rise in elevation. Small ruminant species sheep and goats constitute the major animal species in the higher hills, whereas large ruminants such as cow and buffalo predominate the lower and mid hills. Small ruminants can easily adapt to the harsh climate and water scarcity, tolerate long tortuous hill walks, utilize grazing lands more efficiently, serve as source of wool, meat, and high-quality skin, and thus provide employment to man and women in wool spinning, weaving work, and carpet industry. Unlike large animals, small ruminants could be easily handled and disposed off as cash. Compared to sheep, goat is a precious economical animal for landless and poor section of the rural hill people. The cost of goat rearing is relatively low as it is raised through grazing and on agriculture by-products. Of nearly 135.2 million goats in the country, 23 million belong to well-defined breeds with specific characteristics and the rest are nondescript. The well-known goat breeds of north-western Himalayas are Bakarwal, Gaddi, Changthangi, Chegu, and Kangan with a total population of about 6.1 million.

The Changthangi (Pashmina) and Chegu goats found at elevations of 4000–5500 m in Jammu area and union territory of Ladakh are famous world over for fine warm wool fiber. The body of goats is covered with smooth silky hair about 10–12 cm long below which is present a fur-like undercoat called Pashmina which is used to make Kashmiri shawls. Of different materials, shawl prepared from pashmina is most graceful, warm, soft, and attractive. The goat owners are nomadic and practice migration from as far as Indus river belt to alpine pastures of Changthang region of Ladakh.

1.6 Small Ruminant Genetic Resources in North-Western Indian Hilly States and Their Contributions

Table 1.3 gives the highlight of 20th Livestock Census conducted during the year 2019. Table 1.4 presents a comparative picture of 19th Livestock Census (2012) and 20th Livestock Census (2019) to show a change in sheep and goat population for the three north-western Indian states (GOI 2019). As per the 20th Livestock Census (2019), the total sheep and goat population in the three north-western Himalayan states including union territories of Jammu and Kashmir and Ladakh has been

Table 1.3 Sheep and goat population (thousands) in major north-western Indian states and union territories as per the 20th Livestock Census-2019 (GOI 2019)

Species	Type	Jammu and Kashmir including Ladakh	Himachal Pradesh	Uttarakhand
Sheep	Indigenous	1663.4 (51.2%)	718.5 (90.8%)	208.2 (73.2%)
	Exotic/ crossbred	1584.0 (48.8%)	72.8 (9.2%)	76.3 (26.8%)
Total sheep		3247.5	791.3	284.6
Goats		1730.2	1108.4	1371.9
Total sheep and goats		4977.7	1899.7	1656.5

Table 1.4 Comparative population (thousands) of sheep and goats in the major north-western Indian states and union territories as per the 19th Livestock Census-2012 and after 20th Livestock Census-2019

Species	Type	Jammu and Kashmir including Ladakh		Himachal Pradesh		Uttarakhand	
		2012	2019	2012	2019	2012	2019
Sheep	Indigenous	1074.0	1663.4	499.6	718.5	241.2	208.2
	Exotic/CB	2315.4	1584.0	305.2	72.8	127.5	76.3
Total sheep		3389.4	3247.4	804.8	791.3	368.7	284.5
Goats		2017.9	1730.2	1119.4	1108.4	1367.4	1371.9
Total sheep and goats		5407.3	4977.6	1924.3	1899.7	1736.1	1656.4

estimated at 4323.4 and 4210.5 thousands, respectively. The following are other salient features of the census.

1. In the north-western states of India Himachal Pradesh and Uttarakhand and the UTs of Jammu and Kashmir and Ladakh, there was a clear trend of decline in population of exotic/crossbred sheep and a consistent increase in the number of indigenous sheep except Uttarakhand.
2. The UTs of Jammu and Kashmir and Ladakh reared more sheep than goats, but it was opposite in case of Himachal Pradesh and Uttarakhand.
3. The UTs of Jammu and Kashmir and Ladakh have highest number of sheep and goats followed by Himachal Pradesh and Uttarakhand.
4. Compared to 2012 Census, there was a decline in number of goats in Himachal Pradesh and UTs of Jammu and Kashmir and Ladakh and marginal decrease in the number of sheep in both the states and UTs.

Values in parentheses show percentages of indigenous and exotic/crossbred sheep of the total sheep population.

As compared to large ruminants cattle and buffalo, sheep and goat require smaller investment and there is a quicker return on investment due to their short breeding cycle. Furthermore, the herds of small ruminants recover quickly following the incidence of natural disasters such as drought (Mayberry et al. 2018). The most prominent indigenous sheep breeds of north-western temperate region including UTs of Jammu and Kashmir and Ladakh, Himachal Pradesh, and Uttarakhand and their utility are presented in Table 1.5.

Sheep of Purgi breed are found in the Kargil area of Ladakh and are raised under range system and semi-intensive system in summer and winter seasons, respectively (Hamadani et al. 2022). Sheep breeds of Jammu and Kashmir that have already become extinct include Ghidord Phamphri, Punchi Bakerwali, Bani, and Karnahi (Puri 2007; Hamadani et al. 2022). In addition to Gaddi, Muzzafarnagri breed of sheep is also found in Uttarakhand (Aggarwal et al. 2022).

India accounts for significant proportion to the global goat population, and goats are contributing 27.80% of the total livestock population in India (GOI 2019; Chauhan et al. 2021). Due to their lesser requirement for feed, goats can be reared on marginal lands. The capability of goats is better to digest roughages, and moreover, they can easily browse on shrubs and trees (Mayberry et al. 2018). Goat breeds distributed in Jammu and Kashmir are Gaddi, Bhakarwal, and Kashmiri goat. These breeds have exhibited highest adaptability to the varied climatic situations of the state. The Gaddi breed also known as the "White Himalayan goat" is widely distributed in the Jammu region. Gaddi goats are well adapted to migratory system and are used for meat and fiber production as well as for the purpose of transport of commodities (Hamadani et al. 2022). Bhakarwal goats are found in Poonch, Jammu, Rajouri, Raise, Udhampur, and Kathua districts. This breed is being used for the production of meat and milk (Rathera et al. 2022). Kashmiri goat is another breed of goat found in Kashmir valley, and these animals are mostly raised by grazing and stall feeding during early spring and late autumn, respectively. There are some

Table 1.5 Distribution and utility of various sheep breeds of some north-western states/union territories

State/union territory	Breed	Distribution	Utility	Reference
Jammu and Kashmir	Bhakarwal	Pir Panjal ranges of the Jammu division, Kashmir Valley and other low lying hills	Colored coarse wool	Yadav et al. (2017); Hamadani et al. (2022)
	Poonchi	Poonchi district and adjacent districts Rajouri and Reasi of Jammu	Meat, wool	Taggar et al. (2018); Hamadani et al. (2022)
	Gurez	Gurez and Tulail Tehsils of the Bandipora district	Meat, wool	Ganai et al. (2010); Hamadani et al. (2022)
	Karnah	Karnah Tehsil of Kupwara district	Meat, wool	Khan et al. (2021); Aggarwal et al. (2022)
	Gaddi	Kishtwar and Bhadarwah Tehsils in the Jammu region	Meat, wool	Yadav et al. (2017); Aggarwal et al. (2022)
Ladakh	Changthangi	Changthang region of Leh district	Meat, wool	Malik et al. (2021); Aggarwal et al. (2022)
Himachal Pradesh	Rampur Bushair	North temperate Himalayan region, Lahaul and Spiti, Kinnaur and Shimla	Meat, wool	Sankhyan et al. (2018); Aggarwal et al. (2022)
	Gaddi	Hamirpur, Ramnagar, Udhampur and Kulu and Kangra valleys	Meat, wool	Yadav et al. (2017); Aggarwal et al. (2022)
Uttarakhand	Gaddi	Dehradun, Nainital, Tehri Garhwal and Chamoli districts	Meat, wool	Yadav et al. (2017); Aggarwal et al. (2022)

breeds of goat in Jammu and Kashmir that are either on the verge of extinction or have already become extinct. Kaghani, Lubdi, and Kilan breeds of goats reared by Bhakarwals are on the verge of extinction. Four goat breeds, namely, Gurziya, Belori, Lamdi, and Goodri, have become extinct (Puri 2007; Hamadani et al. 2022).

Goats of Changthangi breed are found in Leh and Kargil districts of Ladakh, which are known for producing world famous Pashmina fiber which is one of its kinds among animal fibers particularly in terms of its fineness, warmth, and softness. Pashmina produced by Changthangi goat is superior to that of Chegu in terms of fiber diameter and staple length (Kour et al. 2018; Aggarwal et al. 2022). Besides Changthangi, goats of Purgi breed are widely distributed In Kargil. This breed is used for both meat and fiber production and plays an important role in providing livelihood support to Kargil tribes (Alam et al. 2021).

In Himachal Pradesh, the total goat population is comprising of two breeds, namely, Chegu and Gaddi. Goats of Chegu breed produce a finest and warmest natural fiber Pashmina, and their true breeding tract is confined to the cold desert region of Lahaul and Spiti and Kinnaur area in Himachal Pradesh (DAH HP). Chegu has adapted to the harsh agro-climatic and agro-pastoral conditions (Katoch et al. 2020). Goats of Gaddi breed are found in Kangra and Kullu valley, Chamba, Sirmur, Simla, and Lahaul and Spiti in Himachal Pradesh (Dhara et al. 2020).

The breeds of goats found in Uttarakhand include Gaddi, Chaugarkha, Black Bengal, Beetal, and Barbari (DAH UK). Udaipuri is yet another goat breed found in the district of Pauri (Bhatt and Padmakumar 2012). Gaddi goats are found in hilly districts of Tehri Garhwal and Chamoli of the state. Chaugarkha has been identified as a unique new breed since 1992 and the breeding tract of Chaugarkha is spread across Almora, Bageshwar, Champawat, and Pithoragarh districts of Kumaon region (Chauhan et al. 2021). Goats of Chaugarkha breed are of small size and raised mainly for meat purpose (Jana et al. 2018). Pantja is another breed of goat raised for meat purpose by the farmers of Tarai region of Uttarakhand (Khadda et al. 2017).

In order to improve sheep wool quality and quantity in these hilly Indian states, attempts have been made to introduce exotic germplasm and switch to crossbreeding. The breeding policy, sheep exotic crosses, and other aspects of sheep and goat development in the north-western Indian Himalayan states and union territories are presented below.

The union territories of Jammu and Kashmir and Ladakh are ideally suited for rearing sheep and goats due to its favorable climate and terrain, and the region is the second-largest producer of wool in the country with around 70 lakh kg of wool per year (Hamid 2021). Before independence, hardly any distinct breed was available except some local nondescripts. Later, proper breeding policies were framed. The current breeding policy in Jammu and Kashmir envisages crossing of local Kashmir valley sheep with Merino or Rambouillet rams to the level of 75%. Kashmir merino breed was, in fact, evolved by crossing local Kashmir ewes with Tasmanian Merino rams. The resulting F1 ewes were then mated to Delaine rams imported from the USA to produce a stock with 75% level of inheritance from two exotic breeds. This was further followed by rigorous selection and upgradation of Changthangi sheep of Ladakh. Besides wool breeds, world famous Pashmina goat breed Changpa found in Changthang subdivision of Leh district of Ladakh are being upgraded by selective breeding. Introduction of crossbreeding has resulted in enhancement of fine wool production significantly (DAH JK).

In Himachal Pradesh, sheep and goat rearing is one of the main occupations and a sustainable source of livelihood for the poor and marginal farmers. Nearly 80% of sheep and 70% of goat population are estimated to be migratory (Sankhyan et al. 2016). Sheep crossbreeding program was started in Himachal Pradesh as early as 1956, and during the subsequent years, several exotic sheep breeds including Spanish Merino from Spain, Soviet Merino from the then USSR, Rambouillet from the USA, and Polwarth and Corriedale from Australia were imported (Chauhan 1990). Most exotic sheep were reared on the State Government Farms. The Subcommittee of Planning Commission on Sheep and Goat Development for North Temperate Zone recommended Rambouillet and Soviet Merino for fine wool production. The main limiting factor for optimum performance of exotic breeds and their crossbreds has been poor feed and fodder resources. Currently, the main indigenous breeds include Gaddi and Rampur Bushehri. The state has five sheep breeding farms at Jeory (District Shimla), Karchham (District Kinnaur), Tal (District Hamirpur), Sarol (District Chamba), and Nagwain (District Mandi). Some of the farms solely maintain pure exotic sheep. The existing breeding policy is to breed indigenous Gaddi and local Rampur bushahari ewes with pure Rambuillet or pure Russian Merino to obtain F1 animals. In subsequent generation, rams with 75% exotic inheritance are to be used for crossbreeding to be able to maintain exotic and indigenous inheritance at 75:25 level (Gupta 2015).

In the state of Uttarakhand, which was created in the year 2000, sheep and goat husbandry is of prime importance and has emerged as a sustainable source of livelihood for the poor and marginal farmers. To promote and encourage scientific outlook towards sheep and goat development, an organization named Sheep and Wool Development Board came into existence in the year 2005 (SWDBU 2005). The main objective of the board is improvement in germplasm of livestock by following a well-defined breeding policy for sheep, goat, and rabbit and to facilitate increase in livelihood of the farmers. The breeding policy of the state lays downupgradation of local sheep breed and Gaddi and Rampur Bushair sheep by crossing with purebred high quality rams of Merino and Rambouillet breeds. It was mentioned that one-third of the total Uttarakhand population of sheep was already crossbreds which needed to be upgraded through crossbreeding with pure breed rams of Merino and Rambouillet breed. For goat improvement in the plain region of the state, local goats are to be up-graded by crossing with Barbari and Jamunapari breeds of goats. In hilly region, mohair fiber would be encouraged by crossbreeding local goats with Angora bucks and to check inbreeding in germplasm at Government Angora Goat Breeding Farm.

1.7 Pastoralism and Historical Background

The term pastoralism has been described in various ways. Shridhar (2021) of *Outlook* magazine refers to pastoralists as "Original Livestock Keepers of Civilization" (Pastoralists, Outlook September, 2021). The magazine describes pastoralism as a living pattern of trading herds of animals in continuing mobility and closeness to the nature and rangelands. According to Dong et al. (2016), pastoralism is the caring and herding of large animals in dry land areas, whereas International Fund for Agricultural Development describes it as a subsistence and successful living through herding livestock in less productive rangelands (IFAD 2012). Pastoralism in the Himalayas has been described as a way of life. Pastoralists make their living in some of the hardest conditions and unprecedented climatic conditions.

As per the available records, grazing lands cover about half of earth's land surface. In India, the rangelands are estimated to be 121 million hectares of which about 18% lie in Hindu Kush Himalayan region (Arjjumend 2018). In China, vast rangelands of the Qinghai-Tibetan Plateau (QTP) comprise one of the most important pastoral areas of the country and are known for supporting the survival and livelihoods of millions of pastoralists and agro-pastoralists (Kreutzmann 2011). Approximately 90% of the global mountain population live in the developing countries including China, India, Nepal, and other Asian countries (Gentle and Thwaites 2016), and an estimated population of around 10 million livestock-dependent people are known to inhabit Himalayan mountain areas (Namgay et al. 2013). Although official data on the number of pastoralists in India are not available, a figure of 35 million is frequently mentioned (Kishore and Köhler-Rollefson 2020).

By raising different livestock species that adapt to variable environments, pastoralism plays a vital role in alleviating poverty and providing food security in these regions. Pastoralists feed their animals through varied landscapes, and while working proficiently with nature, they contribute in productivity, sustainability, and welfare of animals. Besides providing affordable, high-quality proteins and nutrients to meet local demand, pastoralism acts against climate change, functions as an early warning system for pandemics, strengthens the labor force, and offers employment opportunities. Furthermore, it decreases competition between feed for farm animals and food for human consumption, protects animal diversity, and helps in maintaining and renewing the forestlands (FAO 2021).

Migratory pastoralism on the Himalayas especially the hills of Hindu Kush mountainous region has been a very old cultural and economic activity as indicated by archeological excavations in the countries of this region. In Afghanistan, grazing of goats and sheep (nomadism) has been a major source of livelihood for ages. The Kochi nomads were believed to travel thousands of miles every year in Pakistan, Iran, and Afghanistan in search of pasture lands (Rashid 2001). According to one report, pastoralism on Tibetan plateaus was practiced around eighth century, which seems to suggest that migratory animal husbandry was in vogue in that part of Himalayas as early as 2000 year (Negi 1998). It has been suggested that sheep and goats were probably the first animals to be domesticated well before 6000 BC on the highlands of Persia. In the Indian subcontinent, finding of animal bones and parts in the excavations at Mohenjo-Daro and Harappa sites suggests signs of domestication of sheep long time ago. Ingty (2021), who has studied rangelands and grazing systems in the alpine Himalayas, has stated that pastoralism has existed for millennia and impacted the evolution of many rangelands, from the biodiverse rich Africa to the highest elevations in the Tibetan steppe.

In several mountainous countries with common borders, nomads and tribes join together at common geographical location and cooperate for facilities by barter system. For instance, on Pak-Afghan border, China-Nepal border, and Indo-Tibetan border, graziers from respective countries share facilities. In the north-western region of India, the main tribes that have been following the transhumance are Bakarwals and Gujjars in Jammu and Kashmir, Gaddis and Kinnauras in Himachal Pradesh, Van Gujjars and Bhotiyas in Uttarakhand, and Bhutias in the state of Sikkim. In Meghalaya, two principal tribes, namely, Garos and Khasis, have been recognized, but they are hardly involved in annual transhumance as in the north-western states. On the Indo-Tibetan-Nepal borders, Gaddis and Nepali shepherds use pastures of northern highland mountains for grazing their livestock. Transhumance and pastoral life on mid and high hills of north-western Himalayas has been in practice for time immemorial, but its status has been dwindling over the past few decades.

1.8 The Hardy Shepherds/Graziers

It is interesting that the age-old tough profession of transhumance has evolved with it the hardy graziers/shepherds such as the Gaddis of Himachal Pradesh. These and other pastoral communities with distinctive work and cultural characteristics are involved in transhumance activity in different Indian states and neighboring countries. The detailed account of major transhumant pastoralist communities of Himalayan countries including their names, locations, livestock species reared, migratory patterns, and other characteristics have been discussed later in Chap. 3. In the present chapter, we have described the distinctive features and socio-economic status of major Himachal pastoral community known as Gaddis.

1.8.1 The Gaddis of Himachal Pradesh

Gaddis are a major tribe of Himachal Pradesh who are involved in migratory pastoralism or shepherding. The tribe is an amalgam of several castes including Brahman, Rajput, Kharti, Sippi, and Rane. The people love outdoor life, worship God Shiva, and live a simple life. Similar to males, Gaddi women are tough and hard working. During the day, the ladies are busy in cutting wood and grasses and in the evening, they knit and spin wool to make woolen threads. As far as migratory flock is concerned, mostly male members are involved in it. With time, some of the Gaddis have turned semi-sedentary and some sedentary. In spite of abandoning transhumance, even sedentary Gaddis practice livestock rearing as a major activity. There are interesting tales about the origin, characteristics, living habits, and dress code of this famous tribe. In Himachal Pradesh, the tribe is mainly concentrated in four to five districts, namely, Kangra, Kullu, Chamba, Kinnaur, and Sirmor. The origin of Gaddis is traced to Bharmour town of Chamba district of Himachal Pradesh which is also named as Gadiyar or Gaderan (land of Gaddis) after its inhabitants. Gaderan area extends from the bank of river Beas up to Chamba and Kangra districts and encompasses both sides of the Dholadhar range of Himalayas. Bharmpur territory was established by King Meru Varma who took the throne (Gaddi) in the year 550 (Molly 1998). The inhabitants of the area acquired Gaddi name during subsequent years. It is also believed that Gaddis migrated to the hills from plains of Punjab, Rajasthan, West Bengal, and eastern Uttar Pradesh at the time of Muslim invaders. Most Gaddis worship God Shiva—the lord of Dholadhar and Chamunda Mata (mother) (Fig. 1.2).

Story goes that God Shiva created Mani Mahesh (a town near Bharmour) as his abode. But the people believed that lord resided in this town only for a period of 6 months and migrated to some unknown destination during winter. The lord created the Gaddis for the protection of sheep and goats and Gaddis are nomads because

Fig. 1.2 A Gaddi shepherd leading the Gaddi flock in Kullu town in Himachal Pradesh. (Photograph by courtesy of Director ICAR-CSWRI, Avikanagar)

their lord himself was a nomad (Molly 1998). The Gaddis roam from hill to hill, from one pasture land to another because their lord instructed them to do so. There are countless stories about Gaddis' sacrifice and devotion to lord Shiva. According to one of the tales, once lord Shiva tested the faithfulness of Gaddi shepherd to the sheep flock. When the Gaddi was eating his meal, the lord drove his flock away. The Gaddi did not care about the food and ran away to collect his sheep. Similarly, in another story, one night when the Gaddi woke up from sleep under the rock, he found his flock missing. He spent whole night tracing his flock. Lord Shiva was very pleased with the devotion of Gaddi shepherd and he blessed the Gaddi for his love to the flock. Undoubtedly, the Gaddi's love for his flock is unparalleled (Fig. 1.3). He rears these animals like his own child. For this reason, the flock is called by him as his "Dhan" or "Mall" (money or property).

There are two more striking characteristics typical about Gaddis: their dress and housing. The dress is designed such that it is fit to protect him from harshness of climate and rugged hard life. The chola (main dress), the dora (string or rope around waist), and cap are important pieces of the Gaddis' dress. The chola protects him from extreme cold winter and the dora, wrapped around his waist, serves multiple roles. It supports the back during climbing the mountains with heavy loads and serves as a pillow during travel and rest. The cap keeps the head warm during winter

Fig. 1.3 Made for each other: a Gaddi shepherd's total dedication to his flock

and also protects from bright sun. The Pattu or blanket is also commonly used to protect from cold and drizzle. Most Gaddis' houses are made of wood, mud, and slate slabs and have generally three storeys: the obra (ground floor), the mandeli (first floor), and chhappar (second floor). The ground floor is used to keep cattle, sheep, and goats and other animals and the first and second floors are used as family quarters.

In the beginning, large numbers of Gaddi farmers were involved in migratory practice due to availability of plentiful of grazing lands and paucity of government jobs and limited avenues in industry. With dwindling pastures, the Gaddis have switched to farming, horticulture, business, and government jobs. The average land holding reportedly varied from 0.2 to 0.42 hectares per family. According to one estimate, around 30% of the Gaddis are still fully migratory, whereas 70% have turned to semi-sedentary and sedentary life (Misri 2003). During migration, Gaddis are known to follow traditional methods of animal treatment. Most Gaddi children nowadays go to schools. However, earlier reports indicated that compared to the state literacy rate of 60%, only 30% of Gaddis were literate. Another report suggested a literacy rate of 50.7–94.7% in Gaddis settled in Palampur town area of Kangra district of Himachal Pradesh (Misri and Dev 2001). During the past four decades, there has been a distinguishable change in the social makeup of the Gaddi-inhabited region of Himachal Pradesh. Besides increased literacy rates, the population densities have doubled, and joint households have been separated into nuclear households and trends toward emigration, particularly from the core areas. It has been suggested that such transformation is a cause of and a response to changes in human-environment relations (Sharma 2021) (Fig. 1.4).

Fig. 1.4 The Bakarwals with a flock of sheep at alpine pastures in Jammu and Kashmir. (Photograph by courtesy of social activist, Mr Sankar Sridhar with permission)

1.9 Grazing Rights and Forest Act

The use of pasture lands by shepherds and Gujars for grazing of animals has always been under some kind of regulation. It was never a free-for-all type situation. In pre-colonial era, hundreds of years ago, the Rajas or landlords of the area used to control the grazing rights. In Himachal Pradesh, for instance, Rajas of Chamba or Kullu and in Keylang town area of Spiti, Thakur (landlord) of Keylang used to fix the Trini (grazing charges) rates for sheep, goats, and other animals and earmark grazing land (Lyon 1993). Arrival of British in India and spread of their dictate to the mountainous region did not affect the system for some time and the Rajas continued to regulate grazing rights which were inherited from the past. Later, with the commercialization of wood and other forest produce, the grazing rights of shepherds were re-examined.

National Forest Law was enacted in the year 1865, which sought to regulate grazing rights to each shepherd, fix herd/flock size and fee, migration route, and other aspects. In the twentieth century, there was increased shortage of grazing areas, growing conflict among stake holders, degradation of pastures due to overuse, soil erosion, and growth of obnoxious weeds. Goats were blamed as a threat to forest and pastures. Consequently in 1927, the British Government enacted Indian Forest Act 1927 to tap the country's forest wealth. It gave tribes and shepherds rights to use forest land and exploit forest resources (Chauhan 2008).

After India gained independence in 1947, Himachal Pradesh government appointed two commissions in the years 1959 and 1970; they, besides other issues, froze the flock size, but their recommendations were never implemented. Concerned with the continued worsening of forest health due to ever increasing encroachment and over exploitation of grasslands, the Indian Government promulgated another Act in 1980, the Forest Conservation Act (1980) (Ministry of Environment, Forest and Climate Change of India), which took away the rights of tribals and forest dwellers. They were labeled as encroachers on forest land on which they had been living for generations. In pursuance with this approach a step further, the Inspector General of Forests issued a circular in the year 2000, stating that people encroaching the forest land could be evicted. These orders were protested by the tribals and other forest dwellers, and their cause was supported by the local Non-Governmental Organizations (NGOs). To resolve the issue, the Government promulgated Forest Dwellers Act 2007 (Gazette of India 2007) giving tribals and forest dwellers their rights over the land and forest produce if they had been living there for the last 25 years.

1.10 Migratory Practice—A Necessity or an Age Old Social Habit?

Nature seems to have provided a rare gift in the form of highly nutritious grasslands and pastures in high altitude alpine areas. Migration of livestock to these pastures in summer seems to be an efficient way of avoiding fodder scarcity during summer and utilize the nature's precious gift, i.e., the grasslands. This suggests that basically seasonal migration or transhumance is an economic activity in response to ecological needs—a balance between mutual demands of the herders and their animals and judicious use of both resources in the plains and alpine grasslands. It is thus a reality that in some of the north-western states, animal migratory practice is a necessity for subsistence. Seasonal movement is a strategy in the interest of flocks. These animals need to be fed and grazed throughout the year. When the hot summer sets in the plains, there is hardly any grass and the migratory tribes need to take their animals to the cooler areas of alpine pastures where abundant nutritious grasses are available. Cold and snow in the months of September and October force them to leave the highlands for the wet plains. The available vegetation and other facilities control their route and time of migration. Movement of cattle, sheep, and goats during upward and downward journey helps afforestation and manuring of land and fields and the rotational grazing helps regeneration of grasses and plants. The National Commission on Farmers (NCF 2006) headed by the eminent Agricultural Scientist, late Dr M.S. Swaminathan, in its report has emphasized significant role of pastoralists in supplying organic manure to soil, conserving domestic diversity and providing food security on dry lands without depleting ground water resources. But with time, this practice seems to have put tremendous strain on the grasslands and ecology and environment. The ecological balance also seems to have been adversely affected in the last five decades due to interventions such as construction of dams and the industrialization of hills and overgrazing and growth of weeds and harmful plants in the grasslands.

In order to find out the views of shepherds on the practice of seasonal migration and related issues, we directly approached some of them through interviews and questionnaires. Their responses have been presented in Chap. 4 under Sect. 4.5.

Since migration of livestock, especially sheep and goats, has been continuing as an essential activity in the north-western and north-eastern Himalayas, it is desirable to provide insight into various aspects of this farming system and examine its status and constraints with a view to improve animal health and production. The annual pattern of animal migration shows that the animals migrate to high altitude pastures in summer and start backward journey for the base camps located in the plains before the onset of winter. The present and subsequent chapters seek to unravel various aspects of small ruminant migration in the hills of Himalayas. The facts and detailed information on the following aspects would provide better recognition and understanding of the migratory farming system.

- Significance and need for migratory farming system
- Section of the society and community engaged in animal migratory practice
- The extent of animal migration including the numbers of different species
- Problems and constraints faced at the base camps and alpine pastures during ascending and descending journeys
- The mineral and nutritional status of migratory animals as well as pastures and grasses
- High altitude environmental adaptations of the migratory animals
- Parasitic infestations and vaccines for small ruminants
- Impacts of climate change and shepherds' perceptions to climate change on their profession

The present and subsequent chapters have attempted to unravel different aspects of small ruminant migration in the Himalayas.

References

Aggarwal RAK, Narula HK, Dixit SP, Arora R, Yadav DK, Ganguly I (2022) Goat and sheep genetic resources of India. Indian J Plant Genet Resour 35(3):237–241

Alam S, Kaur J, Rather MA, Shanaz S, Mir S, Shah R, Hamdani A (2021) Socio-economic and management features of Purgi goat and their contribution in rural livelihood security. Pharma Innov SP-10(8):1195–1197

Arjjumend H (2018) Review of pastoralists' resilience and adaptation to climate change: can technology help pastoralists mitigate the risks? IJBSM 9(1):118–128. https://www.academia.edu/36889497/Review_of_Pastoralists_Resilience_and_Adaptation_to_Climate_Change_Can_Technology_Help_Pastoralists_Mitigate_The_Risks

Bhat MS, Salam MA (2016) An assessment of nature and extent of crop diversification across agro-climatic zones of Jammu and Kashmir: spatial and temporal analysis. IOSR J Agric Vet Sci 9(11):33–40. www.iosrjournals.org

Bhatt PM, Padmakumar V (2012) Goat-based livelihood improvement in Uttarakhand-recommendations evolved in workshop sponsored by Tata. International Livestock Research Institute (ILRI), New Delhi. https://cgspace.cgiar.org/bitstream/handle/10568/16769/goat-VCD_uttarakhand.pdf

Chauhan SK (1990) Wool production and sheep management practices in the western Himalayas. Wool Wool India 27(4):17–22

Chauhan C (2008) Forest Act comes into force, green activists cry foul. Hindustan Times, January 3, 2008. https://www.hindustantimes.com/india/forest-act-comes-into-force-green-activists-cry-foul/story-olaef0elkijJhWOnonczNK.html

Chauhan I, Meena RK, Gautam S, Kharayat NS, Jana C, Sankar M (2021) Growth and reproduction performances of Chaugarkha goat of Kumaon region of Uttarakhand. IJBSM 12(3):205–210

Department of Animal Husbandry Himachal Pradesh (DAH HP), India. http://hpagrisnet.gov.in/hpagris/AnimalHusbandry/?Language=En

Department of Animal Husbandry Jammu and Kashmir (DAH JK), Srinagar, India. https://jkanimalhusbandry.net/

Department of Animal Husbandry Uttarakhand (DAH UK), India. https://ahd.uk.gov.in/

Dhara KC, Kesh SS, Roy S (2020) Goat breeds of India. IJCRT 8(10):3204–3214

Dong S, Liu S, Wen L (2016) Vulnerability and resilience of human-natural systems of pastoralism worldwide. In: Dong S, Kassam KA, Tourrand J, Boone R (eds) Building resilience of human-natural systems of pastoralism in the developing world. Springer, Cham. https://doi.org/10.1007/978-3-319-30732-9_2

ENVIS Bulletin (2010) Compendium of statistics for the states of Indian Himalayan Region, official web portal of GB Pant National Institute of Himalayan Environment and sustainable development, Kosi-Katamal, Almora, India. http://www.gbpihad.gov.in

FAO (2021) Seven reasons why pastoralism supports a better future. Food and Agriculture Organization of the United Nations. https://www.fao.org/fao-stories/article/en/c/1453839/

Forest Conservation Act (1980) Ministry of Environment, Forest and Climate Change, India. https://www.indiacode.nic.in/bitstream/123456789/10815/1/forest_(conservation)_act,_1980.pdf

Ganai TAS, Misra SS, Sheikh FD (2010) Gurez – a threatened sheep breed of Kashmir. Indian J Anim Sci 80(8):806–808

Gazette of India (2007) The Scheduled Tribes and other Traditional Forest Dwellers Act 2006, Ministry of Law and Justice, Government of India. https://www.indiacode.nic.in/bitstream/123456789/2070/1/200702.pdf

Gentle P, Thwaites R (2016) Transhumant pastoralism in the context of socioeconomic and climate change in the mountains of Nepal. Mt Res Dev 36(2):173–182

Government of India (GOI) (2012) 19th Livestock census: all India report. Department of Animal Husbandry, Dairying and Fisheries, Ministry of Agriculture, Government of India. Available: https://dahd.nic.in/sites/default/filess/Livestock%20%.pdf

Government of India (GOI) (2019) 20th Livestock census: all India report. Department of Animal Husbandry and Dairying, Ministry of Fisheries, Animal Husbandry & Dairying, Government of India. Available: https://dahd.nic.in/sites/default/filess/20thLivestockCensus2019AllIndiaReport.pdf

Gupta BB (2015) Development of small ruminants in Himachal Pradesh. South Asia, Pro-Poor Livestock Policy Programme. http://www.sapplpp.org/files-repository/information-hub/national-concluding-workshop-presentations/himachal-pradesh/at_download/file

Gupta RK, Gupta T, Kaushal PS, Pathania MS (1998) A study on the status of fodder balance in Himachal Pradesh. Oecol Mont 7:39–42

Gupta B, Sarvade S, Singh M (2017) Species composition, biomass production, and carbon storage potential of agroforestry systems in Himachal Pradesh. In: Gupta SK, Panwar P, Kaushal R (eds) Agroforestry for increased production and livelihood security. New India Publishing Agency, New Delhi, pp 242–270

Gurung H (1990) Mitigation of environmental risks in the highlands. Paper presented at IFAD Project implementation workshop for the Asia and Pacific Region, Chengdu, China, October 22, 1990. https://lib.icimod.org/record/4877

Hamadani A, Ganai NA, Rather MA, Shanaz S, Ayaz A, Sheikh M, Nazir S (2022) Livestock and poultry breeds of Jammu and Kashmir and Ladakh. Indian J Anim Sci 92(4):409–416

Hamid A (2021) Sustenance of Kashmir's Wool Industry depends on adapting and adjusting its practices to the changing times. https://www.jkpi.org/sustenance-of-kashmirs-wool-industry-depends-on-adapting-and-adjusting-its-practices-to-the-changing-times/

ICAR-IGFRI (2021) Fodder resources development plan for Uttarakhand. Indian Grassland and Fodder Research Institute, Jhansi. https://igfri.icar.gov.in/wp-content/uploads/2021/09/Uttarakhand-Fodder-Plan.pdf

IFAD (2012) Women and pastoralists. Available: https://www.ifad.org/en/web/knowledge/-/publication/women-and-pastoralism

Ingty T (2021) Pastoralism in the highest peaks: role of the traditional grazing systems in maintaining biodiversity and ecosystem function in the alpine Himalaya. PLoS One 16(1):e0245221. https://doi.org/10.1371/journal.pone.0245221

Jana C, Amol G, Sankar M (2018) Prospects of Chaugarkha as meat goat breed in Kumaon region of Uttarakhand, India. In: 8th Conference of Indian meat science association on technological innovations in muscle food processing for nutritional security, quality and safety. WBUAFS, Kolkata, pp 307–309

Katoch A, Sharma A, Negi M, Mandial RK, Wadhwa DR, Sharma D (2020) Haematobiochemical and mineral status in Chegu Pashmina goats of Himachal Pradesh. Indian J Vet Med 40(2):20–22

Khadda BS, Singh B, Singh DV, Singh SK, Singh CB, Singh JL, Palod J (2017) Growth performance of Pantja goats under field conditions in their home tract. Indian J Anim Res 53(2):264–269

Khan NN, Rather MA, Hamadani A, Bukhari S, Shah R (2021) Wool quality traits of sheep reared in Karnah tehsil of J&K. J Entomol Zool Stud 9(2):255–257

Kishore K, Köhler-Rollefson I (2020) Accounting for pastoralists in India. Ober-Ramstadt, Germany, League for Pastoral Peoples and Endogenous Livestock Development (LPP). Available: http://www.pastoralpeoples.org/wp-content/uploads/2020/09/Accounting4pastoralists-IN.pdf

Kour G, Singh A, Kumar P, Kumar D (2018) An overview of diversified animal genetic resources in the Indian state of Jammu and Kashmir. Int J Curr Microbiol App Sci 7(10):3113–3121

Kreutzmann H (2011) Pastoral practices on the move-recent transformations in mountain pastoralism on the Tibetan Plateau. In: Kreutzmann H, Yong Y, Richter J (eds) Regional workshop in Lhasa, P.R. China. Pastoralism and rangeland management on the Tibetan Plateau in the context of climate and global change. Deutsche Gesellschaft für Internationale Zusammenarbeit (GIZ) GmbH, Bonn, pp 200–224

Lyon F (1993) Semi nomadic pastoralists and environmental degradation in Himachal Pradesh, India. Report summary, Conservation and Development Action Research Institute (CADARI), Report No 1

Malik AA, Khan HM, Sofi AH, Mir MS, Farooq J, Sheikh FD, Mir AQ, Abdullah M (2021) Wool characteristics of Changthangi sheep. Small Rumin Res 199:106364. https://doi.org/10.1016/j.smallrumres.2021.106364

Mayberry D, Ash A, Prestwidge D, Herrero M (2018) Closing yield gaps in smallholder goat production systems in Ethiopia and India. Livest Sci 214:238–224

Mir NH, Suheel A, Verma DK (2016) Livestock rearing – sectorial status and fodder feed strategies in Kashmir Himalaya. Ann Biol 32(2):253–259

Misri B (2003) Migratory goats and sheep rearing in Himachal Pradesh, India. In: Suttie JM, Reynolds SG (eds) Transhumant grazing systems in temperate Asia, Plant production and protection series no. 31. Food and Agriculture Organization of the United Nations, Rome

Misri B, Dev I (2001) Systems study on sedentary Gaddis of Kangra valley. ENVIS Bull Himal Ecol 9(1):14–20

Molly K (1998) Sacred response to environment: the Gaddi and his mountain. In: Saraswati B (ed) The cultural dimension of ecology. IGNCA and D K Printworld Pvt Ltd, New Delhi

Namgay K, Millar J, Black R, Samdup T (2013) Transhumant agro-pastoralism in Bhutan: exploring contemporary practices and socio-cultural traditions. Pastoralism 3:13. https://doi.org/10.1186/2041-7136-3-13

National Commission on Farmers (NCF) (2006) A draft national policy for farmers, 4th report, New Delhi, India. https://www.agriwelfare.gov.in/sites/default/files/NCF5%20Vol.-2%20%281%29.pdf

Negi RS (1998) Symbiotic relationships between man, animal and nature: a study of the Gujar of Garhwal. In: Saraswati B (ed) Life-style and ecology. IGNCA and DK Printworld, New Delhi

Partap T (2011) Hill agriculture: challenges and opportunities. Indian J Agric Econ 66(1):33–52

Puri L (2007) Rare livestock breeds in Himalayan belt under threat. The Hindu. http://www.thehindu.com/todays-paper/rare-livestock-breeds-in-himalayan-belt-under-threat/article1934090.ec

Rashid A (2001) Taliban: the story of Afghan warlords. Pan Books, Pan Macmillon Ltd, London. https://books.google.com/books/about/Taliban.html?id=_GR5tXpppS0C

Rathera MA, Bukhari S, Shanaz S, Alam S, Mir SA, Shah R, Hamdani A (2022) Bakerwal goat: the robust goat breed of Jammu and Kashmir. Sci World 2(3):300–307

Rawat AS (2004) Livestock subsistence of Gujars – the forest dwellers of Tarai – history of nomads struggle for survival. In: Singh V, Gautam PL (eds) Proceedings of the international symposium "Livestock production systems for sustainable food security and livelihood in mountain areas". GB Pant University of Agriculture and Technology, Pantnagar, Uttarakhand

Sandhu JS, Gautam US, Ghosh PK, Dubey SK, Singh A, Kumar RV, Singh SV (2016) Agro-climatic region centered research and development planning (Central Plateau and Hill region). Report presented in the brainstorming session held at ICAR-Indian Grassland and Fodder Research Institute (IGFRI), Jhansi, pp 1–28

Sankhyan V, Dogra PK, Thakur YP (2016) Attributes of migratory goats and sheep farming and impact of some improved management strategies en-route migration in adopted flocks of western – Himalayan region of India. Indian J Anim Sci 86(9):1079–1084

Sankhyan V, Thakur YP, Katoch S, Dogra PK, Thakur R (2018) Morphological structuring using principal component analysis of Rampur-Bushair sheep under transhumance production in western Himalayan region, India. Indian J Anim Res 52(6):917–922

Sharma D (2021) Transformation among the Western Himalayan Gaddi tribe of Himachal Pradesh: question of identity and modernity. "Himachal watcher 25.6.2021". https://himachalwatcher.com/2021/06/25/transformation-among-the-western-himalayan-gaddi-tribe-of-himachal-pradesh-question-of-identity-and-modernity/

Shridhar T (2021) Pastoralists, original livestock keepers of the civilization pushed to margins. "Outlook" of Feb 19, 2023. https://www.outlookindia.com/website/story/pastoralists-original-livestock-keepers-of-civilisation-pushed-to-margins/392261

SWDBU (2005) Breeding policy for sheep, goat and rabbit (2005): Uttarakhand Gazette Notification 249/XV-1/2005. Dated 26th May 2005. https://www.uswdbdehradun.in/about/breeding-policies/BKJKUK…UK1

Taggar RK, Chakraborty D, Kumar D, Vohra V, Tantia MS, Sharma A (2018) Wool characteristics of poonchi sheep. Int J Fauna Biol Stud 5(2):133–135

Uddin ME, Kebreab E (2020) Review: Impact of food and climate change on pastoral industries. Front Sustain Food Syst 4:543403. https://doi.org/10.3389/fsufs.2020.543403

Yadav AK, Singh J, Yadav SK (2017) Characteristic features of registered indigenous sheep breeds of India: a review. Int J Pure App Biosci 5(2):332–353

2 Salient Features of Small Ruminant Migratory Farming System

Abstract

Millions of sheep and goats migrate to the hills of north-western Himalayan states and alpine pastures of other Himalayan countries every year as the summer approaches and the ambient temperature starts soaring high. It is quite a scene on the roads to witness vehicle drivers waiting by the road side to pave the way for waves of ascending animals heading for alpine pastures. Similar is the scenario when these animals return from the high-altitude areas with the onset of winter. This chapter presents salient features of animal migratory farming system including the tough journey, en route problems, the practice of manuring, managing flocks at the alpine grasslands, animal husbandry practices, etc. During ascending, the shepherds could be seen carrying ration or food stuff, medicines, heavy woolens, salt bags, and other essential materials. Some shepherds hire mules or yaks to carry the stuff. The animals thrive on tree leaves, seasonal crop residues, or dry hay at the base camp in the plains, whereas during climbing up, the animals just sustain on tree leaves, shrubs, and en route grazing. Depending on the location of the destination (pastures), the shepherds travel along with their flocks for about 25–45 days and cover a total distance of around 200 to 400 km. An interesting aspect of the migratory journey is what is called "the practice of manuring" wherein the en route farmers invite the shepherds for overnight stay of sheep and goat flocks in the fields to provide their droppings as manure and enrich the soil. Stay in the alpine pastures is a difficult task and the shepherds themselves have to conduct all animal husbandry activities including shearing, lambing, and disease management. A unique feature of tribal pastoralists is ethnoveterinary practice and use of different plant species for treatment of animal diseases. The ethnoveterinary knowledge concerns people's indigenous knowledge and practices and their experience for treating animal diseases that is

S. P. Singh, J. K. Malik, *Migratory Small Ruminant Farming System in the Himalayas-Peculiarities, Problems, Prospects, Climate Effects, Nutrition and Health*,
https://doi.org/10.1007/978-981-96-6691-1_2

carried from one generation to the other. The list of common plants of ethnoveterinary importance in the north-western Indian Himalayan states/union territories and ailments treated have been described.

Keywords

Unique migratory farming system · Shepherding · Manuring · Ethnoveterinary practice · Alpine pasture

Abbreviations

EVK	Ethnoveterinary knowledge
NATP	National Agricultural Technology Project
NTRS	North Temperate Research Station, Garsa, Distt Kullu, Himachal Pradesh

The Himalayan migratory small ruminant farming system is both dynamic and quiescent. In the months of April and May every year, the visitors and tourists heading for mountains in vehicles would have to stop sideways on the highways to pave the way for the ascending waves of flocks of sheep and goats traveling on foot for the alpine peaks. It is quite a scene to see the moving animals accompanied by shepherds with watch dogs and paraphernalia. And similar view is observed when the flocks return to the base camps in mid to late September. As the flocks leave the main roads, they have to negotiate their way through narrow winding hillock footpaths along the river fronts. To take hundreds of animals to distant alpine pastures crossing through on way hazards, braving through cold winds, feeding flocks on scare feed resources, and keeping them healthy for months, performing all the essential animal husbandry practices all by themselves and bringing the animals to the base camps fit and healthy is a story of toughness and hardship by the shepherds.

This chapter presents series of features covering the pattern and extent of migration, preparations, start of journey, animal husbandry practices, management of sick animals in the alpine meadows and actual case studies by shepherds/pastoralists in the north-western Himalayan states of India.

2.1 The Pattern and Case Study of Migratory Flocks

Sheep and goats in the hills are reared in sedentary as well as migratory farming systems. The average number of sheep and goats reared per household has been reported to be about 120 and 49, respectively (Pathania and Dev 2015). The nomads follow an annual migratory cycle especially in Jammu and Kashmir and Ladakh, Himachal Pradesh, and Uttarakhand. In addition, State government farms that have indigenous as well as exotic or crossbred sheep and goats also send their animals to alpine pastures for summer grazing. These farms use vehicles to help cover part of the distance during upward and downward journeys to the alpine meadows. Also the ration, salt bags, and other belongings are transported by the bus or jeep up to the road heads, and then the whole stuff is carried further on the mules' and yaks' back.

Most shepherds and Gujjars have two settlements: permanent house on the lower hills (base camps) and temporary shelter in the alpine grasslands. The base camps also called as Kandi Dhar or Bans (Misri 2003) are located in the plains or foothills (Shivaliks). In Himachal Pradesh, they are located near the towns of Hamirpur, Bilaspur, Nurpur, Una, Dehra, Renuka, Kangra, Jwalamukhi, and Mandi and some parts of Punjab (Pathania and Dev 2015). The forest department allows grazing facilities to the flocks in and around the base camps for which the department levies a nominal fee called as Tarni of Rs 0.20 /sheep and Rs 0.50/goat. The rates keep varying from time to time and depending on the forest policy.

2.1.1 Life at the Base Camps (Permanent Homes)

Most Gaddi nomads spend almost 5–6 months at their permanent homes in the plains/foothills. The major feed sources for sheep and goats include crop residues, dry hay, tree leaves, and scant grazing available in the village fields. Some Gaddis who have turned semi-migratory or sedentary spend most of their time in the plains doing their routine chorus. Misri and Dev (2001) reported the socio-economic pattern, income sources, cropping pattern, animal husbandry practices, feeding pattern, and herd structure of some of the Gaddis. The inhabitants and flocks of eight villages of Kangra valley of Himachal Pradesh were included in the study. The families reared cows, buffaloes, sheep, and goats. The average number of cows, buffaloes, sheep, and goats per family was recorded as 1.32, 0.80, 136, and 79, respectively. Pathania and Dev (2015) reported average number of sheep and goats reared by Gaddis per household as 120 and 49, respectively. Tree leaves constituted the major feed resources although seasonal crop residues (wheat straw) and green grass (freshly conserved hay) were also available. Most Gaddis discussed about the problem of fodder shortage and the financial difficulty in transporting the wheat straw from adjoining states like Punjab and Haryana. According to the authors, the average annual family income varied from Rs 28.4 to Rs 64.2 thousand in the eight villages studied.

2.1.2 Preparations for Upward Journey

It is customary among all the graziers to carry traditional medicines and common salt for the animals, besides the ration and other materials. Medicines are required for first aid and treatment of disease conditions en route and in the pastures. In some cases, the ration and other items are already transported up to the road head by bus or jeep or on animal back, i.e., prior to the arrival of flock there. On difficult routes, the ration and other stuff are transported on the back of mules, horses, bulls, or yaks.

Almost all the graziers have the knowledge that salt is required by sheep and goats in the alpine pastures because they had observed these animals chasing the attendants carrying salt bags at the alpine grasslands. The Veterinary Pharmacist of Government farm Tal, District Hamirpur, told us that the salt-seeking behavior was not observed in sheep flock at the base camp (Tal Farm) but only during stay at the alpine pastures. We also questioned some private shepherds about the requirement of salt by the migratory herd. One of them, Mr. Om Prakash, replied that salt was essentially required but he did not know what for? According to him, sheep would not graze the grasses if salt was not offered. Others said that if salt was not offered, the oral papillae of sheep became large sized and painful making grazing difficult. By giving salt, the papillae return to their normal size. Graziers told us that some sheep demanded more quantity of salt by making particular sounds and chased the attendants for this purpose. A few shepherds narrated the story how salt was most-sought commodity by Chinese graziers on India-China border. The Chinese demanded salt for their flocks and, in return, offered pulses and grains to the Indian shepherds (Fig. 2.1).

Most graziers offer salt to the animals once or twice a week, and it is placed on the rocks as a free lick. We found that this practice approximately provided only 0.2 to 0.5 g salt/sheep/day and was inadequate to alleviate the salt behavior. Further information relating to salt requirement and other associated issues is included in Chap. 8 under Sect. 8.3.2.

Fig. 2.1 Gaddi sheep and goats on way to high altitude pastures in Kinnaur District, Himachal Pradesh. (Photograph by courtesy of social activist, Mr. BB Negi, with permission)

2.1.3 Location of Pastures

In the state of Himachal Pradesh, there are three or four main locations where pastures are commonly used for grazing purpose. Most of the pastures are located along the major rivers, namely, (i) Kinnaur pastures along the river Sutlej and its tributaries; (ii) Chamba and Bharmour located along the river Ravi; and (iii) Lahaul and Spiti valley pastures along the rivers Chandra and Bhaga and some pasture belts in Kullu area along Beas river. Chap. 3 presents details of the migratory patterns and migratory routes in Himalayan states of India and the neighboring countries.

2.1.4 Commencement of Migration to the Alpine Pastures and Management of the Flocks

Graziers start their upward journey from the base camps around March/April every year. If there are no sideways, the flocks travel on the main roads. The average period taken by shepherds to reach pastures is around 25–45 days, whereas the Gujjars who generally rear large ruminants, travel lesser distance, and therefore, complete the journey within 10–20 days. Most migration takes place in the night in order to avoid road traffic and other hassles of day time. The shepherds and flocks follow a fixed cycle of migration and stick to the same migratory routes and passes during ascending and descending journeys. Table 2.1 provides the details of some parameters pertaining to migration of animals.

The duration of stay in alpine meadows varies, but Pathania and Dev (2015) have reported that maximum number of graziers stayed in meadows for 4 months from June to September. (Figure 2.2).

The upward and downward migration is tough, strenuous, and hazardous and the life in the alpine pastures is very hard and adventurous both for animals and shepherds. Sareen (2022) has reported that in many cases, the families of the shepherds travel up to 600 km on foot and journey could take 2–4 weeks. The alpine shelters erected during the preceding visit are generally get destroyed during winter by

Table 2.1 Ascending and descending journeys of shepherds and Gujjars. (Kant et al. 2004; Wani 2015)

Journey	Phase	Time/duration	
		Shepherds	Gujjars
Ascending	Start of upward migration from the base camps in the plains	April/May	April/May
	Time taken to reach alpine pastures	25–60 days	10–20 days
	Duration of stay in the alpine pastures	4–5 months	4.5–5.5 months
Descending	Start of downward migration from the alpine pastures	September/ October	Mid October
	Time taken to reach back to the base camps	25–45 days	10–20 days
	Duration of stay at the base camps in the plains	4–5 months	5–5.5 months
	Length of the migration route	250–300 km	60–150 km

Fig. 2.2 Gaddi sheep grazing in the meadows in Lahaul and Spiti, Himachal Pradesh (3650 m amsl). (By courtesy of Director, ICAR-CSWRI, NTR, Garsa, Kullu)

strong winds and rains and have to be repaired and re-erected year by year. The problems faced by the graziers during upward and downward migration process and the conflict situation in alpine pastures have been described in the succeeding Chap. 4.

2.2 The Practice of Shepherding

Normally the flock owners themselves manage small and medium flocks during ascending or descending journeys, but over the years, a contract of shepherding system (hiring other shepherds) has emerged. This has been developed especially in holders of larger flocks. In Himachal Pradesh, hired shepherds are called Puhals (Misri 2003). The system is also quite prevalent and organized in some areas of Nepal (Joshi et al. 2004). In this system, most flock management is handled by the hired shepherds and not by the owners. The owner provides inputs in the form of grains, pulse, salt, etc., and, in some cases, a minor salary and also gift sheep and goat as an offer for shepherd's personal use. The hired shepherds are supposed to be expert in managing all operations such as lambing, shearing, sale of animal and its pricing, and flock movement including healthcare. The shepherds have developed all kinds of local methods for treating animal ailments (Tables 2.6 and 2.7).

2.3 The Practice of Manuring

During the course of migration, the flocks stay overnights in the fields or farm lands and provide in situ fecal manuring. A unique feature of small ruminant rearing in the hills is their role in fertilizing the soil. Farmers believe that goat and sheep fecal material is superior as a manure compared to the cattle and buffaloes. Similarly, urine of sheep and goat has been found to be very rich in nitrogen and phosphorus. This has been confirmed in the subsequent laboratory tests, which showed that sheep and goat manure material contained 0.61 and 0.83% higher nitrogen, respectively, compared to feces of cattle (0.25%) and buffaloes (0.33%) (Ghmire 1987). In an interesting study on migratory flocks of Himachal Pradesh, Pathania and Dev (2015) also reported that the droppings of sheep and goats make very good manure and during the migration, most of the farmers in plains encourage the graziers for a night halt of the flocks on their fields. In a supporting piece of work, Dhital et al. (1990) have observed that in situ manuring system (burring the crop in the field as such) was found to increase the crop production by 28% compared to the application of compost. For these reasons, this practice has become an attraction and many farmers come forward to allow their farm land for the stay of flocks to get benefit of field manuring. It was observed that in some places, the owners of fields offer food and ration or pay a token fee to the shepherds in return for manuring.

2.4 Monitoring of Flocks Migrating to the Alpine Pastures

There are some published reports providing information on exactly how many sheep and goats cross various passes in different hill states to reach alpine pastures and what problems, disorders, and difficulties the animals come across en route. In Himachal Pradesh, some estimates are available with the forest check posts, which check and issue the grazing permits to the migratory shepherds. During the course of investigation in the ICAR-funded research project at Palampur Agriculture University in Himachal Pradesh, we had the opportunity to interview and scrutinize records of some of these check posts. The case illustrated is an example of migratory pattern and practice. The figures of migrating animals and other parameters would, obviously, keep on changing year by year. The first two case reports presented below are related to the flocks which actually passed through Palampur University town, and in the latter two cases, the information was collected from check posts located in Kinnaur and Lahaul and Spiti districts of Himachal Pradesh.

Case 1 Local shepherds on way to alpine pastures through Palampur town

As stated, the Project Scientists had the opportunity to intercept some local shepherds passing through Dehra and Palampur area, interviewed them, and examined their animals. Summary of these cases is presented in Table 2.2.

In this study, blood and fecal samples were collected from sheep suffering from wool shedding and diarrhea to determine the causes of the ailments.

Further, the shepherds complained that these animals could not be treated as there were no hospitals or dispensaries located en route to the hills.

As regards the composition of migratory flocks, it varied from case to case. In a recent study on migratory sheep and goats in the north-western states, Dogra et al. (2019) have reported that majority of the flocks were medium sized (200–350) and

Table 2.2 Cases of local shepherds on way to the high altitude pastures

S. No	Name and address of shepherd and place of origin	Number of animals in flock		Condition of animals	Alpine destination	Via/mid stations
		Sheep	Goat			
1	Chuda Singh, village & PO Bundla, Dist Kangra (Jwalaji)	50	20 (Gaddi)	Pica & alopecia in sheep	Manali	Utrala
2	Kirat Ram, village & PO Jwalaji, Dist Kangra (Mubarakpur)	60	20 (Gaddi)	Half of the total sheep suffering from diarrhea	Holi Chamba	Utrala
3	Lalchand, village and PO Punnera near Bhawarana Dist Kangra (Gar Shankar near Hoshiarpur (Punjab)	120	180 (Gaddi)	Wool shedding and alopacia in sheep	Bharmour Chamba	Palampur/ Bandla

only 9.2% flocks were larger sized (>350 animals). In another study in Kangra district of Himachal Pradesh, Singh et al. (2006) found that about 46% flocks were small and 54% were larger in size.

Case 2 Flocks migrating from Hamirpur, Nangal, and Una areas of Himachal Pradesh.

A large number of sheep and goats migrate to the alpine pastures from Hamirpur, Nangal, and Una districts. Table 2.3 provides information related to the base camp, routes, and alpine destinations of these flocks. A rough estimate showed that annually approximately 51 thousand sheep and goats migrate from this area to the high mountain pastures.

Case 3 Sheep and goats crossing through forest check post "Wangtu" in Kinnaur district, Himachal Pradesh

The author and his associates from the College of Veterinary and Animal Sciences of Himachal Pradesh Agricultural University, Palampur, surveyed the sheep and goat flocks passing through forest check post at a place named "Wangtu" in Kinnaur district in the north-eastern part of Himachal Pradesh. Wangtu check post is located on the river Satluj. Most of the flocks were heading for pastures located in Kinnaur. The data recorded from the check post during the month of May–June 1996 are presented in Table 2.4. As stated above, the number of animals keeps changing, but the information is important and reveals the pattern and routes of migration.

The total number of sheep and goats migrated from base camp towns through Wangtu forest check post to alpine pastures were 71,567. Most of these animals were owned by the private shepherds who hailed from different parts of Himachal Pradesh. Scrutiny of their addresses indicated that more than 80% of the shepherds who crossed over through Wangtu check post belonged to various villages of

Table 2.3 Migratory routes and alpine destinations of flocks of some districts of Himachal Pradesh. (Chauhan 1990, Report, Forest Department, Government of Himachal Pradesh)

S.No.	Base camp town	Via/mid station	Number of animals in the flock		Alpine destination
			Sheep	Goat	
1	Hamirpur, Nangal, Una, Jwalamukhi	Kangra, Palampur, Baijnath,	14,006	11,391	Bharmour, Holi in Chamba district
2	Jwalamukhi	Baijnath	14,765	10,846	Kugti, Mani Mahesh, Lahual and Spiti

Table 2.4 Sheep and goats crossing through Wangtu forest check post of north-eastern Himachal Pradesh (Forest Department, Himachal Pradesh)

Number of shepherd families	Flock size			Number of sheep	Number of Goats	Base camp town	Alpine destination
	Small[a]	Medium[b]	Large[c]				
234	37	60	137	34,059	37,508	Sirmour, Ponta Sahib, Dehradun, Nahan, Renuka, Bilas-pur, Kota, Sunder Nagar, Karsog	Bhabha Kanda, Sucha Kanda, Lipu Kanda (Kinnaur), Ribba Kanda, Raksham, (Pangi, Pin valley, Spiti)

[a]Less than 300 animals; [b]300–1000 animals; [c]More than 1000 animals

Kinnaur district and most of them were heading to the high altitude pastures located in Kinnaur district (Bhabha Kanda, Sucha Kanda, Lipu Kanda, Ribba Kanda, Raksham). Some shepherds intended to migrate their animals to pastures as far as Pangi in Chamba and Pin valley in Lahaul and Spiti. Each shepherd had specific grazing permit and had paid Treni (fee) at the prevailing rate of Rs 0.30/animal.

Besides the flocks of private shepherds, approximately 2000 sheep and goats belonging to the State Government Sheep Farm in Tal, District Hamirpur, and Jeory, District Kinnaur, also cross the Wangtu check post annually heading for Bhabha Kanda (Kara and Kanta) and other pastures in Kinnaur district.

In another report available with the Forest Department of Lahaul and Spiti and Kinnaur districts, it was shown that nearly 2,34,000 flocks crossed over through different mountain passes.

Case 4 Flocks crossing over to Thanpattan and other pastures in Lahaul and Spiti valley of Himachal Pradesh

Thanpattan pasture with an area of nearly 4000 hectares is located in Lahaul and Spiti at an altitude of 3450 to 4365 m amsl. The Gaddi shepherds have traditional grazing rights and thousands of sheep and goats migrate to graze the pasture every summer. Most shepherds belonged to Districts Kangra and Chamba who reached the high-altitude pasture Thanpathan after crossing Kalicho pass at 4803 m amsl.

Shepherds arriving from lower ranges commonly follow two main routes to reach Thanpattan, viz., Kullu valley and Chamba district. Some flocks also cross-over to Lahaul from Kinnaur area. Chadha (1989) reported that a number of flocks crossover to Lahaul and Spiti through Pir Panjal mountain range. Lyon (1993) conducted a study on conservation and environmental degradation by pastoralists in Himachal Pradesh and reported that more than two lakh (0.2 million) sheep and goats crossover to Thanpattan through Rohtang, Hampta, Kugti, Kalicho, and Chobia passes.

2.5 Animal Husbandry Practices at High-Altitude Pastures

2.5.1 Animal Husbandry Practices at Pasture

The animal husbandry activities included shearing, lambing, traditional methods of disease management, and disposal of flocks, etc. Being tough terrain and harsh climate, the on-site scientific studies on animal husbandry activities at the high alpine pastures are very few. Millions of sheep and goats and other animals move to the alpine pastures every summer and stay there for nearly 4–5 months. The distribution of flocks in different alpine locations, their flock size, breeding practices and lambing pattern, shearing wool yield, and disposal of animals have been studied in the selected pastures (Lyon 1993; Dev et al. 2003). Table 2.5 provides the available details of the flock size and other attributes at these pastures.

The study by Lyon (1993) included about 700 flocks consisting of more than 23 thousand animals. These animals grazed in the upper and lower alpine pastures of Chandra and Bhaga rivers and the Mayar Nallah (rivulet). Dev et al. (2003) conducted the study in Thanpattan pastures (altitude 3450–4365 m amsl) of Lahaul and Spiti which are the largest pastures in Himachal Pradesh having an area of about 4300 hectares, but the author's study area was confined to 16 flocks grazing on the left bank of Mayar Nallah.

Table 2.5 Some attributes of the animal husbandry activities at the alpine pastures of Lahaul and Spiti (Himachal Pradesh)

Attribute	Lyon (1993)	Dev et al. (2003)
Number of flocks grazing	700	16
Flock size		
Average	338	237
Largest	409	413
Number of sheep in the flock	147–232	88–220
Number of goats in the flock	102–208	92–193
Lambing percentage (average)	73%	28.3%
Kidding percentage (average)	–	1.5%

In another study, Sankhyan and Thakur (2019) reported that in general migratory herds are classified into three categories. Small flocks consist of 100–200 animals, medium flocks comprise of 200–350 animals, while the large flocks consist of >350 animals. Survey conducted revealed that majority of the shepherds (50%) keep medium flock size, followed by owners (40.2%) of small flocks, whereas a very few shepherds (9.8%) were raising larger size flocks.

Shearing

In most studies cited above, the shearing operation was undertaken two to three times a year (January/February, April/May, and September/October).

However, at the State Government Farm, Tal, District Hamirpur, Himachal Pradesh, shearing was reported to be undertaken only once a year in the month of October soon after flocks return from alpine pastures. The wool's quality as well as quantity was found to be best and highest in the shearing of September and October. Shepherd, Lalchand, interviewed in Dehra area of Kangra district stated that shearing two times a year was a better proposition. He was also of the opinion that the wool of crossbred sheep was too fine to work. Generally, 20–25 sheep are shorn in 1 day and the average quantity of wool per sheep per annum varied from 1.5 to 3 kg. As the marketing facilities were poor and disorganized, most wool was sold to the middlemen who come to the pastures at the time of shearing. Some wool was always saved for use by the family and most households had an indigenous handloom.

Lambing

Practically there are two breeding seasons annually, viz., September and April–May. The ratio of ram to ewes in Bharmour study was reported to be 1:18. (Chauhan 1990). In case of September mating, lambing takes place in February or March when the flocks are in safe quarters in the plains. The lambs are covered with blankets and kept in protected place which is commonly known as "Dalara" in Bharmour areas. When mating is in April or May possibly during ascending journey, lambing takes place in September or October either at the alpine pastures or during descending journey. The lambs born during descending journey were found to be comparatively weak (Lyon 1993). In Lahaul study, lambing rate was found to be around 64–91% (Lyon 1993). At Government Sheep Farm, Tal, mating is generally done in September/October in the alpine pastures so that the lambing takes place in January every year at the time when animals are safely back at the farm.

2.5.2 A Unique Attribute of Tribal Pastoralists: Ethnoveterinary Practice and Use of Plants for Animal Treatment

Ethnoveterinary practice deals with the use of plants to treat livestock diseases. In fact the word "ethno" signifies race or culture. It is well known that throughout the history, people from different cultures around the world have made use of medicinal plants for treatment of diseases due to their healing properties (Miranda 2021). It is believed that ethanoveterinary knowledge (EVK) concerns people's experience and memory which is stored in the system of the race and carried on from one generation to the other. The pastoralist tribal leaders have a long history of using traditional plant-based medicines. They possess and are endowed with a special attribute to use various parts of the plants as medicine for treating different disease conditions of the migratory animals. Some of these methods are raw and untested, but they still persist in the shepherds community (Verma 1988; Rana et al. 2019).

In spite of questions, the advocates of ethnoveterinary medicine consider that traditional practice of veterinary medicine is legitimate and seeks to validate their results (Köhler-Rollefson and Bräunig 1998, Wiki). Many systems of disease treatment not based on western principles do exist in the third world countries and are doing a great service in the management of diseases. These include technique of acupuncture, use of herbal medicine in China, Tibetan veterinary medicine system, and Ayurveda and naturopathy in India. In fact, India has a long history of the use of plant-based medicines for animal treatment (Sikarwar and Tiwari 2020). The Vedic literature is a store house of indigenous medicines for treatment. The tribal people are well aware of the principles and skills for drug administration. These traditions have written records that go back thousands of years past. A large number of nomadic people use local herbal medicines for treatment of animal diseases and the role of ethnoveterinary medicine in livestock development seems beyond dispute (Tiwari and Pande 2010). These people seldom have access to the modern veterinary treatment methods. Recognizing the importance of traditional methods of animal treatment, FAO in the past had published a number of bulletins on traditional system of Medicine for farmers for Asian countries including India (Beigh et al. 2004).

Although very few ethnoveterinary medicines have been tested clinically in the livestock species using standard laboratory methods and experimental animals, a good attempt has been made to document the plant species having therapeutic uses in the migratory animals. In an important study on ethnobotanical knowledge among the semi-pastoral Gujjar tribe in Chamba district of Himachal Pradesh, Rana et al. (2019) reported 83 plant species to have ethnobotanical use. Leaves, fruits, and roots were the most commonly used parts in various medicinal preparations. The most important plants found in the study were *Pteridium aquilinum, Juglans regia, Corylus jacquemontii, Urtica dioica, Diplazium maximum,* and *Angelica glauca.*

The tribal people of the north-western Indian Himalayan states of India wholly depend on traditional methods of treatment. Tables 2.6 and 2.7 present an overview of commonly used plant species of ethnoveterinary importance being used for treatment of different animal ailments in Jammu and Kashmir, Himachal Pradesh, and Uttarakhand.

Besides India, the tribal people of neighboring countries such as Pakistan have also been using ethnoveterinary plants for treatment of various animal ailments. Rafique Khan et al. (2021), Tariq et al. (2014), and Abbasi et al. (2013) have reported 39, 41, and 89 species of plants are being used extensively for treatment of animal diseases in Neelam valley, Sulaiman range, and Khyber Pakhtunkhwa areas, respectively.

Table 2.6 Plant species of ethnoveterinary (EV) importance used by pastoralists in the north-western Himalayan states of India

State/Union Territory	Place of study	Number of plant species of EV use	Number of important families	Reference
Jammu and Kashmir	Several places in the valley	32	19	Ahmad et al. (2017)
	Poonch district	31	24	Dutta et al. (2021)
	Kulgam district	35	21	Sofi et al. (2019)
	Kupwara district	32	23	Mir et al. (2023)
Ladakh	Cold desert	25	14	Batool et al. (2022)
Himachal Pradesh	Chitkul range, Kinnaur district	33	–	Radha and Puri (2020)
	Simla district	44	–	Radha and Puri (2019)
	Kamru Sangla, Kinnaur district	76	–	Radha et al. (2019)
	Dharampur, Mandi district	88	42	Kumar and Duggal (2019)
Uttarakhand,	Alaknanda area, Chamoli district	73	45	Phondani et al. (2010)

Table 2.7 Some ethnoveterinary plants used by pastoralists to treat various ailments of migratory animals

State/Union territory	Ailment/ activity	Botanical name of plant	Reference
Jammu and Kashmir and Ladakh	Foot and mouth disease (FMD)	*Alnus nitida, Zea mays L, Prunus prsica, salix alba, Cicer microphyllum*	Ahmad et al. (2017), Dutta et al. (2021), Sofi et al. (2019), Mir et al. (2023), and Tariq et al. (2014)
	Diarrhea	*Cedrella toona, Datura stramonium, Bergenia ligulata wall*	Ahmad et al. (2017), Sofi et al. (2019), and Mir et al. (2023)
	Wound healing	*Chenopodium album, Berberis lyceum Royle, Curcuma longa, Galium aparine L, Artemisia absinthium*	Ahmad et al. (2017), Dutta et al. (2021), Sofi et al. (2019), Mir et al. (2023) and Tariq et al. (2014)
	Wormicide	*Cannabis sativa L, Artemisia absinthium, Capsicum annuum L*	Dutta et al. (2021), Sofi et al. (2019), Mir et al. (2023), and Tariq et al. (2014)

(continued)

Table 2.7 (continued)

State/Union territory	Ailment/ activity	Botanical name of plant	Reference
Himachal Pradesh	Skin infection	*Acer pictum, Amaranthus viridis L, Cedrus deodara*	Radha and Puri 2020, Radha and Puri (2019), and Thakur et al. (2020)
	Wound healing	*Aconitum heterophyllum, Urtica dioica L, Butea monosperma seeds*	Radha and Puri (2020), Radha and Puri (2019), and Thakur et al. (2020)
	Indigestion	*Cynodone dactylone, Celtis tetrandra Roxb, Trachyspermum ammi*	Radha and Puri (2020), Radha and Puri (2019), and Thakur et al. (2020)
	FMD	*Asperagus filcinus, Acacia catechu*	Radha and Puri (2020)
Uttarakhand	Skin infection	*Capsicum annuum L, Glycine max*	Tiwari and Pande (2010),
	Diarrhea	*Cariandrum sativum L*	Tiwari and Pande (2010)
	Constipation	*Brassica compestris L, Bombax ceiba, Ocimum tenuiflorum L*	Tiwari and Pande (2010), and Phondani et al. (2010)
	FMD	*Acacia catechu, Mangifera indica, sAllium cepa*	Phondani et al. (2010)

2.5.3 Disposal of Animals in Flock and Financial Status of Shepherds

The graziers sell about 12–15% of animals each year for meat and fattening purpose when they are 6 –12 months of age. Most male lambs/kids are sold to butchers or land owners. The ewes are retained up to fifth to seventh lambing. If some animals are incapacitated due to disease or accident during the journey, the flock owner contacts the local people for auction of the same. Shepherds Chuda Singh, Kirat Ram, and Lalchand interviewed by us in Dehra area also stated a loss of 10–15 sheep due to injury and attack by wild animals every season (Singh et al. 1997, Personal communication). Chauhan (1990) has reported that majority of the animals (88–89%) are sold to butchers and around 5.5–7% are consumed at home. Some animals are also lost during accident and attack by wild animals.

How much the Gaddis earn out of the migratory animal husbandry practices and what has been their economic status over the years? These and other questions have remained unanswered in the absence of suitable scientific studies. Recently, a systematic 3-year study under the National Agricultural Technology Project (NATP 2004) of the ICAR was undertaken to investigate the status of agriculture and animal husbandry in hill and mountain agro-ecosystem covering Jammu and Kashmir, Himachal Pradesh, and Uttarakhand. The salient findings of the study are presented in Table 2.8.

Table 2.8 Estimated income (Rupees) of shepherds from migratory sheep and goat rearing in different states/union territory (NATP hill & mountain Ann Rept)

Attribute	Area (State/Union territory)		
	Kangra and Kinnaur districts (Himachal Pradesh)	Chamoli district (Uttrakhand)	Poonch and Rajouri districts (Jammu and Kashmir)
Pastoralist	Gaddis	Bhotia	Bakarwal
Number of families	30	20	30
Family size of shepherds	6.5	7.8	8.0
Sheep flock size	154	128	52
Goat flock size	133.9	72.2	10
Gross household income (Rs/family/year)	153,293	141,821	40,610
Average net income (Rs/family/Year)	53,000	57,000	35,000

The study revealed that annual gross household income per family of shepherds in Himachal Pradesh was highest (Rs 153,293) of which nearly 44.1% was contributed by rearing of sheep and goats alone. The lowest returns in Jammu and Kashmir were attributable to the small flock size.

References

Abbasi AM, Khan SM, Ahmad M, Khan MA, Quave CL, Pieroni A (2013) Botanical ethnoveterinary therapies in three districts of the lesser Himalayas of Pakistan. J Ethnobiol Ethnomed 9:84. https://doi.org/10.1186/1746-4269-9-84

Ahmad S, Radotra S, Singh JP, Verma DK, Sultan SM (2017) Ethnoveterinary uses of some important plants by pastoralists in Kashmir Himalaya. SKUAST J Res 19(1):121–128

Batool Z, Singh K, Gairola S (2022) Documenting potential ethnoveterinary knowledge from the indigenous communities of Cold Desert of Ladakh: a trans-Himalayan region of India. Ethnobot Res Appl 24:43. https://ethnobotanyjournal.org/index.php/era/article/view/4235

Beigh SY, Nawchoo IA, Iqbal M (2004) Traditional veterinary medicine among the tribes of Kashmir Himalaya. Int J Geogr Inf Syst 10(4):121–127

Chadha SK (1989) Problems of animal grazing in Himalayas. A case study of Himachal Pradesh. In: Semi nomadic pastoralists and environmental degradation in Himachal Pradesh, India. Consevation and development action research institute (CADARI), Report No 1. https://eurekamag.com/research/002/466/002466633.php

Chauhan SK (1990) Wool production and sheep management practices in the western Himalayas. Wool Woollens India 27(4):17–22

Dev I, Singh V, Misri B (2003) Socio-economic profile of migratory graziers and participatory appraisal of forage production and utilization of an alpine pasture in north-west Himalaya. ENVIS Bull Himal Ecol 11(2):52–62

Dhital BK, Sthapit BR, Joshi KD, Pradhanang PM, Subedi KD, Vaidya A, Gurung J, Kadayat KB (1990) A report on maize research at LAC and its off-station research sites, LAC seminar paper 13. Lumle Agricultural Centre, Kaski

Dogra PK, Thakur A, Sankhyan V (2019) Migratory sheep and goat production in North-Western Himalayas: an overview. In: Kumar N, Shekher P, Sinha RRK, Kumar P, Singh PK, Kumar R (eds) 19th ISSGPU national seminar on current scenario and future strategies for augmenting productivity of small ruminants held at BASU, Patna, 14–16 February, pp 258–266

Dutta A, Singh K, Singh B, Sharma YP (2021) Exploring traditional veterinary practices from Gujjar and Bakarwal tribes of district Poonch, Jammu & Kashmir: a boon for animals from our ancestors. Res Sq:1–30. https://doi.org/10.21203/rs.3.rs-255153/v1

Ghmire SC (1987) The role of small ruminants. In: Abington JB (ed) Sustainable livestock production in the mountain agro-ecosystem of Nepal. FAO animal production and healthpaper 105

Joshi HD, Rana RB, Joshi BR, Subedi, A (2004) The role of migratory sheep and goats in the livelihood of communities in the mountains- a case study from Nepal. In: V Singh, PL Gautam (eds) Livestock production for sustainable food security and livelihoods in mountain areas. International symposium, G.B.Pant University of Agriculture and Technology. Pantnagar

Kant N, Chadha KK, Manoj G (2004) Augmenting productivity of migratory herds/Flocks under hill and agro-ecosystem. In: V Singh and PL Gautam (eds) Livestock Production for sustainable Food security and Livelihoods in mountain areas. International Symposium, G.B.Pant University of Agriculture and Technology Pantnagar, p 165–176

Köhler-Rollefson I, Bräunig J (1998) Anthropological veterinary medicine: the need for indigenizing the curriculum. Paper presented at the 9th AITVM conference held at Harare, 14–18 September, 1998. Available from: http://www.vetwork.org.uk/ilse2.htm

Kumar G, Duggal S (2019) Ethnobotanical wisdom among the *Kiratas* and *Hindu-Gujjar* tribes in Dharampur region of Mandi District, Himachal Pradesh, (India). BFAIJ 11(1):156–171

Lyon F (1993) Semi nomadic pastoralists and environmental degradation in Himachal Pradesh, India, Conservation and Development Action Research Institute (CADARI), Report No 1

Mir AY, Riyaz M, Ignacimuthu S (2023) Assessing ethno-veterinary practices in Kashmir Himalayas: traditional knowledge and its role in animal healthcare. Nova Geod 3(2):131. https://doi.org/10.55779/ng32131

Miranda JJM (2021) Medicinal plants and their traditional uses in different locations. In: Bhat RA, Hakeem KR, Dervash MA (eds) Phytomedicine: A treasure of pharmacologically active products from plants. Academic. pp. 207–223

Misri B (2003) Migratory goats and sheep rearing in Himachal Pradesh, India. In: Suttie JM, Reynolds SG (eds) Transhumant grazing systems in temperate Asia, Plant production and protection series no. 31. Food and Agriculture Organization of the United Nations, Rome

Misri B, Dev I (2001) System study on sedentary Gaddis of Kangra valley. ENVIS Bull: Himal Ecol 9(1):14–20

NATP (2004) National Agricultural technology project hill and mountain agro-ecosystem. Final report (2004–2005) Indian Council of Agriculture research (ICAR), Central soil and water Research and Training Institute, Dehradun

Pathania MS, Dev I (2015) Rearing practices of migratory sheep and goats in Himachal Pradesh: a case study of *Gaddi* tribe. Indian J Agrofor 17(1):57–61

Phondani PC, Maikhuri RK, Kala CP (2010) Ethnoveterinary uses of medicinal plants among traditional herbal healers in Alaknanda catchment of Uttarakhand, India. Afr J Tradit Complement Altern Med 7(3):195–206

Radha, Puri S (2019) Study of wild medicinal plants used by tribal migratory shepherds in hills of Shimla district, Himachal Pradesh. Plant Arch 19(1):785–790

Radha, Puri S (2020) Ethnoveterinary medicinal uses of some plant species by the migratory shepherds of the western Himalaya. Plant Arch 20(SI-AIAAS-2020):29–31

Radha, Puri S, Kumar S (2019) An ethnobotanical study of wild medicinal plants used by migratory shepherds—a tribal community of western Himalayas. Asian J Pharm Clin Res 12(4):137–144

Rafique Khan SM, Akhter T, Hussain M (2021) Ethno-veterinary practice for the treatment of animal diseases in Neelum Valley, Kashmir Himalaya, Pakistan. PLoS One 16(4):e0250114. https://doi.org/10.1371/journal.pone.0250114

Rana D, Bhatt A, Lal B (2019) Ethnobotanical knowledge among the semi-pastoral Gujjar tribe in the high altitude (Adhwari's) of Churah subdivision, district Chamba, Western Himalaya. J Ethnobiol Ethnomed 15(1):10. https://doi.org/10.1186/s13002-019-0286-3

Sankhyan V, Thakur Y (2019) Migratory sheep husbandry in Himalayan state of Himachal Pradesh, India: status and headway options for improvement and sustainability. Int J Livest Res 9(8):71–77

Sareen P (2022) Impact of climate change on nomads in Jammu and Kashmir with a view of public health facilities. Policies and practices, Issue No. 132, Mahanirban Calcutta Research Group, Kolkata. http://www.mcrg.ac.in/PP132.pdf

Sikarwar RLS, Tiwari AP (2020) A review of plants used in ethnoveterinary medicine in Central India. Indian J Tradit Knowl 19(3):617–634

Singh C, Ram K, Lalchand (1997) Personal communication by shepherds

Singh DR, Kaul S, Sivaramane N (2006) Migratory sheep and goat production system: the mainstay of tribal hill economy in Himachal Pradesh. Agric Econ Res Rev 19(2):387–398

Sofi SA, Hakeem R, Manzoor MA, Wath MR, Sofi KA (2019) Ethno- veterinary practices performed for animal care in Kulgam district, (Jammu and Kashmir). Int J Res Anal Rev 6(1):i657–i663

Tariq A, Mussarat S, Adnan M, AbdElsalam NM, Ullah R, Khan AL (2014) Ethnoveterinary study of medicinal plants in a tribal society of Sulaiman range. Sci World J 2014:127526. https://doi.org/10.1155/2014/127526

Thakur A, Sharma M, Thakur R, Sharma A, Suman M, Verma N, Vanita B (2020) Ethnoveterinary practices among Gaddi pastoralists of North-Western Himalayan region, Himachal Pradesh. Indian J Anim Prod Mgmt 36(3–4):97–100

Tiwari L, Pande PC (2010) Ethoveterinary medicines in Indian perspective: reference to Uttarakhand. Indian J Tradit Knowl 9(3):611–617

Verma LR (1988) Livestock and fodder management. In: Negi A, Sharma PN (eds) Indigenous technology knowledge for watershed management in upper north-West Himalayas of India, participatory watershed management training in Asia, (PWMTA) program GCP/RAS/161/NET. FAO (UN), Kathmandu. https://www.fao.org/4/x5672e/x5672e00.htm#Content

Wani SJ (2015) Small ruminant rearing among Bakerwal pastoralists of Jammu and Kashmir- an exploratory study. (PhD thesis). National Dairy Research Institute, Karnal

Migratory Patterns and Migratory Routes of Pastoral Small Ruminants in the Himalayas

3

Abstract

Pastoralism has served as a key livelihood strategy of people living in the high mountains and plays a vital role in caring for animals, alleviating poverty, providing food security, and maintaining fertility of soil by providing manure from migratory livestock. The major pastoral communities of Himalayan countries, namely, India, Nepal, Bhutan, Pakistan, and China, rear small ruminants and other domestic animals. Some of the major communities of the Indian Himalayan regions practicing transhumance are Bakarwal and Gujjar in Jammu and Kashmir, Changpa in Ladakh, Gaddi, Kinnaura, and Kanet in Himachal Pradesh, Bhotiya in Uttarakhand, Bhutia in Sikkim, and Monpa in Arunachal Pradesh. In Nepal, a number of ethnic groups practicing pastoralism are Thakuri, Sherpa, Rai, Limbu, Tamang, Kirats, Gurung, Thakali, Chettri, and Magar. The communities in the Himalayan region of Gilgit-Baltistan of Pakistan practicing pastoralism include Balti, Gujjar, Wakhi, Uygur, Shina, Yashkun, Kirghiz, and Bakkarwal. On the basis of migration types, the major pastoral groups in the Himalayan region have been broadly classified into three categories, viz., nomadic pastoralism, semi-nomadic pastoralism, and transhumant herders. The nomads differ depending upon the type of animals they raise. For instance, Gujjars raise buffaloes, cattle, and sheep, whereas Bakarwals and Gaddis generally raise sheep and goats. Changpa nomads raise yaks, sheep, goats, and horses. The Changthangi (Pashmina) and Chegu goats are famous world over for fine warm wool fiber. Bhutias of Sikkim raise sheep, goats, yaks, cattle, dzow (cross of yak and cattle), and horses. Pastoral groups of the Himalayan region of Gilgit-Baltistan of Pakistan raise sheep and goats besides other livestock species and nomadic pastoralists of Tibet raise sheep, goats, yaks, yak–cattle hybrids, and horses.

S. P. Singh, J. K. Malik, *Migratory Small Ruminant Farming System in the Himalayas-Peculiarities, Problems, Prospects, Climate Effects, Nutrition and Health*,
https://doi.org/10.1007/978-981-96-6691-1_3

Seasonal migration by pastoralists to the alpine pastures and back to the plains annually is a hard task involving long walking hours, accidents, attacks by wild animals, and other hazards. In general, uphill ascending and descending migratory routes, places of enroute stay, destination of alpine pastures, and time of departure and arrival in alpine pastures are almost fixed and identical by all categories of herders. Weather or climate change is the major factor which governs the departure during ascending and descending journeys. Other distinctive features of the patterns of migration by different communities have been described. In India while ascending, the Bakarwals and Gujjars pass through seven major tribal migratory routes and majority of the undertaken routes pass through different regions of Pir Panjal range. Their migration has been categorized as intra-district, inter-district, inter-province, and inter-state migration. Gaddis of Himachal Pradesh rear sheep and goats by following vertical and horizontal transhumance pattern. Kinners migrate to higher reaches of Kinnaur hills of Himachal Pradesh and the Kumaun and Garhwal hills of Uttarakhand. During the winter season, they move with their animals to the plains of Kangra and Sirmaur districts of Himachal Pradesh. Bhotiyas of Uttarakhand occupy the entire alpine regions of western Himalayas in the state. The Changpa nomads have fixed routes and keep sheep and goats together and restrict grazing of yaks and horses to separate pastures.

The nomadic pastoralists of Tibet/China have several distinctive features. The Qinghai-Tibet Plateau (QTP) has the largest area of continuous grazing grassland in the world and maintains approximately 20 million yaks and 50 million Tibetan sheep. The Tibetan sheep grazes on natural pastures with no additional feeding and is considered as a special breed that has well adapted to alpine harsh environments including severe cold, low oxygen content of the air in high altitude and high solar radiation. The low temperature- and hypoxia-resistant Tibetan sheep are the main source of earnings for pastoral nomads, providing them with valuable resources including milk, meat, and fur and wool clothing, and dung as fertilizer and fuel. Nomads move with their animals to different grasslands at different seasons of the year and are in search of suitable forages. In Nepal, pastoral communities raise sheep, goats, yaks, and cattle under the transhumant production system. The downward journey commences as temperature starts to fall and their animals migrate through naked hills towards the land covered by forest and vines. The mobile pastoral herds supply manure to the uncultivated agricultural fields, and this practice of manuring is quite common in Nepal. During the winter season, Nepal pastoralists use castrated sheep and goats as pack animals. In Bhutan, similar to other Himalayan countries, pastoralism is a vital component of the agricultural farming system. Transhumant agro-pastoralism (TAP) involves vertical transhumance and a so-called alpine yak system. In this system, when yaks move down to temperate pastures, cattle move further south to subtropical pastures and then migrate upward to temperate pastures as yaks migrate back to alpine pastures in summer season. Another distinctive feature of the traditional transhumant production system of Bhutan is what is called "Tsamdro resource-sharing system" in which multiple households gain grazing access to the same tsamdro (rangeland) during different seasons. The current chapter mainly focuses on the principal nomadic Himalayan communities, the way they have been classified, the patterns of their migration in major Himalayan countries including India, Nepal, Bhutan, and China.

Keywords

Transhumance · Migratory patterns and routes · Pastoral communities · Sheep/goats · Himalayan regions

Abbreviations

QTP Qinghai-Tibet plateau
TAP Transhumant Agro-pastoralism

Pastoralism is a traditional and extensive form of rearing farm animals. By raising different livestock species that adapt to variable environments, pastoralism plays a vital role in alleviating poverty and providing food security. Pastoralists feed their animals through varied landscapes, and while working proficiently with nature, they contribute in productivity, sustainability, and welfare of animals. Furthermore, it decreases competition between feed for farm animals and food for human consumption, protects animal diversity, and helps in maintaining and renewing the forest lands (FAO 2021).

Himalayan mountain pastoralism is an efficient way of utilizing marginal environments. The continued existence of pastoralism in the Himalayan regions is interlinked with numerous aspects of sustainable use of land. Livestock species generally reared by the Himalayan pastoralists are sheep, goats, cattle, buffaloes, horses, yak, and yak-cattle hybrids. The migratory farm animals provide commodities such as, milk, meat, blood, wool, hair, leather, horn, and/or manure. The food obtained from animals is key components of the diet in mountain communities. Other benefits provided to humans are agricultural labor and transportation. It has been recognized that pastoralism supports a population with the limited resources of available land. The key mechanisms of pastoral adaptation are low population density, mobility, and multifarious information systems. In addition to breeding their livestock, pastoralists invest in caring for animals and thus increase their reproduction and survival rates, and it makes pastoralism the most efficient way of using resources in marginal areas. Pastoral livestock production is an environmentally sustainable system which provides a range of services including conservation and restoration of biodiversity in the rangelands, agricultural intensification in mountain regions, and sources of income by selling livestock products. Pastoralists are known to preplan their actions and continuously adjust to changing conditions. It is documented that all pastoralist groups do not exist at the basic subsistence level, but there are pastoralists who are wealthy through their economic activities. Furthermore, it is uncommon that any pastoral group lives solely with the products of their domesticated farm animals (Bishop 2009; Bhasin 2013).

Pastoralism seems to be much nearer to the grazing patterns of wild animal species, which mimics natural ecosystem interactions and functional roles. Since rangeland ecosystems are dependent on grazing, pastoralism plays an important role in maintaining fertility of soil, water regulation, pest and disease regulation, conservation of biodiversity, and fire management (McGahey et al. 2014). A recent study on the impact of the traditional pastoral system on high elevation plant species in Lachen valley, Sikkim, has shown that grazing significantly contributes to greater plant species diversity such as species richness, Shannon index, Simpson diversity index, and Pielou evenness index and ecosystem function (Ingty 2021).

3.1 Classification of Pastoralism

Pastoralism can be categorized in a number of ways. The most important way of classification is based on the degree of movement, species, management strategy, geography, and ecology. The most common categorization is by degree of movement, from highly nomadic through transhumant to agro-pastoral (Blench 2001; Sharma et al. 2003).

The major pastoral groups in the Himalayan region categorized based on the migration types include the following:

3.1.1 Nomadic Pastoralism

Nomadic herders migrate in an irregular manner to search new pastures for grazing their livestock. The movements of nomads are opportunistic and depend almost directly on the availability of forage resources. They grow no crops and their livelihood is based entirely on pasturing of animals. Since nomadic herders do not cultivate land, they mostly carry out extensive economic exchange with their neighboring agricultural communities for cultivable goods. Nomadic pastoralists comprising of Van Gujjars of Uttarakhand and Himachal Pradesh and Changpas in the union territory of Ladakh move with their whole families in search of pasture resources (Sharma et al. 2003).

3.1.2 Semi-nomadic Pastoralism

Semi-nomadic pastoralists migrate seasonally to higher pastures with their animals. These pastoralists have their own cultivable land, and during half of the year, they are engaged in agricultural practices. Examples of semi-nomadic pastoralists include Gaddis and Bhotiyas of the north-western Himalayas as well as Bhutias living in the Lachen and Lachung valleys of Sikkim and Monpas of Arunachal Pradesh (Sharma et al. 2003; Bhasin 2013; Simmy 2019).

3.1.3 Transhumant Herders

Transhumant pastoralists make the regular movement of livestock between fixed points to take advantage of seasonality of pastures. In the Himalayas, this is often vertical movement between established points and may have a regular and ancient pedigree. Pastoralists living in villages practice long-distance herding of livestock. In transhumance, animals are generally accompanied by hired men but also by livestock owners and their relatives, but seldom by a whole family, on a long-distance movement or journey between two seasonal ranges (Blench 2001; Sharma et al. 2003).

3.2 Pastoral Nomadic Communities of the Himalayas

Some of the well known pastoral communities of Himalayas practicing transhumance include Gaddis, Gujjars, Bakarwals, Kinnauras, Kaulis and Kanets of the north Indian Himalayas, Bhotias of Garhwal Himalayas, Monpa of Arunachal Pradesh, Bhutias of Lachen and Lachung, Sikkim, Changpas of Changthang, Ladakh and Bhotias and Sherpas of Khumbu valley of Nepal, Kirats of eastern Nepal (Bhasin 2013). Recently, Singh and Kerven (2023) documented various pastoral communities of the Himalayan regions of India, Gilgit-Baltistan of Pakistan, Bhutan, and Nepal on the basis of the secondary data analysis and primary data collection by the lead author with resource mapping and interviews with the regional experts. Of pastoral groups in Bhutan, Ghalley, and Gurung are sheep herders, Brokpa (Merak), Brokpa (Saktenpa), and Bumtap keep sheep and yak. Sephup community raises yak, cattle, and sheep. Lunap, Baylagap, and Dagalagap keep both yak and cattle, whereas Layap, Lingzhip, Haap, Nubri, and Soeyaksa rear only Yaks. Among pastoralist communities of the Himalayan region of Nepal, Magar and Chettri raise sheep and goats and Gurungs are sheep herders. Thakali rear yak, goats, and sheep, whereas Thakuri keep a mix of animals including yak, horses, sheep, and goats. Rai and Limbu communities rear yak and goats. Sherpa and Tamang are Yak and cattle herders. Pastoralists of the Pakistan Himalayan region of Gilgit-Baltistan include Balti, Gujjar, Wakhi, Uygur, Shina, Yashkun, Kirghiz, and Bakkarwal, who rear sheep and goats besides other livestock species (Singh and Kerven 2023). Tibetan nomads raise several livestock species including yaks, yak-cattle hybrids, sheep, goats, and horses (Isom 2009). Table 3.1 presents the pastoral communities of the Indian Himalayas.

Table 3.1 Pastoralists of the Indian Himalayas

Pastoral community	Livestock species	Location	Reference
Bakarwal	Sheep and goats	Jammu and Kashmir	Simmy (2019)
Gujjar/ Gujjar-Bakarwal/Van Gujjar	Buffaloes, cattle, sheep, and goats	Sirmaur, Shimla, Mandi and Chamba of Himachal Pradesh, Jammu and Kashmir, and Uttarakhand	Ahmad et al. (2017), Simmy (2019), and Singh and Kerven (2023)[a]
Changpa	Sheep, goats, and yaks	Ladakh	Namgail et al. (2007)
Brokpa	Yak, sheep, goats, and horses	Ladakh	Singh and Kerven (2023)[a]
Gaddi	Sheep and goats	Himachal Pradesh and Jammu and Kashmir	Bhasin (2013), Ahmad et al. (2017), and Simmy (2019)
Kinnaura	Sheep and goats	Kinnaur district of Himachal Pradesh	Simmy (2019)
Lahuala	Goats, sheep, yak, dzo, cattle, horse, and donkeys	Lahual and Spiti region of Himachal Pradesh	Singh and Kerven (2023)[a]
Spitian	Goats, sheep, yak, dzo/dzomo, cattle, horse, and donkeys	Lahual and Spiti region of Himachal Pradesh	
Kanet	Sheep and goats	Himachal Pradesh	Simmy (2019)
Brokpa and Sherpa	Yak	Arunachal Pradesh	Singh and Kerven (2023)[a]
Mishmi	Mithun	Arunachal Pradesh	
Bhotiya	Sheep, goats, and cattle	Upper regions of Garhwal and Kumaon (Uttarakhand)	Mitra et al. (2013) and Simmy (2019)
Bhutia	Sheep, goats, yaks, and cattle	North Sikkim	Bhasin (2013)
Gurung	Sheep	Sikkim	Singh and Kerven (2023)[a]
Chettri	Cattle		
Dokpa	Yak		
Monpa	Yaks and cattle	Tawang and west Kemeng districts of Arunachal Pradesh	Pandey et al. (2020)

[a]Based on the secondary data analysis and primary data collection by the lead author with resource mapping and interviews with the regional experts

3.3 Migration Patterns by Pastoralists

3.3.1 Indian States/Union Territories

3.3.1.1 Gujjars and Bakarwals (Union Territory of Jammu and Kashmir)

Gujjars and Bakarwals are the tribal communities and have been ranked as the third largest community in the union territory of Jammu and Kashmir. Bakarwals are nomadic pastoralists and raise livestock for their livelihood. Gujjars and Bakarwals in Jammu and Kashmir are predominantly nomadic community who raise buffaloes, sheep, and goats etc. Horses play an important role in the migration process and are utilized as pack animals. Gujjars are concentrated in Poonch, Rajouri, Anantnag, Udhampur, and Doda districts. As compared to Bakarwals, Gujjars have a low tendency for seasonal migration and so at times, they are also known as Semi-Nomads. Since Gujjars own their permanent houses and are relatively economically sound and educated in comparison to Bakarwals of Poonch district, they are less likely to participate in the seasonal migration of animals. Nevertheless, only a few percent of Gujjars called Dhokes migrate seasonally to the hilly areas, as they are weak socially as well as economically. Bakarwals are generally dependent on naturally occurring resources and follow seasonal migration in order to utilize pasture lands for nourishing their animals (Kheraj et al. 2017; Kirmani et al. 2020).

Migration journey of the Gujjars and Bakarwals from plains towards upper reaches of Himalayas begins with the onset of summer season and they migrate back to the lowlands in winters. They pass through seven major tribal migration routes and all the major routes and sub-routes pass through various mountain passes. The starting points for the seasonal migration are Rajouri, Poonch, Reasi, Marhot, Surankote, Mendhar, Udhampur, Thanamandi, and Jajerkotlietc and majority of the undertaken routes pass through different regions of Pir Panjal range. The migration route is fixed, and each year after crossing Pir Panjal, Bakarwals move towards a particular region which has many pastures. A certain route to a particular pasture land is followed by various herders and they generally adhere to their traditional route. The migration route is only changed in the event of harsh weather or if the terrain is blocked due to some natural calamity. Different herders cross through different mountain pass routes. Bakarwals generally travel for approximately 6 to 8 h each day. At times, halts are taken due to harsh environmental conditions in order to minimize casualties. These emergency halts are for about 2 to 3 days duration until the environmental conditions get normalized. It takes about 40 to 55 days to complete the journey (Kheraj et al. 2017; Kirmani et al. 2020).

In general, Bakarwals stay 6 months each on plains and on highland pastures of mountains during the year. In winter season, they reside in plains and in summers they live on the mountainous part of Jammu region around Kathua, Doda, Udhampur, or in around pastures in Ganderbal, Daksum, Uri, Sonmarg, Anantnag, etc., located in Kashmir. In addition, numerous Bakarwals also reside in and around Poonch and Rajouri (Kirmani et al. 2020). Approximately 20,000 herders migrate to summer pasture annually and travel back at the end of grazing season. The seasonal movement of Gujjars and Bakarwals is dependent on climate and location of the area. As per their requirement, some herders migrate in winter time and some others in summer time. Gujjars and Bakarwals follow different migration routes including jamiyagali, gorabatta, Nanansar, RopadiDharal Pass, Mughal road, etc. In Jammu and Kashmir, migration generally takes place from Poonch, Rajouri, and Kishtwar regions and all these routes lead to different passes in the Pir-Panjal ranges. Of the three regions, the Poonch and Rajouri migration makes the heaviest route (Kheraj et al. 2017).

In a recent study, Kirmani et al. (2020) identified various migration routes adopted by Bakarwals in Anantnag and Ganderbal districts of Jammu and Kashmir region. Bakarwals usually crossed Pir Panjal, Rupari pass, and Budhil pass to reach Sonmarg and allied areas and for reaching Pahalgam and allied areas, they crossed Zojimarg and Banihal pass. On arrival to their routine pasture, Bakarwals set up kutcha shelters (Dheras) to protect themselves from harsh weather. Often Dheras set up during the previous migration are utilized after minor repairs in case damaged by extreme climatic conditions. During summers, Bakarwals migrate to different pastures located in the district Ganderbal, and many of these pastures are located around Sonmarg including Thajwas valley, Baltal, Lidderwat, etc. The most common routes being utilized to reach the district included: (1) R.S Pura (ChakRoi) → Banihal (Jammu-Srinagar road) → Ganerbal → Sonmarg; (2) Reasi → Bodal → Pir Panjal → Shopian—Srinagar → Kangan → Sonmarg; (3) Rajouri → Thanamandi → Mohra → Shopian → Wayil → Sonmarg; (4) Poonch → Shopian → Wayil → Kangan → Sonmarg; (5) Nagrota → Kalakot → Shopian → Kangan → Sonmarg; (6) Katra → Reasi → Rajouri → Shopian → Ganderbal → Sonmarg; and (7) Rajouri → Thanamarg → Sathri → Pampore → Shalimar → Ganderbal → Sonmarg. On the other hand, the most common routes followed by Bakarwals to pasture lands situated on the upper reaches such as Aru valley, Lidderwas valley etc. in Pahalgam included: (1) Rajouri →Thanamandi → Shopian → Anantnag → Pahalgam; (2) Reasi → Kalakot → Rajouri → Shopian → Pahalgam; (3) Nagrota → Galai → Rajouri → Shopian → Pahalgam; (4) Udhampur → Patnitop → Banihal → Anantnag → Pahalgam; (5) Reasi → Katra → Udhampur → Ramban → Banihal → Anantnag → Pahalgam; (6) Udhampur → Doda → Kishtwar → Kokernag → Anantnag → Pahalgam; and (7) Katra → Ramban → Verinag → Anantnag → Pahalgam. Both Sonmarg and Pahalgam receive huge number of nomads every year.

More recently, migration patterns of Gujjar-Bakarwals in Jammu and Kashmir were categorized as: (1) *Intra-district migration:* This type of migration includes movement of people from villages in plains or mountain bases to highland pastures or alpine meadows within the district; (2) *Inter-district migration:* It involves the migration of nomads from plains of Jammu, Samba and Kathua districts of Jammu division to highland pastures in Udhampur, Chenab region, and adjoining mountain ranges; (3) *Inter-province migration*: It is adopted by the majority of the Bakarwal pastoralists who migrate from most of the districts in Jammu region to Kashmir region during summer season and travel back during winter season; and (4) *Inter-state migration:* It is followed by less than one-tenth of this transhumant population who migrate to Ladakh, Himachal Pradesh, and Punjab. In numerous cases, families of shepherds travel up to 600 km on foot, along with their animals. During migration, they have halts at various transit points and the journey can take from 2 to 4 weeks or 6 weeks (Sareen 2022).

3.3.1.2 Changpas (Union Territory of Ladakh)

Rupshu-Kharnak in Changthang provides sustenance to a large number of domestic animals. As the environmental conditions are not favorable to any other form of land use, grazing land is not sown fertilized or irrigated. Changpa nomads raise sheep, goats, yak, and horses. They migrate with their herds across the Changthang. Changpas in the high altitude areas such as Changthang in Ladakh have relatively regulated grazing pattern by dividing the season into two parts and have assigned the names to their pastures as winter pasture and summer pasture. They utilize these two season pastures in a rotational grazing system. The migration between summer pastures and winter pastures is cyclic and on regular basis. Moreover, they have fixed routes and pastures and schedules of movement. They keep sheep and goats together and restrict grazing of yaks and horses to separate pastures and in doing so changpas avoid concurrent disturbance of pastures. Their indigenous practices of maintaining the livestock ratio and proper use of pastures and avoidance of excessive use of some pastures have played pivotal role in continuance of grazing of animals in the Changthang region of Ladakh (Bhasin 2013).

The animals raised by Changpas fulfill their many needs. Sheep, goats, and yaks provide them with wool required for ropes, tents, clothing beddings, as well as milk, meat and transportation and dog protects the domestic animals and belongings. The domestic goats raised by herders in Changthang produce one of the finest cashmere wool or pashmina. Changthangi goats are raised under harsh environmental conditions at temperatures ranging from −40 °C to +40 °C. Being the mainstay of economy of Changpa nomads, these animals are reared in large numbers. It is estimated that annual production of raw pashmina fiber in Changthang region alone is approximately 35,000 kg from around 0.15 million of Changthangi goats, which constitutes 80% of the total pashmina produced in the country (Bashir Beigh and Bashir 2020). Changpas compose their herds not randomly, but it is based on adaptive response to environment and the available resources. Rearing of a mix of yaks, goats, sheep, and horses by changpas has some advantage to reduce threat of losses due to disease or harsh weather conditions as not all the animals will be lost and herds can be again restored. Different animals consume different plants in the same pasture which enables proficient use of available resources (Miller 2000; Bhasin 2013). Since there are a few areas where grasses stop growing in the winter season, changpas do not migrate to distant places.

3.3.1.3 Gaddis (State of Himachal Pradesh)

In the state of Himachal Pradesh, Gaddi is the most prominent tribe involved in transhumant pastoralism for hundreds of years. Originally (Fig. 3.1), they belong to Bharmour town of Chamba district, but over the years many of them have settled in Kangra district and graze their animals and follow migration at north and south sides of Dhauladhar mountain range. Transhumance is manifested in two main forms, vertical transhumance in which shepherds with their livestock migrate from lowland winter pastures to high-altitude summer pastures and horizontal transhumance wherein herders move from dry season pastures to rainy season pastures. Vertical transhumance is predominantly practiced in high mountain ranges such as the Himalaya. Majority of Gaddis from Kangra valley are semi-nomadic as they possess permanent houses in the valley and claim inheritance to the property. (Sharma et al. 2003; Bhasin 2013; Sharma et al. 2022).

Fig. 3.1 A Gaddi shepherd with his flock of sheep and goats on way to alpine pastures in Kinnaur district of Himachal Pradesh. (Photo, courtesy of Mr. BB Negi with permission)

Gaddi pastoralists have reared sheep and goats for generations, and they have deep religious links and long-standing association with pastoral practices. As per their traditions, men are involved in seasonal migration of sheep and goats, whereas women stay back at their permanent settlements and remain actively engaged in agricultural production during the whole year. In the Gaddi pastoral system, there has been a customary practice for ages of hiring herders who in the local dialect are called as puhals. In the recent years, the need to employ puhals has increased manifold that help in the movement of sheep and goat flocks across the mountains and in performing various day-to-day activities (Malhotra et al. 2022).

Broadly, the pattern of animal migration followed by all the Gaddis in the area is almost identical, but it does differ depending on the travel route taken, weather conditions, place of enroute stay, destination of alpine pasture, number of people accompanying the flocks, and many other factors. In most cases, Gaddis start their uphill journey from the lower hills of Shivaliks in mid to end of April, traverse through the lower ranges of Dhauladhar, cross the mountain passes, and arrive at the high altitude meadows by mid to end of June covering long distances on foot. They start downward journey with their livestock from higher hills in the month of September. In a way the animals migrate for grazing in three ecological zones, namely, subtropical pastures, temperate-temperate pastures, and alpine pastures. In Kangra valley and some other areas of Himachal Pradesh, Gaddis are involved in the practice of camping their flocks in the farmers' fields with a view to enhance soil fertility by providing natural manure to the fields for which Gaddis receive payments from the land owners. In addition, they are well informed regarding ecosystems, the spatial distribution of resources, access, as well as the use of resources (Bhasin 2013). They also have knowledge of species used as firewood and fodder (Ramprasad et al. 2020).

Recently, Sharma et al. (2022) documented the seasonal migration trends of Gaddis in the Bharmour region of Chamba, Himachal Pradesh. The migratory routes followed by Gaddis are well-defined. Since Gaddis repeatedly follow the same migratory route across years, they hold vast knowledge of route and the available resources. During the winter season from December to February, they graze their animals in the Shivalik foothills at an elevation of about 300 m. During the summer and rainy seasons from July to September, Gaddis camp with their livestock for grazing in the alpine meadows (dhars) at an elevation of approximately 4000 m. They strictly have to stick to grazing their livestock in their respective dhars which are assigned to an individual shepherd or a group of shepherds. Increase in the temperature in the lowlands is an indicator for the initiation of upward journey to alpine meadows which begins during March, while the downward journey starts during September when winter begins to set in. During their migration, Gaddis use four types of lands, namely, forests, high altitude pastures, villages areas, and private lands. During their uphill journey (during March) and downhill migration (during November), Gaddis camp near to their permanent houses and at this time they keep their animals in the adjoining forests. They have permanent settlements in between and close to the migration routes. One-way journey on an average takes 49 ± 6.9 days and they cover a distance of approximately 400 km. They often move with their livestock through populated areas in the late hours.

3.3.1.4 Kinnauras (State of Himachal Pradesh)

The inhabitants of Kinnaur in Himachal Pradesh are known as Kinners. They are more often referred as Kinnauras and reside under very harsh conditions, created by the extreme weather and climate conditions. They inhabit valleys between high mountains. Similar to Gaddis, Kinnauras are agro-pastoralist, semi-nomadic people who are engaged in pastoralism as well as agricultural activities during varying seasons. For grazing their animals in summers, the Kinnaura herders migrate with their flocks of sheep and goats to higher reaches of Kinnaur district of Himachal Pradesh and the Kumaun and Garhwal hills of Uttarakhand. During the winter season, they move with their animals to the plains of Kangra and Sirmaur districts of Himachal Pradesh. As per their tradition, Kinnauras employ shepherds from different parts of Himachal Pradesh and disburse them clothes, shoes, basic foods including maize, salt, spices, etc., as well as 7–8 animals depending upon size of the flock (Sharma et al. 2003; Simmy 2019).

3.3.1.5 Bhotiyas (State of Uttarakhand)

In Uttarakhand, Bhotiyas are scattered over the seven main river valleys, namely, Jadh, Mana, Niti, Johar, Darma, Byans, and Chaudans. Bhotiyas constitute a considerable portion of the state's total population. They are mostly found in five districts of the state, namely, Almora, Chamoli, Pithoragarh, Bageshwar, and Uttarkashi. They are scattered over 291 villages (Farswan 2017; Sharma et al. 2019). In

Uttarakhand, Bhotiyas migrate with their sheep and goats from foothill forests to alpine pastures and vise versa for grazing the animals on naturally available forage resources. Approximately 90% sheep in Uttarakhand are migratory (Das et al. 2015). In the summer season, the transhumant pastoralists occupy the entire alpine regions of the western Himalayas including Bedni, Ali, Dyara, Gidara, Sahstratal, Auli, Rudranath, Pindari, and Chhiplakedar in Uttarakhand state, apart from for the alpine regions of Protected Areas (Mitra et al. 2013).

Mitra et al. (2013) reported baseline information regarding the transhumant pastoralists in Niti valley which is situated in the extended buffer zone of Nanda Devi Biosphere Reserve (NDBR) in the western Himalaya in Uttarakhand. It has seven migratory villages, namely, Kailashpur, Mahergaon, Gurgutti, Farkia, Bampa, Gamsali, and Niti. The settled rural people in the NDBR belong to the Bhotiya tribe. The migratory people own two permanent settlements: one summer settlement in the higher elevation, i.e., Niti valley, and one lower-altitude winter settlement. In general, people belonging to the lower elevations of the Himalayas are the transhumant herders. They are mainly involved in small-scale agriculture and raise livestock. During summer season, they migrate with their herds to higher alpine meadows for grazing. Kuari pass and Tapovan, the last permanent village on the route to Niti valley, are preferred halts for transhumant pastoralist groups travelling with their livestock to the Niti valley during summer season. Kuari pass at an elevation of 3640 m is a venue where majority of the migratory herders moving from lower altitudes camp for a few days to rest and then scatter to different alpine grasslands.

Each pastoralist group consists of 3–5 male shepherds, 3–4 horses or mules, and 3–4 dogs. Sheep are raised for wool and meat, while goats are reared for meat and to a lesser extent for obtaining rough wool which is utilized to make rope and handicrafts. Horses and mules are used chiefly for transporting loads. Dogs are kept for the safety of livestock and protection from wild animals in lower and upper regions. One or two male members of the family participate in the upward journey from the village. Shepherds from the same village take their herds together for grazing on the summer pastures. Area for summer grazing is mainly selected on the basis of their evaluation of the quality of fodder and grazing area availability. Every year the same group of shepherds may visit the same area. One alpine meadow is generally occupied by one group of shepherds so that overcrowding is avoided. Each shepherd does not stay continuously for the entire duration of grazing season of 4–5 months on the alpine pasture. Two or three persons take turns to graze the animals, while others return to the village to join their families. At times, shepherds also bring animals to the summer pastures from other people who are not the part of livestock grazing and in return they get money. Seventy percent shepherds halt at Niti before moving on to the Kalajowar, Gothing, or Geldung area. On this route, Niti is the last village on the Indian side. Geldung area, with an elevation from 4400 to 5500 m, has three mountain passes, namely, Niti pass, Lepcha pass, and Selakang pass where these shepherds take their animals for grazing. Apart from these passes, a few shepherds also tend to stay in adjoining areas such as Laldhang and Keolangmandi. The downward journey to their villages commences in late August to early September and shepherds take the same route as on the way of upward journey. The availability of fodder and largely the climatic conditions are the determining factors for duration of their stay in Niti valley. The shepherds usually stay in the valley for 3–4 months, beginning from late June to August (Mitra et al. 2013).

In district Uttarkashi of Uttarakhand, Bhotiya tribe is known as Jaad who raise sheep and goats and practice transhumant pastoralism. Jaads live at Dunda in winter season, while during summers they migrate to higher places such as Bagori and Harsil. These herders migrate with their livestock from lower altitudes of the sivalik Himalaya to higher Bugyal (alpine) regions in summers and then travel back to lower altitudes with the onset of autumn (Singh and Goswami 2020).

3.3.1.6 Bhuttias of Lachen and Lachung (State of Sikkim)

The Bhuttias are the most prominent agro-pastoralist community of the alpine regions of Lachung and Lachen valleys of the northern districts of Sikkim. They raise sheep, yaks, *dzow* (a breed of yak and common cow), horses, and mules and practice marginal agriculture. The Bhuttias own permanent houses at middle altitude and migrate with their animals to high altitude pasture in summer and to lower altitude pasture in winter. Lachenpas have seasonal migration pattern and have encampments at different places during migrations to high altitude pastures by using *byah* (tents) made of yak hair. Lachenpas occupy Lachen from mid-February to mid-May and then migrate towards Menshithang. Lachungpas migrate with their animals in the month of May to high-altitude pasture stopping at Yumthang, Yume Samdong, and Chholhamo. They travel 6 km up the Lachung to Chholhamo and 6 km down Lachung to Khedum, Leema Lothen (Bhasin 2013).

3.3.2 Migration Patterns in Tibet (China)

The Qinghai-Tibet Plateau (QTP) has the largest area of continuous grazing grassland in the world and maintains approximately 20 million yaks (Fan et al. 2020) and 50 million of Tibetan sheep (Langda et al. 2020). The Tibetan steppe extends approximately 3000 km from east to west and almost 1500 km from north to south. Besides yaks and sheep, nomadic pastoralists of Tibet raise goats, yak-cattle hybrids, and horses. Sheep and goats are considered important in many areas of the Tibetan Plateau, particularly in western Tibet, where available pastures are better suited to small ruminant species. Tibetan sheep extensively grazes in alpine meadows, at an altitude ranging between 3000 and 5000 m (Zhou et al. 2015; Cui et al. 2021). Tibetan sheep grazes on natural pastures with no additional feeding and is considered as a special breed that has well adapted to the alpine harsh environments including severe cold, low oxygen content of the air in high altitude, and high solar radiation (Ding et al. 2020; Langda et al. 2020).The low temperature- and hypoxia-resistant Tibetan sheep are the main source of earnings for pastoral nomads, providing them with valuable resources including milk, meat, and fur and wool clothing and dung as fertilizer and fuel (Zhao et al. 2011; Cui et al. 2021).

Tibetan pastoral production system is a unique system with mixed different livestock species. Each of the animal species including yak, sheep, goat and horse, etc., has its own specific characteristics and adaptations to the alpine harsh environments. Although yak characterizes best Tibetan pastoralism, sheep are usually more important from the economic point of view. As compared to yaks, sheep can browse shorter grass and are more common in the western regions of Qinghai-Tibetan Plateau. Moreover, sheep meat is preferred by nomads and in agricultural and urban areas, and also they provide one of the best carpet wools in the world (Ding et al. 2020). On the other hand, Tibetan goats provide cashmere which is one of the best cashmeres in the world. Goats are also used for the purpose of meat and milk. Since different animal species graze on different plants, herding different small and large ruminants and horses together on the same rangeland make more efficacious use of available forage. Herding of different animal species together is also beneficial to lessen the risk of losses due to disease or severe winters. All round the year, animals live almost entirely by grazing in Tibetan pastoral areas. Increasing number of pastoralists are planting sown pastures for either winter–spring grazing or for hay. They cut some hay for feeding weak animals and horses in the winter and spring seasons (Miller 2000).

In Tibetan nomadic pastoral system, livestock migration is still a very important part in production practices for most nomadic pastoralists. But with increasing settlement of nomads, animal migration patterns are being restricted. Nomads move with their animals to different grasslands at different seasons of the year and are in search of suitable forage conditions. Movement between different pastures enables pastoralists to graze their animals on growing forage during summer and to keep grass growth preserved for fall and for utilizing in the early winter season. Most of the nomads own houses and have built animal sheds for use during the winter and migrate with their animals to distant grasslands during the year. They keep permanent camps even at the height of 5100 m. Tibetan pastoralists have also developed management skills that help them in balancing livestock with the pastureland resources. Furthermore, the ratio of different animal species reared and herd size vary in the pastoral region of Tibet depending upon pastureland factors and the suitability of the landscape for different livestock species. As compared to Hongyuan County, Shuanghu County is drier and the dominant alpine steppe vegetation is more suitable to small ruminant species. In Luqu County, due to active encouragement by the government, pastoralists raise much higher percentage of sheep than the neighboring counties (Miller 2000).

The pastoral practices on the Tibetan Plateau have been categorized into three utilization strategies (Kreutzmann 2011). The first utilization strategy is combined mountain agriculture. Shepherds from their permanent habitations on average altitudes ranging between 3300 and 4400 m migrate with their herds to higher elevations for seasonal natural grazing during summer season. The natural grazing provides fodder for a period of 3 months annually. The additional fodder requirement of animals for up to 9 months that are grazed in the high-lying pastures during the summers has to be produced on private or common property village lands. Harvested hay and grass storage may enhance the fodder capacity. Shorter migrations are needed to access the pastures. The combined mountain agriculture is carried out largely in the western Tibetan river valleys of the Yarlong-Tsangpo valley and its tributaries at an altitude up to 4400 m, the Qaidam Basin in the north and in the eastern lower valleys down to 3300 m.

The second utilization strategy is nomadism which includes migration of shepherds with their herds staying in high-altitude pastures during summer and subsequently grazing in winter pastures in low-lying basins in the northern foothills or on the plateaux within the Tibetan steppe. At times herds graze in spring and/or autumn pastures depending upon availability of suitable forage. Detached mountain pastoralism is the third utilization strategy which is relatively a more recent one that includes forced sedentarization and permanent settlement in high-altitude grazing lands. This strategy has replaced long-distance migrations by short-distance migrations at a high altitude of permanent houses, generally at the elevation above 3000–4000 m (Kreutzmann 2011).

3.3.3 Migration Patterns in Nepal

The mountain areas of Nepal are inhabited by a number of ethnic groups such as Sherpa, Rai, Limbus, Tamang, Jirel, Gurung, Thakali, and Magar. They are primarily engaged in animal husbandry practices and are predominantly involved in the rearing of livestock (Tiwari et al. 2020). Sheep, goats, and yaks are major livestock species raised under transhumant production system in Nepal. Sheep is a multipurpose small ruminant and is one of the most accepted meat animal species. Besides meat, wool is another valuable product obtained from sheep and the total sheep population in Nepal is 0.8 million (Sapkota et al. 2020). Similar to sheep, goat is also one of the most populous, accepted, and easily available meat animal in Nepal with a population of 11.16 million (DLS 2018; Sapkota et al. 2020). Grazing in high altitude alpine pasturelands in summer season and progressive downward migration through mixed forest areas and staying at open lands and crop aftermaths in winter season is a distinctive feature of transhumant production system in Nepal (Parajuli et al. 2013; Tiwari et al. 2020).

In the Annapurna Conservation Area, Chauri, sheep and goats are major livestock species raised under the transhumance system. In general, yaks are kept free in the high pasturelands, while goats and sheep are allowed to graze during the day in the grasslands and brought back around the tent-house during the night. Horses and guard dogs are normally used by herders while shepherding (Pandey and Chhetri 2005). The route of migration is fixed and it depends on factors such as season including temperature, humidity and rainfall, and availability of fodder in the region. In summer season, livestock owners migrate with their animals up towards the naked hills and the region that are covered with snow whole year. The downward journey commences as temperature starts to fall and their animals migrate through naked hills towards the land covered by forest and vines (Sapkota et al. 2020).

Keeping goats and sheep in the same herd is advantageous as goat is a selective feeder and browses the top nutritious part of the pasture, whereas sheep is nonselective feeder and graze the whole tiller, even uprooting them. The hill is left naked at the end of grazing by goats and sheep together. In turn, mobile pastoral herds contribute to the supply of manure to the pasturelands. The grasses above 4000 m grow very rapidly and provide nutritious feed to the grazing animals during summer season. Depending upon the availability of grasses and ambient temperature, the animals are maintained in this type of pasture for few weeks in summer. Following stay for few weeks at temperate pasturelands and forests, herds mostly reach valley and lowland villages during late autumn and winter. Animals feed on aftermath in the rice-fields and also graze on nearby forest. The local farmers in these villages permit herders to stay and graze stocks on their land (Sapkota et al. 2020).

In Ghermu Village Development Committee of Lamjung district of Nepal, transhumant pastoralism ranges from the lower grounds along the floors of valley and residential areas at the height of about 800 m to the high-elevation pasturelands (kharkas) at the height of 4500–5000 m. For a period of about 6 months each year from September/October to March/April, the herds of sheep and goats are kept in the lower grounds and are grazed in community forests, farmlands, and other grazing lands along rivers, roads, and forests. From April/May to September/October, the herds are taken to the kharkas, movement of the animals between grazing grounds takes time and to reach at the final destination at the highest elevation requires 1–2 months. Herders while travelling graze their animals in more than 20 kharkas. During the 6-month period at higher elevations, animals can graze on highly nutritious kharkas with ample of grass, whereas at lower elevations there is shortage of grazing lands and nutritious grasses. On their downward journey, herders are invited by the local farmers to graze the herds of sheep and goats on the uncultivated fields and manure the soil (Gentle and Thwaites 2016).

Ichikawa (2012) documented agro-pastoral production systems practiced by Sherpa communities in the north-eastern mountains of Nepal. The Solu-Khumbu district is located in the northernmost part of the Sagarmatha zone in Nepal having an area of approximately 3300 km^2 (Kano 2001; Ichikawa 2012). Besides practicing agriculture, some of the farmers of the Junibesi Valley of the Beni Village in the Solu-Khumbu district also keep female yaks and a single crossbreed between yaks and cows (zoms). The method followed to raise livestock takes the form of transhumance, in which animals move around the rangelands at different elevations in accordance with the natural climate cycles. In the Junbesi Valley, the transhumance is characterized by vertical movement of livestock over a relatively short distance in spite of the significant difference in the altitude from the residential area at an elevation of 2500–3000 m to the pasturelands at about 4500 m (Iwata 2000; Ichikawa 2012). During the winter season, animals are grazed in forests located near agricultural fields close to settlements and are slowly migrated to the upper reaches of valley streams between late March and early April. The cattle are grazed from August to September in summer camp site pastures set up at an elevation of about 4500 m and are then taken back to the residential areas. Animals on their way back are maintained by feeding in the form of glade grazing in the month of October and at this time ambient temperature drops sharply (Kumagai 2005).

The Sherpas in the Khumbu region in the Solu-Khumbu district practice a form of transhumance with multiple plots of arable land, grasslands, and houses on sites at different elevations in specific valleys (Kano 2001; Ichikawa 2012). Based on the climatic cycle, they also practice transhumance by migrating up to the highland in summer and descend to areas at a lower altitude in the autumn. They again practice transhumance in the highlands during the period between late autumn and early spring. The hay that was cut in the autumn and stored in highland houses is used to feed animals to deal with forage scarcity in winter season (Inamura et al. 2000; Ichikawa 2012).

Majority of the households in Saipal Rural Municipality of Bajhang district in the western Nepal are dependent on agriculture and raising of goats and sheep for their livelihoods (SRM 2020; Pasakhala et al. 2021). The transhumant herding was practiced in Saipal by more than one-third of the total households in 2015 (KVDC 2016; Pasakhala et al. 2021). In Saipal, shepherds reared two types of small ruminants: (1) male and female sheep and goats that were either castrated and raised for meat or uncastrated and reared for breeding (paltey) and (2) castrated sheep and goats used as pack animals (jakhna) during the winter season. In addition to raising sheep and goats, transhumant herders practiced agriculture, made woolen carpets and clothes either for personal use or business purpose, and engaged in the business of goods and medicinal plants (Pasakhala et al. 2021).

From April to September, the transhumant herders with their sheep and goats migrate towards north to the mountain regions bordering TAR, China. This enables them to find forage for animals and to get away from the heat and flies during summer months. In the month of September, as ambient temperatures fall and the availability of livestock fodder decreases, herders move down to their villages, where they stay until October/November. Before commencing their succeeding migration towards south, animals are grouped into two categories namely paltey and jakhna. The herds of paltey are moved southwards to the lower belts of Bajhang and adjoining Doti district, where the animals are kept until March. The herds of castrated sheep and goats used as pack animals move many times from the nearest market back to Saipal transporting food and goods, mainly rice. Both the paltey and jakhna herds move back to the summer pastures in north in the month of April. Furthermore, migration of the animals through different routes between summer and winter pastures permits regeneration of the already grazed pastures (Pasakhala et al. 2021).

Before commencement of migration, 2 or 3 herding households join together their labor and sheep and goats to form a collective herd known as a gol. The number of animals in each gol ranges between 250 and 700 and it is managed by 3–5 transhumant herders. The herders who give their animals to the gol continue to reside in the villages. They either take care of the agricultural lands of the migrating herders or share the progeny of the animals lent to the gol. The migrating herders on the way are invited by the sedentary households to stall sheep and goats on their uncultivated agricultural fields to supply with manure. In turn, the sedentary households offer food and other commodities to the migrating herders (Pasakhala et al. 2021).

3.3.4 Migration Patterns in Bhutan

Similar to other Himalayan countries, pastoralism is a vital component of the agricultural farming system of Bhutan. Transhumant agro-pastoralism (TAP) is a widespread practice in Bhutan. Although yak herding is the main pastoral practice in all mountain communities of Bhutan, they also raise other animals including dzo-dzom, cattle, equine, sheep, and goats (Wangdi and Norbu 2018). TAP involves vertical transhumance synchronized with the alpine yak system in such a way that in winter season when yaks move down to temperate pastures, cattle move further south to subtropical pastures and then migrate upward to temperate pastures as yaks migrate back to alpine pastures in summer season (Ura 2002; Namgay et al. 2014).

The pattern of traditional transhumant production system in Bhutan has a distinctive feature of tsamdro resource-sharing system; multiple households, villages, and districts gain grazing access to the same tsamdro during different seasons (Moktan et al. 2008; Wangdi and Norbu 2018). The nature of the contemporary practices and associated socio-cultural traditions of TAP in Bhutan have been reported by Namgay et al. (2013). They employed qualitative and quantitative research methods to study the contemporary practices and socio-cultural traditions of TAP in the country. They gathered background quantitative data based on a structured survey of 75 TAP households in six different villages, four in the west and two in the central-east region of Bhutan. It was reported that the key reasons for migration of animals are avoidance of production loss, reducing mortality of migratory animals due to cold, scarcity of forage, better income opportunities by staying away from farm, parasite infestation avoidance in the south and leaving the grasslands for yak grazing during the winter season. The established migration routes are followed regularly by the transhumant nomads. The camp sites utilized by them during migration and the final camps where they erect semi-permanent sheds in the winter season are also established. The journey in a single direction could range between 4–5 days and more than a month. The Jabana (Tshebji and neighboring areas)-Dungna route is the shortest migration route, which could be reached within 4–5 days. The longest route is the Urchi (Bumthang)-Doban (Sarpang) route that takes over a month to arrive at the final destination in one direction. Herders with their animals cover shorter distances on a daily basis. They halt for several days along the way for allowing animals to graze the pasture. In addition, a variety of other reasons force them to halt such as adverse weather conditions, occurrence of natural calamities, poor health of the migrating animals, and attacks by the wild animals

The timing of migration is similar in seasons across the west and central regions of Bhutan. Herders move to south between September and October while return to north for grazing at temperate pasture from May to June. Besides using the established schedule, other factors taken into consideration by the pastoral nomads to start the migration include indicators relating to climate, village location, animals etc. An important climatic indicator to begin migration is decrease or increase in temperatures. As autumn sets in, the ambient temperature at the villages starts dropping and appearance of frosts on the ground in the morning time are considered as indicators to commence journey down to the southern grazing resources. On the other hand, as spring sets in and the temperature starts rising in the southern grazing resources and animals begin to sweat, the transhumant nomads commence their north-bound return migration. The falling of leaves from deciduous trees and the drying up of grasses around their villages ultimately leading to feed shortage for the animals are also indicators for initiating the movement. For keeping their livestock healthy and achieving efficient production, pastoral nomads closely observe their animals and the environment prior to making strategic movements. There are several established transit camp sites along the migratory routes, and it is vital for the herders to plan before moving as to which camp to reach for the day. It is necessary to avoid being caught up in the heavy snow fall on mountains particularly in the autumn season. Other factors considered by the herders for migration of animals are the availability of forage resources along the way and the risk of attacks on livestock by carnivore predators (Namgay et al. 2013).

Transhumant nomads in Bhutan use grazing resources usually known as tsamdros, with usufructory rights; but they do not own the land. Tsamdro user rights exist in three forms, namely, private use, joint ownership between a couple of households and communal holdings. Herders possessing usufructory rights for private use of tsamdros have full rights to graze animals in tsamdro registered in their households' names. Private use and joint ownership types of tsamdros are more prominent in the western Bhutan. Conversely, only a few household herders had private tsamdro rights in the central-east Bhutan, and most usufructory rights for tsamdros in this region are held in common. Although the tsamdro was nationalized in the recent past, the pastoral nomads still continue to hold tsamdro ownership rights. Building tsamdro border structures such as walls and fences by the pastoral nomads of Merak and Sakteng indirectly assisted in management of the tsamdro and ensured harmony within the nomadic herder society by decreasing conflicts arising due to trespassing of domestic animals (Namgay et al. 2013; Wangdi and Norbu 2018).

References

Ahmad S, Radotra S, Singh JP, Verma DK, Sultan SM (2017) Ethnoveterinary uses of some important plants by pastoralists in Kashmir Himalaya. SKUAST J Res 19(1):121–128

Bashir Beigh A, Bashir S (2020) Pashmina goat farming in cold arid desert of Changthang and its impact on economy of Changpa nomads of this region, Goats (Capra) – from ancient to modern. IntechOpen. https://doi.org/10.5772/intechopen.84523

Bhasin V (2013) Pastoralists of Himalayas. J Biodivers 4(2):83–113

Bishop NH (2009) Himalayan mountain pastoralism. Mt Forum Bull 9(1):28–29. https://lib.icimod.org/record/14054

Blench R (2001) 'You can't go home again': pastoralism in the new millennium. ODI, London. https://odi.org/en/publications/you-cant-go-home-again-pastoralism-in-the-new-millennium/

Cui X, Wang Z, Tan Y, Chang S, Zheng H, Wang H, Yan T, Guru T, Hou F (2021) Selenium yeast dietary supplement affects rumen bacterial population dynamics and fermentation parameters of Tibetan sheep (*Ovis aries*) in alpine meadow. Front Microbiol 12:663945. https://doi.org/10.3389/fmicb.2021.663945

Das D, Jha NK, Maikhuri RK (2015) Fragmentation of pastoral grazing landscape and herd migratory routes: a case study from Indian Central Himalaya. Int J Life Sci 9(3):18–23

Ding L, Long R, Guo X, Bayer W (2020) Sheep on the Tibetan Plateau flock structure and breeding strategies. IGC Proceedings (1993–2023). p 30. https://uknowledge.uky.edu/igc/21/9-2/30

DLS (Department of livestock Services) (2018) Livestock statistics of Nepal. Department of livestock services, Lalitpur

Fan Q, Wanapat M, Hou F (2020) Chemical composition of milk and rumen microbiome diversity of yak, impacting by herbage grown at different phenological periods on the Qinghai-Tibet Plateau. Animals (Basel) 10(6):1030. https://doi.org/10.3390/ani10061030

FAO (2021) Seven reasons why pastoralism supports a better future. Food and Agriculture Organization of the United Nations. https://www.fao.org/fao-stories/article/en/c/1453839/

Farswan DS (2017) Tribes in Uttarakhand: status and diversity. Int J Multidiscip Res Dev 4(1):89–93

Gentle P, Thwaites R (2016) Transhumant pastoralism in the context of socioeconomic and climate change in the mountains of Nepal. Mt Res Dev 36(2):173–182

Ichikawa K (2012) Socio-ecological production landscapes in Asia. United Nations University Institute for the Advanced Study of Sustainability (UNU-IAS), Tokyo, Japan. https://www.semanticscholar.org/paper/Socio-ecological-Production-Landscapes-in-Asia-Ichikawa/fbe765a439bc45435fdce0b4a800d932dd02d8b3

Inamura T, Furukawa A, Yuuki F (2000) People and the livestock moving up and down in the Junvesi Valley. In: Yamamato N, Inamura T (eds) Nature of the Himalayas -the world of Sherpa and the nature of the mountainous areas. Yasaka Shobo, Tokyo, pp 182–198

Ingty T (2021) Pastoralism in the highest peaks: role of the traditional grazing systems in maintaining biodiversity and ecosystem function in the alpine Himalaya. PLoS One 16(1):e0245221. https://doi.org/10.1371/journal.pone.0245221

Isom J (2009) Tibet's nomadic pastoralists: tradition, transformation and prospects. Indigenous Affairs, no. 3–4, pp 6–13. https://www.iwgia.org/images/publications/IA_3-09_tibet.pdf

Iwata S (2000) The Hymalayas, the roof of the world-the geographical formation and the impact. In: Yamamato N, Inamura T (eds) Nature of the Himalayas -the world of Sherpa and the nature of the mountainous areas. Yasaka Shobo, Tokyo, pp 3–22

Kano K (2001) The Sherpa-the 20th century of the mountain people. Meikeido, Tokyo, p 277

Kheraj, Ahmed A, Ahmed P, Meenaxy (2017) Seasonal migration of Gujjar and Bakarwal in Poonch district of Jammu and Kashmir. Int J Acad Res Dev 2(4):364–368

Kirmani NR, Banday MT, Wani A, Pampori ZA, Adil S (2020) Routes adopted by Bakarwals during migration of livestock and its constraints. J Krishi Vigyan 8(2):242–246

Kreutzmann H (2011) Pastoral practices on the move-recent transformations in mountain pastoralism on the Tibetan Plateau. In: Kreutzmann H, Yong Y, Richter J (eds) Regional workshop in Lhasa, P.R. China. Pastoralism and rangeland management on the Tibetan Plateau in the context of climate and global change. Deutsche Gesellschaft für Internationale Zusammenarbeit (GIZ) GmbH, Bonn, pp 200–224

Kumagai H (2005) Natural resources management in mountainous areas of northern Nepal- literature survey on the management of pastures, farmlands and forests with the transhumance systems. Institute for Peace Science Center Hiroshima University (IPSHU). Research Report 35, pp 91–106

KVDC (Kanda Village Development Committee) (2016) Kanda village profile (in Nepali). KVDC, Kanda

Langda S, Zhang C, Zhang K, Gui B, Ji D, Deji C, Cuoji A, Wang X, Wu Y (2020) Diversity and composition of rumen bacteria, fungi, and protozoa in goats and sheep living in the same high-altitude pasture. Animals (Basel) 10(2):186. https://doi.org/10.3390/ani10020186

Malhotra A, Nandigama S, Bhattacharya KS (2022) *Puhals*: outlining the dynamics of labour and hired herding among the Gaddi pastoralists of India. Pastoralism 12:30. https://doi.org/10.1186/s13570-022-00237-5

McGahey D, Davies J, Hagelberg N, Ouedraogo R (2014) Pastoralism and the green economy – a natural nexus? International Union for Conservation of Nature and Natural Resources and

United Nations Environment Programme, Nairobi, pp 1–58. https://portals.iucn.org/library/sites/library/files/documents/2014-034.pdf

Miller D (2000) Tough times for Tibetan nomads in Western China: snowstorms, settling down, fences, and the demise of traditional nomadic pastoralism. Nomadic Peoples 4(1):83–109

Mitra M, Kumar A, Adhikari BS, Rawat GS (2013) A note on transhumant pastoralism in Niti valley, Western Himalaya, India. Pastoralism 3:29. https://doi.org/10.1186/2041-7136-3-29

Moktan MR, Norbu L, Nirola H, Dukpa K, Rai TB, Dorji R (2008) Ecological and social aspects of transhumant herding in Bhutan. Mt Res Dev 28(1):41–48

Namgail T, Bhatnagar YV, Mishra C, Bagchi S (2007) Pastoral nomads of the Indian Changthang: production system, landuse and socioeconomic changes. Hum Ecol 35(4):497–504

Namgay K, Millar J, Black R, Samdup T (2013) Transhumant agro-pastoralism in Bhutan: exploring contemporary practices and socio-cultural traditions. Pastoralism 3:13. https://doi.org/10.1186/2041-7136-3-13

Namgay K, Millar JE, Black RS, Samdup T (2014) Changes in transhumant agro-pastoralism in Bhutan: a disappearing livelihood? Hum Ecol 42(5):779–792

Pandey MR, Chhetri M (2005) Nomads and pastoralism: linkage with biodiversity conservation in upper mustang. Nepal Our Nat 3(1):42–49

Pandey NK, Somvanshi SPS, Kumar S, Prakash B, Singh CK (2020) Yak rearing practices by (Brokpa) pastoralist of Tawang Arunachal Pradesh. J Entomol Zool Stud 8(3):1067–1071

Parajuli DP, Paudel LN, Gyawali RR (2013) Changes in pastoral production systems in high-altitude village-rangeland interfaces in Nepal. In: Ning W, Rawat GS, Joshi S, Ismail M, Sharma E (eds) High-altitude rangelands and their interfaces in the Hindu Kush Himalayas. ICIMOD, Kathmandu, pp 48–54

Pasakhala B, Ghate R, Phuntsho K, Gentle P, Gurung J, Shrestha A, Gurung K, Thapa S (2021) Against the tide: the future of transhumant herders in the Kailash sacred landscape of Nepal. Mt Res Dev 41(4):R8–R15

Ramprasad V, Joglekar A, Fleischman F (2020) Plantations and pastoralists: afforestation activities make pastoralists in the Indian Himalaya vulnerable. Ecol Soc 25(4):1. https://doi.org/10.5751/ES-11810-250401

Sapkota A, Sapkota B, Khatiwada S, Neupane G (2020) Transhumant goat and sheep husbandry practices in high hills of Annapurna conservation area. IJEAB 5(5):1209–1214

Sareen P (2022) Impact of climate change on nomads in Jammu and Kashmir with a view of public health facilities, Policies and practices, Issue No. 132. Mahanirban Calcutta Research Group, Kolkata. http://www.mcrg.ac.in/PP132.pdf

Sharma VP, Köhler-Rollefson I, Morton J (2003) Pastoralism in India. A scoping study. DFID, New Delhi. https://assets.publishing.service.gov.uk/media/57a08ce2e5274a31e00014fa/ZC0181b.pdf

Sharma S, Sharma KK, Patel S (2019) Geographical analysis of internal migration and livelihood options for mountain sustainability along Indo-China border in Niti valley of Uttarakhand, Western Himalaya. J Himalayan Ecol Sustain Dev 14:49–65

Sharma A, Parkash O, Uniyal SK (2022) Moving away from transhumance: the case of Gaddis. Trees For People 7:100193. https://doi.org/10.1016/j.tfp.2022.100193

Simmy (2019) Pastoralists of the Himalayas: Gaddis of Bharmaur tribal region in Himachal Pradesh. South Asian Anthropol 19(2):99–107

Singh V, Goswami DC (2020) Changing pattern of transhumance pastoralism in upper Bhagirathi Basin: a case study of Jaad Bhotiya community. IJIRAS 7(4):174–176

Singh R, Kerven C (2023) Pastoralism in South Asia: contemporary stresses and adaptations of Himalayan pastoralists. Pastoralism 13:21. https://doi.org/10.1186/s13570-023-00283-7

SRM (Saipal Rural Municipality) (2020) Mainstreaming of gender and social inclusion into natural resources management plan for rural municipality level (2019–2023) (in Nepali). SRM, Saipal

Tiwari KR, Sitaula BK, Bajracharya RM, Raut N, Bhusal P, Sengel M (2020) Vulnerability of pastoralism: a case study from the high mountains of Nepal. Sustainability 12(7):2737. https://doi.org/10.3390/su12072737

Ura K (2002) The herdsmen's dilemma. J Bhutan Stud 7:1–43

Wangdi S, Norbu N (2018) Good fences are key to sustainable pasture management and harmonious pastoral society of Merak and Sakteng in Bhutan. Pastoralism 8:4. https://doi.org/10.1186/s13570-017-0106-0

Zhao Y, Zhao E, Zhang N, Duan C (2011) Mitochondrial DNA diversity, origin, and phylogenic relationships of three Chinese large-fat-tailed sheep breeds. Trop Anim Health Prod 43:1405. https://doi.org/10.1007/s11250-011-9869-2

Zhou JW, Guo XS, Degen AA, Zhang Y, Liu H, Mi JD, Ding LM, Wang HC, Qiu Q, Long RJ (2015) Urea kinetics and nitrogen balance and requirements for maintenance in Tibetan sheep when fed oat hay. Small Rumin Res 129:60–68. https://doi.org/10.1016/j.smallrumres.2015.05.009

4 Problems/Constraints Faced by the Migratory Animals and Pastoralists, Strategies for Their Amelioration, and Current Status of Migratory Pastoralism

Abstract

Sheep and goat pastoralism has been a unique feature of mountain societies wherein the animals exploit the seasonal abundance of alpine grazing areas during summers. However, the life for both the shepherds and their animals is tough and hazardous during upward and downward journeys and stay in the meadows. The en route problems (road blocks, snow, wind, animal sickness, loss of animals by accident, and wild animals attack), heavy parasitism in migratory animals, lack of marketing facilities for the produce, and deteriorating quality of pastures including overgrazing and contamination with weeds and poisonous plants constitute some of the constraints. A serious case of death of thousands of sheep and goats due to *Rhododendron* poisoning has been documented from Bara Bhangal area of Himachal Pradesh. Globalization and commercialization of range lands, establishment of national parks, sanctuaries, forests, and other land uses have seriously affected and put a question on the very future of transhumance and migratory livestock system. This chapter unravels the problems and constraints of the graziers and migratory small ruminants and presents the potential strategies for their amelioration and also throws a light on its current status. Are the migratory livestock system activities causing environmental degradation? Is the migratory practice declining? Most reports suggest that despite its importance, transhumant pastoralism was declining. The National Commission on Farmers (NCF) has also observed that livelihood of pastoralists and small holders has been threatened. The feedback taken from the shepherds through a questionnaire and our own views based on the facts provide answers to these and other related questions.

S. P. Singh, J. K. Malik, *Migratory Small Ruminant Farming System in the Himalayas-Peculiarities, Problems, Prospects, Climate Effects, Nutrition and Health*, https://doi.org/10.1007/978-981-96-6691-1_4

Keywords

Migratory animals constraints · En route problems · Inadequate marketing facilities · Degradation of pastures · Inadequate veterinary services · Status of migratory practice

Abbreviations

Amsl	Above mean sea level
CWDB	Central Wool Development Board
MSP	Minimum support Price
NCF	National Commission on Farmers
NGO	Non-governmental Organization
SAU/ha	Standard Animal Unit per hectare
USWDB	Uttarakhand Sheep and Wool Development Board
WOOLFED	Wool Procurement and Marketing Federation

Summer migration of the domestic livestock to the alpine meadows and back to the plains with the onset of winter every year has been a need based activity to feed the animals in the face of feed and fodder scarcity and escape the harshness of the climate, but the transhumant livestock system is full of difficulties and constraints. Recent studies from the north-western Indian states have shown that climate change and global warming have started affecting the annual migratory cycle of shepherds. In a case in the state of Jammu and Kashmir, it was observed that unusual early warm weather forced the Bakerwal and Gujjar community shepherds to start the upward summer journey in the month of March instead of April (Tufail 2014). Recently, attention of the animal husbandry experts has been drawn to look at the constraints from a scientific angle with a view to ameliorate them. In this chapter, we have endeavored to systematically review and analyze the main constraints and suggest the possible remedial measures to address them.

4.1 The Constraints/Problems

As stated earlier, migratory sheep and goats rearing on the Himalayas has been going on since time immemorial and is very common among weaker sections of the society in tribal areas of India, Nepal, and other neighboring countries. In spite of changing times and availability of diverse avenues for occupation, the system is still in vogue in pastoral communities. According to some estimates, 65% of sheep and 35% of goats in Nepal (Pandey and Gyawali 2012) and 90% of sheep and 70% of goats in Himachal Pradesh, India, (Sankhyan et al. 2016) are still reared under migratory system, and the system is very likely to continue. Over the years, several investigators have pointed out the problems and constraints in the system and suggested measures and strategies for improvement (Wani 2015; Nepali et al. 2007; Kheraj et al. 2017; Pandey and Gyawali 2012).

Based on our own research work experience in the ICAR Project on migratory small ruminants, we offer the following problems and constraints that we came across in the system and possible ameliorating measures.

4.1.1 Problems Concerning Graziers and Animals

At the base camp, the shepherds quite often face conflict situation with the farmers or owners of the land where the flocks stay during non-migratory period in winter. Sheep and goats indulge in grazing in the fields of neighboring farmers resulting in conflicts. The graziers also face host of problems and odd situations during ascending and descending journeys when their flocks pass through different villages and city roads. Sheep and goats wander on the roads and meet accidents or stray into sown or ripened crops inviting wrath from the land owners. People en route do not cooperate and do not allow flocks to graze or stay in their fields, although during earlier years, they used to invite shepherds to halt their moving flocks in their fields for manuring the soil.

A peculiar kind of conflict situation has been reported from pastures of Garhwal hills of the north-western state of Uttarakhand (Negi 1998). It was observed that petty day-to-day disputes arise in alpine pastures between shepherds manning the flocks of sheep and goats and Gujjar farmers rearing buffalo herds. For some years in the past, most shepherds used to raise sheep and goats which alone used to utilize high-altitude pastures. However, in the later years, Gujjars also started taking large ruminants to the same higher reaches of hills. The presence of Gujjars in these pastures (Bugyals) created a situation of conflict. It is not only that the shepherds did not want Gujjars at the alpine pastures, but the presence of two types of livestock species also created situation of competition for limited grass resources. Further, it was observed that sheep did not touch the grass which was already grazed by buffaloes and cattle. Consequently, the shepherds were forced to take away their flocks to further heights under protest. Dispute and conflict situations over grazing territory and area of temporary settlement in the forests have also been reported. In another case in Himachal Pradesh, the shepherds were reportedly attacked by forest officials over petty matters as reported in the local print media (Dainik Jagran 2009). Wani et al. (2017) have recently reported many conflict situations for Bakerwal shepherds of Jammu and Kashmir.

4.1.1.1 En Route Problems

(a) Road Blocks

The shepherd's migratory life is rather tough and hazardous. The ascending and descending journeys last several weeks during which large- and small-sized flocks with guarding dogs travel on roads avoiding regular traffic and vehicles. Some shepherds choose to travel during nights to avoid traffic congestion of the busy roads. The shepherds walk several miles carrying loads of salt, medicines, groceries/rations, heavy woolen clothes, and other essential stuff. Sometimes they have to face road blocks due to traffic congestion, landslides, tree felling, and road repairs and construction.

(b) Snows, Glaciers, and Winds

In the higher reaches of mountains, icy roads and rains are frequent and quite often landslides and boulders block the roads leaving no option before the shepherds but to face uncertain halts with their animals, dogs, and gadgets. At still higher elevations, the flocks are trapped in snow and rains with piercing cold winds especially while crossing the mountain passes. As mentioned in Chap. 7, such cases are frequent and increasing during the past few years due to changing climate. There is no provision of permanent shelter for shepherds and migratory animals, and heavy snow and rains destroy the temporary make shift shelters erected in the previous season. Incidences of unpredictable torrential rains, hail storm, and cloud bursting have become frequent due to climate change (Wani et al. 2017).

(c) Loss of Animals due to Attack by Wild Animals, Theft, and Accidents/Falls

At the present time, there is no authentic data on the loss of migrating sheep and goats due to attack by wild animals, thefts, and accidents, but feedback from local shepherds have revealed that the problem was of considerable magnitude. One press report from Palampur area of Himachal Pradesh claimed the loss of thirty lambs and kids due to attack by wild dogs and leopards during night halt (Dainik Jagran 2012). The livestock depredation losses due to attacks by snow leopards, common leopards, wolves, and wild dogs have been reported in the Nepalese Himalayas. Singh et al. (2006) reported a loss of 8–10% of the flocks due to wild animals' attack. Pandey and Gyawali (2012) have also reported substantial animal losses due to predatory and poisoning problems. Accidental deaths due to slip and falls while walking the narrow hill paths are quite common. Loss of animals by theft has also been reported in the local press. Such cases have come to the notice at the borders of Himachal Pradesh and Jammu and Kashmir in which case animals crossed over from one side to the other side and vice versa (Pritam and Prakash 1996). The shepherds on the other side lured the animals to their side. In addition, cases of loss of animals by injury from fall or accidents are also of common occurrence but have not been authentically recorded.

(d) Lack of Veterinary Health Services

In the absence of organized and timely veterinary help along the migratory routes, there are tremendous losses of animals in moving flocks due to severe diarrhea, skin infestations, infectious diseases, and untimely abortions. Most morbidities are due to PPR, parasites, foot rot, pneumonia, and poisoning (Wani 2015). It becomes difficult to carry the sick animas to distantly located veterinary dispensaries which are mostly available at low and mid elevations. Want et al. (2020) who studied the constraints of sheep breeders in union territory of Jammu and Kashmir have pointed out to the non-availability of quality medicines, lack of dipping tanks, and shortage of feed and fodder to the extent of 35.1%,30.8%, and 36.1% respectively.

4.1.2 Problems Related to Pastures and Grasslands

(a) Shortage of Land Area Under Fodder Crops and Consequent Shortage of Feeds and Fodder

The cultivable land area in the hills is under considerable strain due to difficult terrain and overgrowing population, and whatever land is available, it is being generally used to grow food and commercial crops. In this situation, the area under fodder crops is consistently on decline leading to ever increasing shortage of fodder. The problem of shrinking grasslands for grazing flocks in the union territory of Jammu and Kashmir has been pointed out by a few investigators (Kheraj et al. 2017). The availability of green fodder is scarce and digestibility of crop residues like straws and stoves was very low. In general, majority of the farmers were unaware about the importance of mineral mixture and common salt in the feed of domestic animals. A few shepherds offered salt to the animals only once a week. They were ignorant of the fact that quantity of salt requirement varies in the lactating and non-lactating animals.

(b) Problem of Overgrazing of Pastures and Contamination by Weeds and Poisonous Plants

Constant overgrazing of pastures has caused serious consequences on the health of alpine grasslands. Almost similar scenario has been reported from other neighboring highland Asian countries like Nepal and Pakistan. Large populations of small and large ruminants visit native grasslands every summer with no regulatory control on their movement and activities. Since migration is a regular feature, year after year, the pasture vegetation does not get sufficient time to regenerate. A need has been felt for the past sometime for an appropriate parameter which suitably reflects pressure on grazing land and whether a grassland was overgrazed. Scientists from Himachal Pradesh Agriculture University, Palampur, devised a unit called standard Animal Unit per hectare (SAU/ha) to indicate carrying or grazing capacity of the pasture (Singh 2000). A local non-descript cow weighing about 150 kg was taken as one SAU. The author recommended 2 SAU/ha as normal grazing capacity of a grassland. It was considered that the higher the value of SAU, the more the capacity of grassland. Singh (2000) reported an average SAU value of 5.67/ha in seven villages of Changer area of Kangra district of Himachal Pradesh. A clear and more specific unit to indicate grazing capacity of pastures is still required in order

Fig. 4.1 Flock of local goats and sheep grazing in Chopta area of District Rudraprayag, Uttarakhand (2608 m amsl). (Photograph by the courtesy of Mr. Ashutosh)

to assess the carrying capacity of natural grasslands located in the north-western Himalayan region.

Overgrazing and gradual degradation of forests and pastures (Figs. 4.1 and 4.2) have also affected the erosion of soil which has been increasing with time. Studies have shown that soil loss rates depended on sloppy nature of terraces and the forest cover. The erosion intensity was greater than 40 tonnes/ha/yr. in rain-fed overgrazed mountain slopes. In normal grazing lands, the intensity was recorded as 8 tonnes/ha/yr. (Shrestha 1997). Lack of awareness of harmful effects of overgrazing among the shepherds and confusion on grazing rights leading to unlawful uses of pasture have also contributed to the problem of overgrazing.

Pastures have also been found infested with poisonous plants such as *Lantana, Eupatorium*, and *Ageratum* species (Jithendran et al. 1998; Shrestha and Pakhrin 1989). Many plant species containing hydrocyanic acid and tannins have been found to invade pasture areas which may be unfit for livestock grazing (Pandey 1981) (Fig. 4.3).

4.1.3 Native Breed Versus Crossbred Sheep and Goats in Pasture

As stated earlier (Chap. 1), a lot of efforts have been made to improve the quality and quantity of sheep wool in the north-western Indian Himalayan states by adopting a scientific breeding policy and introducing exotic blood through crossbreeding. The indigenous sheep breeds produce small quantity of coarse wool and have poor growth rate. Adoption of crossbreeding has resulted in quick genetic gain to produce more wool of fine quality. For instance, in the union territories of Jammu and

Fig. 4.2 Degradation of pastures in Garhwal (Uttarakhand) Himalayas due to neglect and soil erosion. (Photograph by the courtesy of, Mr. Ashutosh)

Kashmir and Ladakh, the Kashmir Merino sheep produce 3–4 times more wool than the local sheep. All these efforts resulted in tremendous boost in wool production from the crossbred stock initially, but it was not always possible to maintain the production level in the fields due to inability to timely supply quality rams and semen and other inputs and maintain the right health status of crossbred animals. Adaptive problems to new terrain, variable elevations, and feed resources were other constraints.

According to local shepherds experience of introducing exotic breeds and improving the productivity has not been very rewarding in hilly terrains both in large and small ruminants. In most hilly regions, majority of the livestock owners prefer indigenous cattle, sheep, and goat breeds because the native animals were less demanding, less prone to problems, and provided practical means of using little available feed and other resources. Exotic animals and their crosses did not always perform well because of higher susceptibility to diseases and improper adaptation to the environmental and grazing conditions. This has also been substantiated by some of the foreign agencies which gathered data from tropical countries such as Nepal (Devendra and Gardiner 1995). Feedback from shepherds and employees of the Government Sheep Farm accompanying the flocks during migration revealed many peculiar problems of crossbred sheep. During our investigative work in the ICAR research project on migratory small ruminants, we also witnessed some of the problems while following the migratory flocks in the pastures of Kinnaur district of

Fig. 4.3 A pasture in Kashmir degraded due to road making. (Photograph by the courtesy of social activist Mr. Sankar Sridhar)

Himachal Pradesh. Purebred Rambuillet and Polworth sheep, and even the crossbreds, were not able to compete with the indigenous sheep in climbing up and grazing. The indigenous animals grazed speedily, raced, and climbed along the mountains, whereas the crossbred and purebred animals were sluggish or often hesitant in ascending the slopes (Pritam 1996, personal communication). The indigenous Gaddi sheep walked miles and miles continuously, whereas the crossbred and purebred animals looked tired and unwilling to walk while ascending. Based on the studies in Nepal, Shrestha and Pradhan (1995) have also reported that native breeds had better capability to survive in the harsh mountain environment, as well as in making use of the poor feed resources and management systems.

4.1.4 Inadequate Marketing Facilities

Availability of remunerative prices for the produce from sheep and goats, especially wool, has been a major impediment for the shepherds of all the three north-western Himalayan regions of India, viz., union territory of Jammu and Kashmir and states of Himachal Pradesh and Uttarakhand. Want et al. (2020) recently studied the constraints faced by the sheep farmers in Srinagar area of Jammu and Kashmir. Results showed that inadequate finance (97.8%) and lack of labor (12%), respectively, were among the major and minor constraints. Lack of proper marketing availability was a problem to the extent of 26.5%. These constraints have also been reported from neighboring country Nepal (Nepali et al. 2007). Private breeders and middlemen always had the upper hand and exploited the sheep breeders by negotiating with them during migration and even approaching the shepherds directly in the pastures (Wani 2015). The state-wise picture of the steps taken up by respective north-western Indian states to ensure right remunerative wool prices to the shepherds and take other welfare measures is presented below.

Although the union territory of Jammu and Kashmir is the second largest producer of wool (70 lakh kg of wool/year) in the country, proper infrastructure has not been created for its processing. As per the existing practice, raw wool is transported to the neighboring state of Punjab for processing and, subsequently, the finished products like blanket, apparel, and yarn are supplied to Kashmir valley. Unlike other states, Jammu and Kashmir procures the wool from shepherds at the minimum support price (MSP), but for want of processing facilities, it is not always possible to sell its raw wool stock. Merino wool is the finest wool and most in demand. The average yield of wool/sheep is around 2.28 kg as against the national average of 1.39 kg (Hameed 2021). Besides the unavailability of processing facilities and low remunerative price, there is no proper wool storage facility in the union territory, and as a result, a lot of wool get damaged. For all these reasons, the wool buyers prefer to purchase imported processed wool rather than the coarse wool from the Kashmir valley. The Government of Jammu and Kashmir has taken a number of steps to provide a push to the growth of wool industry. It established Marketing Cooperatives to ensure fair price for pashmina, raised genetic quality of pashmina goats, introduced machine shearing in place of hand shearing, and offered shearing machines to the shepherds at nominal price. The other steps required to be taken up include:

1. Establishment of wool processing plants, for scouring and combing; creation of adequate testing facilities, quality control measures, marketing services, infrastructure, and trained manpower.
2. Helping wool growers in securing proper economic status while being made aware of traditional management practices.
3. Fixing a minimum support price (MSP) fixed for wool growers, workers, shepherds, weavers, and middlemen involved in the wool trade to sustain their daily wages.
4. Promotion, production, shearing, procurement, and processing of wool.
5. Establishment of wool research and quality control laboratories.
6. Imparting training and technical know-how to sheep breeders to increase their skills in shearing and grading.
7. Initiation of other welfare measures for the sheep breeders.

In the state of Himachal Pradesh, most of the wool produced in Kinnaur district is reportedly sold to outsiders, whereas in Bharmour area, 78% of the produce was sold to local traders. Around that time, the average price of the wool in the two areas varied from Rs12 to 30/kg (summer shearing) and Rs 25 to 45/kg (winter shearing). This had always been the situation until the state government interfered at various levels. The Himachal Pradesh government established HP State Cooperative Wool Procurement and Marketing Federation Ltd. (Woolfed) in the year 1988, which came into operation in the year 1991–1992. Initially, the Himachal Pradesh government faced problem of substandard wool due to frequent shearing by the shepherds (2 or 3 times/annum instead of single shearing). But over the years, the situation has improved with the establishment of Wool Procurement and Marketing Federation. Woolfed directly procures the wool from shepherds and sells to agencies at the most reasonable prices.

In the newly formed state of Uttarakhand (created in the year 2000), the work pertaining to sheep and goat husbandry development has been assigned to "Uttarakhand Sheep and Wool Development Board" (USWDB) which came into existence in the year 2005 as a registered body under the Society Registration Act. The main objective of USWDB is improvement in the germplasm of livestock by adopting appropriate breeding policy for sheep, goat, and rabbit and thus facilitating an increase in livelihood of the farmers through skill development. The USWDB distributes elite germplasm rams and bucks produced at 11 Government Sheep Breeding Farms, 3 Goat Breeding Farms, and 5 Ram Raising Units. The USWDB is responsible for strengthening all State Animal Breeding Farms, introduction of latest technology and timely purchase, and market farmers' produce and to make farmers self-reliant through formation of self-help and producer groups.

The Central Government under the Ministry of Commerce and Industry and Ministry of Textiles has been helping the states and has established a "Central Wool Development Board (CWDB)" with the objective of setting up a common facility for wool scouring, dyeing, carding, carbonizing, spinning, and allied activities. The facilities are available to the government or non-governmental organization (NGO) and private entrepreneurs. The CWDB provides cost of machineries or 50% of the total project.

All these efforts have provided some impetus to sheep breeders, but there are still insurmountable problems whose solution is required to sustain sheep breeding as a profitable business in the Himalayan hills. Constant feedback from sheep and goat breeders and their unions is a must to tackle day-to-day problems. In Himachal Pradesh, Gaddi sheep breeders have formed a "Gaddi Union," which has elected office bearers and has held annual meetings to discuss their problems.

4.2 Strategies to Improve Migratory Livestock System and Recommendations to Ameliorate the Existing Problems

Since transhumance is a major activity in the Himalayas and it would continue for a long time to come, it is appropriate on the part of the administration of each state to sort out various issues concerning the shepherds/graziers, en route problems, problems affecting the health of the pastures and grasslands, introduction of fodder trees, and overcome the fodder shortage. The government should encourage the formation of village-level committees to be supervised by their representatives and district officials to sort out various issues. If possible, panchayat land should be made available to the migratory flocks at the base camps. Gujjars rearing buffaloes should be allowed to restrict their animals to lower heights and in no case they should disturb the grazing sheep and goat flocks at higher riches. The forest officials should be advised not to enter in any kind of conflict situation with the shepherds. Specific suggestions for improvement of the migratory livestock system and amelioration of problems being currently encountered are discussed below.

4.2.1 En Route Problems

As already described, there are multi-faceted problems faced by shepherds and flocks, all of which perhaps cannot be addressed. It is not feasible to arrange or provide separate and independent migratory routes for large number of shepherds. The present routes have been existing from time immemorial and the same has been followed year after year.

Based on our own experience (first author) of investigative work on migratory flocks in the ICAR project and the feedback obtained from frequent discussions with the animal husbandry officials, animal attendants, and private shepherds, we offer the following recommendations to improve the migratory rearing system in the mountain regions.

(a) Setting Up of Mobile/Temporary Dispensaries/Hospitals Along the Migratory Routes

During the process of interaction, most shepherds expressed the opinion that during the course of migration, the sick animals cannot be taken to hospitals and dispensaries located far away from the migratory routes. In the flocks heading to alpine pastures during summer, our scientists observed that a large number of migratory sheep were suffering from skin diseases, parasitic infestations, severe anemia, diarrhea, and pyrexia. Extent of parasitic infestations and other disease conditions affecting migratory sheep and goats has been discussed in Chap. 10. We noticed several shepherds carrying syringes and needles and some age-old antibiotics which they use in all kinds of diseases. Of antimicrobial drugs, Terramycin (oxytetracycline) was a familiar name among most shepherds. It is, therefore, recommended that during the migratory season, the Department of Animal Husbandry should set up mobile dispensaries around the main migratory routes so that the diseased animals could be attended to on the road side itself and and medication is prescribed.

(b) Access to Medicines and Dietary Supplements

Veterinary drugs, mineral mixture/supplements, and salt blocks should be made available to the registered shepherds on subsidized rates. Problems relating to mineral deficiency especially salt hunger in migratory sheep and goats have been discussed in Chap. 8.

(c) Training to the Shepherds

Animal Husbandry Department and State Agriculture University should regularly conduct free short-term skill development and training courses for the shepherds/animal graziers in all aspects of management of the flock, treatment of diseases, nursing, and first aid. This would enable to create an army of "Flock Friends" who will be registered and will be available as and when required. It is not always possible to depute veterinary doctors or experts in the alpine pastures or to accompany every migratory flock. The government should constitute an independent standing body comprising of officials from the departments of animal husbandry, revenue, and administration to monitor disease status and keep accurate records of the flocks/animals migrating to the alpine pastures in every season.

(d) Disease Diagnosis Center

An independent disease diagnosis laboratory with a separate section for migratory animals should be set up and entrusted work for monitoring the disease problems and history of the migratory animals.

(e) Laboratory to Handle Toxic Plant Ailments

Toxicity in migratory animals due to poisonous plants and weeds during ascending and descending journeys is quite common (Uniyal et al. 2005), but no analytical facilities exist for the identification of the specific toxic principles. Therefore, it is vital to establish an independent laboratory in every state/union territory to diagnose such cases and quickly provide appropriate treatment.

(f) Setting Up of Animal Check Posts

It is highly desirable to set up animal check posts at strategic points with facility for vaccination, deworming, and treatment as well as to check possible disease carriers. We were told that the dipping tanks for skin infestations were provided at some places, but their number were too inadequate and most of them were not in use for one or the other reasons. We also found that animals from neighboring states cross through inhabited areas of Himachal Pradesh. For instance, sheep and goats from as far as Garishankar area of Punjab were found entering in the village of Gagret on way to crossover the Dholadhar mountain ranges to reach the pastures of Bharmour in Chamba district without passing through any check posts. Such incidences should be avoided and precautionary measures taken.

4.2.2 Degradation of Pastures

Fodder production availability is the most important limiting factor for livestock rearing in the north-western Himalayan states. The current fodder production output is too meager to meet the requirement of farm animals (Mir et al. 2016; Gupta et al. 1998). Although the region is endowed with natural grasslands and alpine meadows and trees, their overexploitation, poor management, weed infestation, and uncontrolled grazing have deteriorated their health deepening the problem further. There is an urgent need to check further damage to the pastures. For this, the state governments have to take stringent steps taking all stake holders into confidence. An awareness campaign has to be launched among farmers and shepherds about the critical issues involved in deterioration of pastures and of the dangers of not acting timely to restore their health. It is absolutely essential to formulate a comprehensive, mandated, and holistic grazing policy requiring periodical shutting down of pastures and grasslands to avoid their overexploitation. Simultaneously, alternate arrangements would be required to acquire fodder from other states for the lean season. In addition, the shepherds and graziers should be given subsidies to enable them to purchase feeds and fodders. Also steps should be initiated to improve the pastures by adopting innovative scientific methods. We intend to present a frame work of the suggested steps without giving specific details.

4.2.3 Strategies for Improvement of Pastures

(a) Long-Term Measures
   (i) Formulate a long-term grazing policy for respective state. Grazing should be permitted based on carrying capacity of the pastures.
   (ii) Change the age-old custom of yearly grazing cycle by the local nomads which does not give regeneration time to the grasses. Shut down the pastures for some time, and the graziers may be explained the need for such action. In addition, there is a need for integrated nutrient management for the grasslands.
   (iii) Establish fodder banks in surplus zones by arranging the supplies from outside the states and make the feed available to the shepherds during the time they are forbidden from using the alpine pastures.
   (iv) Provide subsidies to the graziers and nomads on soft loans to enable them to procure fodder from sources of their liking.
   (v) Make available quality seeds of improved grasses, legumes, and fodder trees to the local farmers for improvement in fodder production locally.

(b) Other Measures
   (i) Control of Weeds and Poisonous Plants

Weeds in the pastures are a great nuisance as they compete with original grasses for light and nutrients. Common weed species in Himalayan grasslands are *Lantana*, *Ageratum*, and *Eupatorium* and some species of *Rhododendron*. A serious incidence of *Rhododendron* poisoning leading to death of 3500 sheep and goats has been reported in Bara Bhangal area of Himachal Pradesh in the year 2005 (Uniyal et al. 2005). On the other hand, Lamsal et al. (2018) have reported noxious plants such as *Acconitum, Anemone, Adonis, Aquilegia, Cincifuga, Clematis, Lepidium, Artimisia, Ranunclus, Stipa sorbinia,* and *Sambucus* as of great consequence to cause deterioration of grass cover. Recently, Jamloki et al. (2022) have documented a number of poisonous plants of the Indian Himalayas which often affect grazing livestock. Prominent of these comprise *Abrus precatorius* Linn, *Ageratum conyzoides* Linn, *Aloe vera* (L.) Burm.f, *Cannabis sativa* Linn, *Capsicum chinense* Jacq, *Datura innoxia* Mill, *Digitalis purpurea* Linn, *Eupatorium adenophorum* Spreng., *Lantana camara* Linn., and *Rhododendron campanulatum* D. Don.

The main technique adopted for eradication of weeds include uprooting or cutting them and application of chemical weedicides (Dev 2001). It is desirable to develop low-cost technology for weed control in the alpine pastures.

(ii) Introduction of Legumes/Grasses

Forage legumes have a high nutritive value and are important because they enrich the nitrogen content of the soil and improve the pasture quality. For various regions, different legume varieties have been identified and recommended for improving the quality of pastures (Hazra 1998).

(iii) Fertilizer Application

Most natural Himalayan pastures and rangelands have been thriving without application of any fertilizers except the organic matter added by way of excreta from large and small ruminants and decay of vegetation. However, it is now generally believed that overexploitation of the range lands may result in deficiency of micro- and macro-mineral elements in soil causing their reduced fertility. Judicious use of fertilizers was found to restore vegetative growth of grasslands. Sharma and Ghosh (1997) reported significant increase in productivity and carrying capacity of grasslands in Shimla district of Himachal Pradesh by application of nitrogen alone.

4.3 Status of Migratory Livestock System—The Declining Trend

As already stated, transhumance of small ruminants in Hindu Kush Himalayan region countries has been going on for ages. Bhasin (2011) has reported that there may be about 200 million nomads in the world. In India, more than a dozen tribes including Gaddis, Gujjars, Kinnauras, Bakarwals, Bhutias, and Changpas have been engaged in pastoralism for centuries. Many of these tribes have got scheduled tribe status by the government of India which carries many privileges.

Although no accurate statistics is available region or country-wise, there is no doubt that the practice of pastoralism is declining gradually due to various reasons in spite of government efforts. Most studies on transhumance in India and neighboring Himalayan countries have revealed that the age-old practice of migration is gradually dwindling. Misri and Dev (2001) reported that in Himachal Pradesh, many Gaddi families have turned from migratory to sedentary pattern and switched to other economic activities in Bharmour area. Choudhary and Garkoti (2021) who studied the transhumance practice of kinnaur community in three villages of Kinnaur districts of Himachal Pradesh observed that, after the year 1980, the number of herds of sheep and goats participating in migration have declined by 44%, and the proportion of households practicing transhumance declined from 67 to 42%. In another but very recent work from Himachal Pradesh, Sharma et al. (2022) from Kangra district have reported that the population of sheep and goats reared by Gaddis was declining year by year as reflected by the lower number of grazing permits being issued annually by the Forest Department. Ladon et al. (2023) who studied the current state of agro pastoralism in the Trans-Himalaya region of Union territories of Ladakh reported abandonment of regular seasonal migration by many shepherds and a decrease in the number of households practicing animal herding. Pastoral practices were also threatened by the wild predators as well as by global warming-induced occurrence of extreme climatic events. Lack of interest by the younger generation also contributed to the decline of traditional practices. Cursing the state polices and anti-pastoral agenda by the governments, Singh and Kerven (2023) in a recent article have brought into light various stresses and adaptations of Himalayan pastoralists. The authors have discussed in depth the factors that are leading to sedentarization of pastoralists including absence or shying away of the young generation from this traditional profession. The authors have rightly given the example of Dokpas of Sikkim at the Indo-Tibetan borders.

In a recent study in Nepal entitled "Against the Tide: The future of Transhumant herders," Pasakhala et al. (2021) reported that despite its importance, transhumant pastoralism was declining in the Hindu Kush Himalayan region. It was happening due to various reasons including socioeconomic, political, institutional, ecological, and climate change. In yet another study from Nepal covering 3 protected areas of high mountain region, Tiwari et al. (2020) observed a decreasing trend of livestock farming mainly due to severe labor shortage, disinterest of younger generation, hardship involved in herding, climate change, and several other reasons.

The practice of transhumant pastoralism has also been in operation in Bhutan for centuries, but it is currently on decline owing to modern changes in life style, children's education, industrial revolution, etc. In order to find out the current status of transhumance, Namgay et al. (2014) conducted an interview and a structured survey of 106 households in villages of Bhutan and reported 31% decline in herders pursuing transhumance pastoralism during a period of 20 years (1990 to 2010). Bhasin (2011) has reported that globalization and commercialization of pastoral lands have created hurdles for the pastoralists. The change in policies of Forest Department has been depriving them from their traditional rights. Many shepherds have been forced to leave their transhumance way of life. The National Commission on Farmers (NCF 2006) has also observed that livelihood of pastoralists and small holder farmers have been threatened by the progressive loss of their grazing land and denial of grazing permits and their nonparticipation in forestry programs. The Commission has recommended restoration of traditional grazing land to the pastoralists.

Contrary to these prevailing views, researchers from Himachal Pradesh have expressed the optimism that the migratory system would remain as an integral part of the agro-economy of hill farming for two reasons (Sankhyan et al. 2016). Firstly, the system was a low input activity which counterbalances its constraints and makes it profitable venture. Secondly, the system being tagged with cultural heritage of tribal population will ensure its propagation. Effective institutional extension support and development of solutions to the constraints would help in its sustainability. In places where migratory practice has been abandoned, the shepherds have switched to a variety of other professions such as taxi driving, sedentary farming, hotel business, and other jobs. The gradual decline in transhumance activity by shepherds and underutilization of vast natural alpine resources are becoming a cause of worry for planners.

In summary, the big question is what is leading to the constant decline in pastoralism? To answer this question, we will have to turn to the question why the practice of animal migration started centuries ago and became a stable feature of hill societies. The answer is obvious and simple. Transhumance was a need-based economic activity for marginal land holders to be able to feed large herds of domesticated cattle, sheep, goats, and other animals and escape from climate stresses. The idea was novel to exploit the gift of nature available in the form of alpine grasslands. This was the best way to convert low-quality plant resources into high quality animal food. Nomadism seems to have taken birth centuries ago as a means of survival both of human beings and of domestic animals in the face of limited land, food, feed, and housing resources. There was perhaps no other way to sustain the life and domestication.

The decline in pastoralism in the Himalayas seems to be directly related to the effect of globalization, industrialization, commercialization, and the faulty government policies. The challenges of declining transhumance pastoralism and future strategies have been under discussion by several investigators (Namgay et al. 2013; Namgay et al. 2014; Namgay et al. 2021; Sharma et al. 2022). Specifically, the following factors seem to be responsible for a decline in transhumance/nomadic way of shepherds' life.

(a) *Hardships to the shepherds*

Limited remuneration, tough life involving long distance travel with flocks, en route problems during ascending and descending journeys, unprotected stay in harsh terrain of alpine during rain and snow.

(b) *Flock/herd Health*

Heavy parasitic infestation and other diseases coupled with non- availability of veterinary care on the migratory routes have been the major obstacles. Setting up of helplines for the migratory flocks would be of a great help (Sharma et al. 2022).

(c) *Problems in the Disposal of produce*

Problems relating to the disposal of produce such as wool and meat at lower remunerations, dependence on middleman, lack of facility to collect the produce from the alpine areas.

(d) *Status of grasslands*

Deterioration of pasture lands due to overcrowding and overgrazing, presence of weeds, invasive plant species and poisonous plants, shrinkage of grasslands and forests due to expending agricultural land, construction of national parks, wild life sanctuaries, dams and other infra-structural facilities.

(e) Inadequate government action plans for sustainable pasture and forest management, forcible evictions, predation and loss of animals, depriving the shepherds of grazing rights ownership and imposing heavy taxes.

(f) *Climate change*

Climate change has been playing havoc with transhumant pastoralists and their animals. Abrupt shift and changes in weather conditions especially unpredictable rains, snow, or thunderstorm have become common for the past few years and have made the life of shepherds more troublesome affecting the transhumance practice more difficult. There have been innumerable cases of shepherds and the animals having been trapped in erratic and unpredictable weather leading to heavy mortality of animals (Sharma et al. 2022).

(g) *Changing the role of society due to globalization*

Rapid urbanization and changes in values of the society, breakdown of village customs and institutions, expanding literacy rate/education level, shrinkage of family size, vast and better opportunities for jobs to the youth and their distant postings in urban cities, and their lack of interest in farm and livestock activities at home.

4.4 The Question of Environmental Degradation

Lately environmentalists and other agencies have often blamed that the Himalayan tribes and grazing pressure by pastoralists were responsible for deterioration and degradation of natural resources and the general environment. Recently, Haq et al. (2021) studied the long-term impact of transhumance pastoralism and associated disturbance on forests of the western Himalayas. It was reported that in recent decades, due to constant anthropogenic pressure and considerable changes witnessed in the climate of the region, many plant species of this region have been threatened. The authors also assessed the impact of nomadic settlement and associated disturbances on ecosystem and fire incidence in high-altitude forests of the western Himalaya, India. They observed a significant difference in plant diversity and higher values of fire incidences. Climate change and overgrazing of Himalayan pastures have certainly contributed to the environment degradation (Escarcha et al. 2018; Saikawa et al. 2019). However, the real culprits of environmental degradation are industrialization and expanding infrastructure like roads, tunnels, dams, factories, and other interventions. Unfortunately, an elaborate and holistic scientific study on the impact of migratory livestock system on environmental damage has not been conducted to unravel the causes. There is no evidence to suggest that Gaddi graziers and other tribesman were responsible for degradation of environment in any way (Glatzle 2019; Mitloehners 2020).

4.5 Feedback from the Shepherds on Migratory Practices and Related Issues in Himachal Pradesh

In order to ascertain the views of local shepherds on issues concerning migratory practices and constraints, we directly approached 11 shepherds/grazier families, out of which seven responded. A detailed 37-point questionnaire was circulated to them for comments individually. Table 4.1 presents the details of the feedback on the 37 points obtained from the shepherds'/graziers' families.

Table 4.1 Responses from seven private shepherds/graziers on issues related to migratory practices

Item		Questions asked	Responses
A	*General information*		
	A1	Family size	4–8
	A2	Area of land in the village	1–8 acres
	A3	Number of members who accompany the flock	1–3
	A4	Breed of migratory animals	Goats: Gaddi(G): 35–250,Rampur Bushair(RB):170, Bakarwal: 45 Sheep: Gaddi 200–350, RB 130–600, crossbred(CB) 100–200(Gx R)
	A5	Vaccination status against diseases	All shepherds except one responded in affirmative that animals were vaccinated against foot and mouth disease (FMD) and PPR
	A6	Facilities of feed in the village (base camp)	Most replied that local grass was the main source of feed, but all said that concentrate is not given to animals. The response about offering of salt to animals varied from once a week (two respondents) to once a fortnight (two respondents)
	A7	Period of involvement in migratory practice	No idea but for long time (three respondents); from time immemorial as it is a family tradition (three respondents); About 7 years (one respondent)
	A8	Purpose of migration	There were varied answers including: (a) In search of fodder as alpine grasses were highly nutritious and to maintain the age old tradition (all respondents) (b) To escape from the hot environment and for selling at higher prices (three respondents) (c) Also to collect medicinal herbs (one respondent)

(continued)

Table 4.1 (continued)

Item		Questions asked	Responses
B	*Upward journey*		
	B1	Commencement of journey	Between first week of April to first week of May (most respondents). May last to first week of June every year (two respondents)
	B2	Route of upward journey	Same every year
	B3	Sources of feed for animals during ascending journey	Feed from road side or grass available on the fields, but sometimes, there is resistance from the local people
	B 4	Management of sick animals	Local veterinarians help if available en route in the lower heights; otherwise self-help or medicines brought from home are used. Some animals cannot make it to the heights
	B 5	Fate of sick animal if not cured	No option, but to sell them or dispose off otherwise
	B 6	Were there ever heavy animal losses due to disease outbreak or attacks by wild animals?	Heavy losses of flock during 1991 when 290 sheep and goats died due to PPR. Also heavy losses due to attacks by wild animals (six of seven respondents)
	B 7	Did you ever insure your animals?	All respondents replied negative
	B 8	Did you ever get your animal vaccinated?	Against FMD and PPR after heavy losses were reported (all respondents replied affirmative)
	B 9	Did you ever get compensation for loss of animals during migration?	Received Rs. 13,000 from the state administration long ago (five respondents). Two said they could not apply for the losses in time
	B 10	How do you manage if animals come in estrus during migration?	There was no option but to have natural mating from males in the flock (most respondents)
	B 11	How do you manage if a pregnant animal gives birth during migration?	The situation is managed at our own (all respondents)
	B 12	Number of animals sold during upward journey	5–15 (three respondents), 25–30 (three respondents), 70–80 (one respondent)
	B 13	How do you manage your food during migration?	Ration brought from home and, if possible, some items bought locally (most respondents)
	B 14	Where do you stay during migration?	Stay in open or in a temporary shed or take shelter under rock as we have to take care of the flocks
	B 15	Other problems encountered during migration	1. Lack of fodder and grazing areas to the flock and non-cooperation and hostility from the local people 2. Sometimes source of drinking water is not available 3. Lot of problems with Gujjars in some areas who don't abide by rules 4. Losses to the flock due to attacks by wild animals 5. In the absence of veterinary aid en route, sick animals cannot be properly looked after

(continued)

Table 4.1 (continued)

Item		Questions asked	Responses
C	*Information about high-altitude pasture*		
	C1	Name of high-altitude pasture commonly visited	4 respondents: Marhod Dhar/ Kalota Dhar tehsil churah, Bhaqnu Dhar Himgiri range Chamba/Radol Dhar Himgiri block Salooni Fifth respondent: Huden Dhar (Pangi) Sixth respondent: Dande pastures Seventh respondent: Spon-Dande Dist Simla
	C2	Period of stay?	June to August. Sometime June to mid-October (most respondents)
	C3	How do you manage your food at high-altitude pastures?	Ration brought from home and, if possible, some items bought from the local
	C4	Where do you stay?	Temporary sheds are constructed which need repair every year. Sometimes support of rocks serve as shed
	C5	What are the main problems/diseases at high altitude?	Respiratory distress, indigestion/gastroenteritis, and diarrhea
	C6	How do you treat sick animals?	Self-help as there is no other option
	C7	Fate of sick animals that do not recover	Leave them on their fate or sometimes slaughter and dry the carcass and sell/consume them
	C8	Have there been heavy losses of your flock at high-altitude pastures?	Heavy losses were incurred in the past but no compensation was granted (3 respondents). Never incurred any major losses (other respondents)
D	*Downward journey*		
	D1	Route	Same routes as in upward journey are followed
	D2	Commencement of downward journey	Late August (5 respondents); first week of September (2 respondents)
E	*Would your new generation/ children continue the age-old practice of migration?*		Majority said no and wanted their children to get good education, be graduate, and join government service. Others replied if children wanted, they could continue
F	*How do you manage education of your children?*		Children not accompanying the flocks go to school
G	*Would you like the government to tamper with high-altitude pastures?*		Majority replied negative, but one shepherd family told that earlier attempts by government to improve pastures *have failed*
H	*Satisfaction with your legal rights/ land deals in plains/high altitude*		Majority of shepherds replied affirmative

The feedback from the shepherds on different issues is self explanatory. It appears that flock health is of major concern for which clear remedial measures are required. Foot and mouth disease (FMD), Peste des petits ruminants (PPR), and heavy parasitism in the flocks require special attention. Setting up of animal check posts along the migratory routes during ascending and descending journeys, mass vaccinations against important diseases, and mandatory deworming of animals would be the right strategies for improvement.

References

Bhasin V (2011) Pastoralists of Himalayas. J Hum Ecol 33(3):147–177

Choudhary DK, Garkoti SC (2021) Weakening socio-cultural linkages imperils the sustainability of the transhumant pastoralism: a case study of *Kinnaura* community of western Himalaya, India. Int J Ecol Env Sci 47(2):91–100

Dainik Jagran May 4 (2009). www.jagran.yahoo.com

Dainik Jagran December 11(2012). www.jagran.yahoo.com,

Dev I (2001) Problems and prospects of forage production and utilization of Indian Himalaya. ENVIS Bull: Himal Ecol 9(2):11–18

Devendra C, Gardiner P (1995) Global Agenda for livestock research- proceedings of consultation for South Asia Region, International Crop Research Institute for Semi Arid Tropics(ICRISAT) Asia centre, Patencheru. https://cgspace.cgiar.org/bitstreams/9f2721ab-7880-4979-ae7a-7f78f39c5100/download

Escarcha JF, Lassa JA, Zander KK (2018) Livestock under climate change: a systematic review of impacts and adaptation. Climate 6(3):54. https://doi.org/10.3390/cli6030054

Glatzle A (2019) Domestic livestock and its alleged role in climate change. IntechOpen. https://doi.org/10.5772/intechopen.80389

Gupta RK, Gupta T, Kaushal PS, Pathania MS (1998) A study on the status of fodder balance in Himachal Pradesh. Oecologia Montana 7:39–42

Hameed A (2021) Sustenance of Kashmir's wool industry depends on adapting and adjusting its practices to the changing times. 13 Dec 2021. https://www.jkpi.org/sustenance-of-kashmirs-wool-industry-depends-on-adapting-and-adjusting-its-practices-to-the-changing-times/

Haq SM, Yaqoob U, Calixto ES, Kumar M, Rahman IU, Hashem A, Abd Allah EF, Alakeel MA, Alqarawi AA, Abdalla M, Lone FA, Khan MA, Khan U, Ijaz F (2021) Long-term impact of transhumance pastoralism and associated disturbances in high-altitude forests of Indian Western Himalaya. Sustain For 13:12497. https://doi.org/10.3390/su132212497

Hazra CR (1998) Integrated nutrient management for sustainable forage production in pasture and silvi-pastures. Fertilizer News 43(3):33–45

Jamloki A, Trivedi VL, Nautiyal MC, Semwal P, Cruz-Martins N (2022) Poisonous plants of the Indian Himalaya: an overview. Metabolites 12(6):540. https://doi.org/10.3390/metabo12060540

Jithendran KP, Sharma OP, Dawra RK, Makkar HPS, Singh B (1998) Survey on animal husbandry practices, bottlenecks in animal production and strategies for improvement: a profile of selected villages in Himachal Pradesh. ENVIS Bull: Himal Ecol 6(2):6–13

Kheraj, Ahmed A, Ahmed P, Meenaxy (2017) Seasonal migration of Gujjar and Bakarwal in Poonch district of Jammu and Kashmir. Int J Acad Res Dev 2(4):364–368

Ladon P, Nüsser M, Garkoti SC (2023) Mountain agropastoralism: traditional practices, institutions and pressures in the Indian trans-Himalaya of Ladakh. Pastoralism 13:30. https://doi.org/10.1186/s13570-023-00289-1

Lamsal P, Kumar L, Aryal A, Atreya K (2018) Invasive alien plant species dynamics in the Himalayan region under climate change. Ambio 47(6):697–710

Mir NH, Ahmad S, Verma DK (2016) Livestock rearing- sectorial status and fodder-feed strategies in Kashmir Himalaya. Ann Biol 32(2):253–259

Misri B, Dev I (2001) System study on sedentary Gaddis of Kangra valley. ENVIS Bull: Himal Ecol 9(1):14–20

Mitloehners F (2020) The scientist setting the record straight on cows and climate change. https://www.darigold.com/the-scientist-setting-the-record-straight-on-(cows-and-climate-change/

Namgay K, Millar J, Black R, Samdup T (2013) Transhumant agro-pastoralism in Bhutan: exploring contemporary practices and socio-cultural traditions. Pastoralism 3:13. https://doi.org/10.1186/2041-7136-3-13

Namgay K, Millar JE, Black RS, Samdup T (2014) Changes in transhumant agro-pastoralism in Bhutan: a disappearing livelihood? Hum Ecol 42(5):779–792

Namgay K, Millar JE, Black RS (2021) The future of transhumants' sustainable resource use in Bhutan: pressures and policies. Front Sustain Food Syst 5:618351. https://doi.org/10.3389/fsufs.2021.618351

National Commission on Farmers(NCF) (2006) A draft national policy for farmers, 4rth report. New Delhi, India. https://www.agriwelfare.gov.in/sites/default/files/NCF5%20Vol.-2%20%281%29.pdf

Negi RS (1998) Symbiotic relationships between man, animal and nature: a study of the Gujar of Garhwal. In: Saraswati B (ed) Life-style and ecology. IGNCA and DK Printworld, New Delhi

Nepali MB, Tiwari MR, Sapkota S, Poudel HP, Acharya BR, Gautam S (2007) Marketing constraints to goats in the western hill of Nepal. Nepal Agric Res J 8:95–102

Pandey KK (1981) Tree and conservation of cultivated lands in Nepal. In: Tingsanchali T, Eggers, H (eds) Proceedings of the South East Asian region symposium on problems of soil erosion and sedimentation. Asian Institute of Technology, Bangkok, p 233–240

Pandey LN, Gyawali R (2012) Constraints and potential of goat and sheep production under transhumance management system in the high mountainous. Paper presented at the conference "Research and Development strategies for goat enterprises in Nepal". https://www.researchgate.net/publication/237054450

Pasakhala B, Ghate R, Phuntsho K, Gentle P, Gurung J, Shrestha A, Gurung K, Thapa S (2021) Against the tide: the future of transhumant herders in the Kailash sacred landscape of Nepal. Mt Res Dev 41(4):R8–R15

Pritam (1996) (Personal communication) Loss of animals by Theft

Pritam, Om Prakash (1996) (Personal communication) Farmers perception about migratory small ruminants farming system

Saikawa E, Panday A, Kang S, Gautam R, Zusman E, Cong Z, Somanathan E, Adhikary B (2019) Air pollution in the Hindu Kush Himalaya. In: Wester P, Mishra A, Mukherji A, Shrestha BA (eds) The Hindu Kush Himalaya assessment—mountains, climate change, sustainability and people springer nature, Cham, pp 340–353

Sankhyan V, Dogra PK, Thakur YP (2016) Attributes of migratory goats and sheep farming and impact of some improved management strategies en-route migration in adopted flocks of Western- Himalayan region of India. Indian J Anim Sci 86(9):1079–1084

Sharma JR, Ghosh AN (1997) Grassland and productivity in the Indian Himalayas and performance of introduced forage species in cold and semi-arid environment. IGC Proc 2024(1993–2023):20. https://ukknowledge.uky.edu/igc/1997/session21/20

Sharma A, Parkash O, Uniyal SK (2022) Moving away from transhumance: the case of Gaddis. Trees For People 7:100193. https://doi.org/10.1016/j.tfp.2022.100193

Shrestha DP (1997) Assessment of soil erosion in the Nepalese Himalaya: a case study in Likhu Khola Valley, Middle Mountain Region. Land Husbandry 2(1):59–80

Shrestha NP, Pakhrin B (1989) Effect of different tree fodders on milk production in buffaloes. PAC Tech Paper 117,. Pakhribas Agricultural Centre, Dhankuta District

Shrestha HR, Pradhan DR (1995) Research priorities for animal agriculture by agro-ecological zone in Nepal. In: Devendra C, Gardiner P (eds) Global agenda for livestock research-proceedings of the consultation for the South Asia region, international crop research Institute for Semi Arid Tropics (ICRISAT) Asia center. International Livestock Research Institute, Patancheru

Singh G (2000) Trends and constraints in livestock rearing in Changar areas of Kangra district of Himachal Pradesh. ENVIS Bull: Himal Ecol 8(2):23–26

Singh R, Kerven C (2023) Pastoralism in South Asia: contemporary stresses and adaptations of Himalayan pastoralists. Pastoralism 13:21. https://doi.org/10.1186/s13570-023-00283-7

Singh DR, Kaul S, Sivaramane N (2006) Migratory sheep and goat production system: the mainstay of tribal hill economy in Himachal Pradesh. Agric Econ Res Rev 19(2):387–398

Tiwari KR, Sitaula BK, Bajracharya RM, Raut N, Bhusal P, Sengel M (2020) Vulnerability of pastoralism: a case study from the high mountains of Nepal. Sustainability 12(7):2737. https://doi.org/10.3390/su12072737

Tufail M (2014) Impact of climate change on the seasonal movement of the Gujjar and Bakerwals: community perceptions. IOSR J Environ Sci Toxicol Food Technol 8(1):41–46

Uniyal SK, Lal B, Kumar A (2005) Mass mortality of livestock in the high altitude areas of western Himalaya. Curr Sci 88(4):534–535

Wani SA (2015) Small ruminant rearing among Bakerwal pastoralists of Jammu and Kashmir- an exploratory study. (PhD thesis). National Dairy Research Institute, Karnal

Wani SA, Meena BS, Ganaie AH, Mir NA (2017) Problem identification and climate change perception of Bakerwal pastoralists of Jammu & Kashmir. J Anim Res 7(1):99–105

Want QH, Banday MT, Adil S (2020) A study on constraints perceived by sheep farmers of district Srinagar, Jammu and Kashmir. Anim Sci Q 1(1):26–30

Part II

High Altitude Acclimatization and Adaptation of Small Ruminants

5 Adaptive Responses to Altitude and Walking Stress in Small Ruminants

Abstract

Domestic and wild animals have lived and adapted to the high-altitude environments much earlier than humans, and extensive research work has been undertaken on the adaptive processes in primates. However, the migratory farming system of goat and sheep-rearing in the Himalayas is unique and different. During ascending to alpine pastures, the animals are certainly vulnerable to cold, hypoxia, and walking stresses which have not been adequately addressed scientifically. Most studies suggest that exposure to cold and hypoxia follows a sequence of physiological, biochemical, and behavioral changes. The most characteristic adaptive changes following exposure to hypoxic high-altitude environment were found to be an increase in the values of hemoglobin (Hb), hematocrit or packed cell volume (PCV), and total erythrocyte count (TEC) and a significant decline in the number of lymphocytes and neutrophils leading to leukopenia. It was also shown in human subjects that change in the hematocrit values accurately reflected as a measure of altitude adaptation. Prolonged stay at the high altitude also resulted in a significant rise in plasma cholesterol, albumin, lactate dehydrogenase (LDH), and glucose-6-phosphate dehydrogenase. Malate dehydrogenase level was decreased. The native mammals such as yak, Tibetan sheep, and goats are believed to adapt to very low partial pressure of oxygen by developing anatomical and physiological alterations that equip them for survival. These include large heart and lungs, high level of arterial saturation, modification of hemoglobin structure, high energy metabolism, larger lung volume, and lack of pulmonary vasoconstriction.

In spite of known facts about mechanism of adaptation, question have always remained how some small ruminants like Tibetan sheep adapt to extreme cold climate in regions like Qinghai-Tibet Plateau of China which is characterized by hypoxia and very low temperature throughout the year. Scientists have marveled

S. P. Singh, J. K. Malik, *Migratory Small Ruminant Farming System in the Himalayas-Peculiarities, Problems, Prospects, Climate Effects, Nutrition and Health*, https://doi.org/10.1007/978-981-96-6691-1_5

at the mechanism of adaptation to such challenging conditions and have turned to the genetic basis of adaptation and the role played by the genes. Whole-genome sequencing has been carried out in different domestic animal species, and various candidate genes responsible for adaptation have been identified. It was shown that rumen microbes develop a versatile and flexible role and modulate their functions according to the season.

This chapter provides a comprehensive overview and new insights into the mechanism of adaptation of small ruminants to cold and hypoxia including genetic basis of adaptation and the role played by the genes. Behavioral changes included development of salt hunger (Sect. 8.3.2 of Chap. 8), geophagia, and changed grazing pattern. Loading the animals, their transport by road in vehicles, and long-distance walking were found to serve as stressful factors for animals and were accompanied by significant rise in plasma cortisol and plasma prolactin levels and increase in circulating neutrophils and a decline in lymphocyte resulting in increased neutrophil-to-lymphocyte ratio.

Keywords

High-altitude acclimatization · Adaptive responses to high altitude · Genomics of adaptation · Effect of loading and transportation · Long distance walking stress

Abbreviations

Amsl	above mean sea level
CO_2	Carbon dioxide
ft.	feet
GxM	Gaddix merino
GxR	Gaddix Rambouillet
H_2O	water
KVK	Krishi Vigyan Kendra
Mm	millimeter
mmHg	millimeter of Mercury
Msl	mean sea level
NDRI	National Dairy Research Institute
O_2	Oxygen
PCV	packed cell volume
PO_2	Partial pressure of oxygen
TEC	Total erythrocyte count
TLC	Total leukocyte count

Unlike the climate at sea level or plains, the high-altitude environment is characterized by distinct changes in environmental attributes, especially composition of inspired and alveolar air. The atmospheric oxygen, essential for animal life, proportionally decreases with the rise in elevation. For instance, at sea level, the concentration of oxygen is about 21%, atmospheric pressure is 760 mm Hg, and partial pressure of oxygen in air (pO_2) stands at 159 mm Hg. As the altitude increases, the concentration of oxygen remains the same (approximately 21%), but because air molecules are more dispersed, each breath delivers less oxygen to the body due to lowering of air pressure. At around 3000 m (10,000 ft), the atmospheric pressure drops to only 523 mm Hg resulting in around 40% fewer oxygen molecules per breath, and pO_2 in air is reduced to 110 mm Hg. At the height of the Mount Everest (8848 m, 29,029 ft), the barometric pressure falls to about 226 mm Hg with pO_2 in air at only 47 mm Hg (Hall et al. 2013).

Other major changes comprised of decreased alveolar partial pressure of oxygen and carbon dioxide and consequent decrease in arterial oxygen saturation. The presence of carbon dioxide (CO_2) and water vapor (H_2O) in the alveoli becomes important at high altitude as the total barometric pressure falls to low levels, while the water vapor pressure does not fall proportionally. The water vapor pressure remains at 47 mm Hg as long as the body temperature is normal and the pressure of CO_2 falls from about 40 mm Hg at sea level to approximately 24 mm Hg at extreme high altitudes due to increased respiration. The animals migrating to alpine pastures have to confront the changing high-altitude environment as well as the physical exertion.

Mainly two types of animal migration pattern are seen in the hills of north-western Himalayas which are as follows:

(a) Sheep and goats walk all the way from the base camps to the alpine pastures. The flocks accompanied by shepherds carrying medicines, salt and other belongings and, often, hunting dogs, travel to pastures on foot along the narrow footpaths, the roads, and river fronts taking several days and months to reach the destination.
(b) Animals are transported in trucks for some distance and then they walk to the alpine pastures. This practice is being generally followed for the animals of the state government farms.

During the long treacherous journey to the alpine pastures, the flocks face transport stress, walking stress, and feed and water starvation. The animals respond to these stresses and try to adapt or acclimatize in a variety of ways as discussed in the following sections. Unfortunately, the work on adaptation of migrating hill animals is very scanty.

5.1 High Altitude Defined

High altitude has been defined by various authors in different ways. An altitude in the range of 1500 to 3500 m (4900 to 11,500 ft) is considered to be a high altitude (Paralikar and Paralikar 2010). International Society for Mountain Medicine

Table 5.1 Changes in environmental parameters with rise in altitude. (Adapted from Johnston 2015 with permission)

Altitude (m)	Barometric pressure (mm Hg)	PO_2 in air (mm Hg)	PO_2 in alveoli (mm Hg)	Arterial oxygen saturation (%)
0	760	159	104	97
1500 (4920)	632	132	78.0	94
2100 (6888)	586	123	70.0	92
3000 (9840)	523	110	67	90
4500 (14760)	429	91	52	74

Values given in parentheses are feet

recognizes that altitude above 1500 m (4900 ft) starts to affect humans (Luks 2017). Table 5.1 summarizes the effects of rise in altitude on various environmental parameters (Johnston 2015).

5.2 Domestic Animals on High Mountains

Besides humans, a large number of domestic animals including sheep, goats, yaks, Ibex, Tibetan gazelle, Llamas, and birds of all kinds have been residing on the high mountains for generations in India, China, Nepal, Tibet, Afghanistan, Pakistan, South America, Ethiopia, Europe, and other countries. In fact, it is believed that animal species have lived at the high-altitude mountains for much longer time than human beings. Over the years, they have adapted to the high-altitude conditions and achieved irreversible, long-term behavioral and genetic changes (Barsila et al. 2014). As mentioned earlier, owing to abundant availability of natural pastures in countries located in the Himalayan belt, a different animal farming system has been evolved for small ruminants. The animals temporarily arrive for grazing at alpine pastures during summer and return to the base camps in the plains during winter. Animals migrating to these heights temporarily undergo short-term adaptation or acclimatization which is basically transitory physiological change to the new environment.

5.2.1 Inbuilt Adaptation Features of Mountain Animals

It is well documented that humans and animals native to highlands have inbuilt physiological features in body systems which make them adaptable to the high-altitude environment. In the native inhabitants, massive alterations have taken place in the oxygen transport system with molecular changes in hemoglobin structure and function. The endogenous secretion of nitric oxide has been shown to play an active role in quick adaptation of Tibetan sheep to high altitude (Koizumi et al. 2004; Jin et al. 2020). Some of these features are visible even in animals which are reared at comparatively lower heights. We observed these attributes in goats which were being reared in the vicinity of the university town Palampur (Himachal Pradesh), where the current research work was in operation. This town is located at the height of about 1450 m (4760 ft) above mean sea level and the adjoining mountain ranges rise quite high.

We estimated the hematological parameters and thyroid status of Gaddi goats of the area which were farm managed (sedentary) and those migrating to the alpine pastures during summer and returning to the base station during winter. Although the elevations were not truly high altitude, it was found that local Gaddi goats had some distinct hematological features such as very high total erythrocyte count (20.4 ± 31 × 0^6/ul), hemoglobin (11.1 ± 0.72 g/dl), and hematocrit (36.5 ± 3.1%). Katoch et al. (2020) have also reported higher levels of RBC count (× 10^6/ cu mm), hemoglobin (g%), and hematocrit (%) to the extent of 21.57 ± 0.35, 11.46 ± 0.27, and 36.30 ± 0.14, respectively, in Chegu pashmina goats of Himachal Pradesh. In contrast to these goats belonging to semi high altitudes, the goats reared in plains were found to have significantly lower values of hematological attributes. For instance, Beetal, a plain region goat, was found to have RBC count, hemoglobin, and hematocrit in the order of 12.23 ± 0.51 × 10^6/mm^3, 7.76 ± 0.21 g/dl, and 30.33 ± 0.80%, respectively (Shah et al. 2010). Of blood analytes, the total protein values ware also of considerable interest. The plasma protein value in Gaddi goats (9.41 ± 0.22 g/dl) was comparatively higher than those reported in Pashmina goats (6.58 ± 0.78 g/dl) (Somvanshi et al. 1987).

We also compared the thyroid status of sedentary (farm-managed) and migratory goats (Kumar et al. 1994). The migratory goats were found to have significantly higher plasma (ng/ml) thyroxine (61.6 ± 3.7) and tri-idothyronine (1.62 ± 0.10) than the corresponding values in farm-managed goats (49.6 ± 2.8 and 1.44 ± 0.10 ng/ml, respectively). The higher thyroxine level in migratory goats was obviously related to their habitat. The migratory animals undertake strenuous physical exercise during migration and grazing, especially when they are on the sloppy mountain ranges under cold climate. The higher plasma levels of T_4 and T_3 in these animals seem to have an adaptation significance as both the hormones are known to be thermogenic.

In addition to Gaddi goats, inbuilt adaptation characteristics were also observed in our study on yaks (Kumar et al. 1992). Yak is a normal habitant of highlands and commonly found in high hill areas of India, Tibet, and China. The yaks under study belonged to the University Yak Farm located at the high-altitude station Sangla in Kinnaur district of Himachal Pradesh. To study the influence of transportation to lower heights on their physiological parameters, six yaks were transported from Sangla (mean height around 2700 m amsl) to the University Campus at Palampur (mean height 1450 m amsl) in vehicles. The most interesting aspect of the study was a gradual declining trend in the values of hemoglobin from 11.0 ± 0.75 g/dl to 8.0 ± 1.69 g/dl, hematocrit from 35.1 ± 3.71% to 27.3 ± 5.36%, and total erythrocyte count from 6.2 ± 0.75 × 10^6/ul to 4.6 ± 1.11 × 10^6/ul in the yaks 3 weeks after their arrival at the lower heights. The decline in hemoglobin concentration and erythrocyte count indicated physiological adjustment of the yaks to the new environment. The requirement for a higher erythrocyte count and higher hemoglobin is obviously much more at the higher altitude than at lower heights, and the decline in the values of these parameters in yaks on arrival at lower altitude is an adaptation change.

The influence of higher and lower altitudes on hematological and biochemical variables of sheep has also been reported from other countries. Titaouine and Meziane (2015) determined hematological and biochemical parameters in Quled Djellal ewes residing at plain region (150 m msl) and mountainous region (1000 m msl) in South East Algeria. The values of hemoglobin and hematocrit and plasma levels of creatinine, total protein, cholesterol, triglycerides, urea, calcium, and phosphorus in sheep residing at the plain region were 73.2 ± 25.4 g/L and 31.0 ± 4.0%, and 105.8 ± 18.74 umol/L, 68.1 ± 15.89 g/L, 1.29 ± 0.75 mmol/L, 0.25 ± 0.01 g/L, 9.63 ± 3.57 mmol /L, 1.12 ± 0.35 mmol/L, and 0.93 ± 0.42 mmol/L, respectively. It was shown that sheep based in the mountainous region had significantly higher values of hemoglobin (120.6 ± 33.0 g /L) and hematocrit (40.0 ± 1.0%) and also had higher plasma creatinine level (119.6 ± 35.71 umol/L) compared to sheep residing at the plain region. On the contrary, sheep staying at 1000 m msl were shown to have significantly lower plasma levels of total protein (59.1 ± 23.5 g/L), cholesterol (0.93 ± 0.31 mmol /L), triglycerides (0.22 ± 0.07 g/L), urea (7.14 ± 2.49 mmol /L), calcium (0.52 ± 0.03 mmol /L), and phosphorus (0.23 ± 0.01 mmol /L) in comparison to ewes based at 150 m msl. The study showed that magnitude of elevation is likely to induce variations in the concentrations of various hematological and biochemical parameters in sheep.

5.2.2 Adaptive Responses to High Altitude

Migratory animals are subjected to stress of transportation, long walking, shortage of feed and fodder, and exposure to extreme climates. Although volumes of system-wise information are available in human subjects and other primates including molecular mechanism of response to hypoxia at high altitude, there is very scanty literature available concerning large or small ruminants, especially the migratory animals. The following discussion unravels how different body systems in migratory sheep and goats react to these stresses and attain adaptation or acclimatization.

5.2.2.1 Hematological and Biochemical Changes

Irrespective of the species, the most characteristic hematological feature after exposure to hypoxic high-altitude environment is an increase in hematological indices like total erythrocyte count (TEC), hemoglobin concentration (Hb), and hematocrit. The mechanism of such hematological changes is well documented in primates. The carotid bodies sense the lowering of arterial oxygen partial pressure and the red blood cells, which are mainly stored in spleen, are released into circulation (Milledge and West 2007). This ultimately increases the hematocrit and enhances the capacity of blood to carry more oxygen. Later, the erythropoietin level increases which stimulates the cell division in hematopoietic stem cells of bone marrow. These changes, considered as adaptive response to hypoxia, have been observed in migratory sheep (Nazki et al. 1988, Shah et al.1991), Quled Djettal ewes (Titaouine and Meziane (2015), crossbred sheep (Kumar et al. 2002a), and goats (Rastogi and Singh 1990) as well as in human subjects. Some of the parameters alter so precisely that Zubieta-Calleja et al. (2007) recommended change in hematocrit value in human subjects as a measure of altitude adaptation. They advocated that adaptation was altitude and time dependent and proposed an equation:

Adaptation = time/altitude. The authors precisely calculated that a complete hematocrit adaptation was achieved in around 40 days in persons climbing from Copenhagen (sea level) to La Paz (3510 m) and staying over there for about 2–3 months at each place (Fig. 5.1).

Hematocrit has also been found associated with erythropoiesis and the process of acclimatization during transport from high to low altitude and vice versa in ewes (Moneva et al. 2016). The effect of migration to alpine pastures and high altitude on sheep and goats has not been adequately investigated in the Indian Himalayan region, and only a few studies have been reported from Himachal Pradesh and union territory of Jammu and Kashmir, India, and Nepal. A few studies have revealed marked alterations in hematological parameters of sheep after exposure to high altitude in Kashmir valley (Nazki et al. 1988; Shah et al. 1991). In these experiments the animals comprised of 100 adult sheep and 10 adult Corriedale rams. The animals were sampled at the base farm before migrating to high altitude and, thereafter, 4–5 weeks subsequent to their stay at the alpine pastures. Migration of sheep from Base camp, Shuhama (1586 m), to highland pasture, Lenderwas (3500 m), was

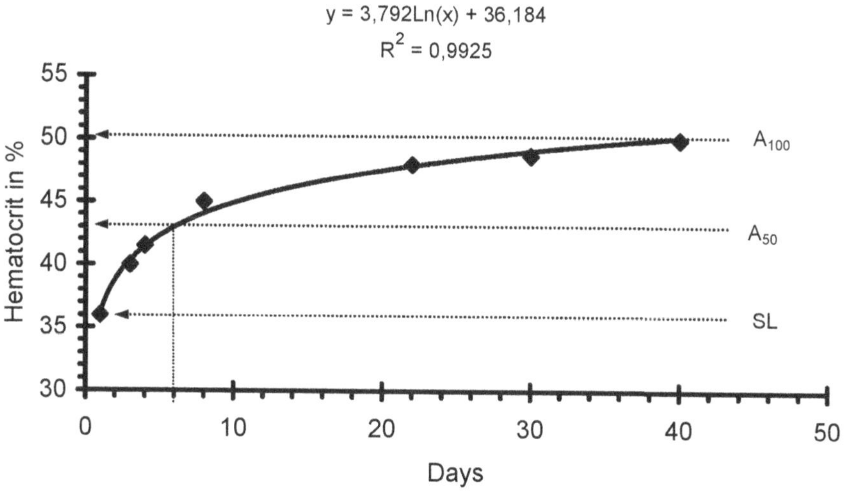

Fig. 5.1 Hematocrit changes in human subjects after high-altitude travel from sea level (35 m) to high altitude (3510 m). (Adapted from Zubieta-Calleja et al. 2007 with permission)

shown to significantly increase hemoglobin (10.5%), TEC (18.7%), and hematocrit (7.23%). Conversely, total leukocyte count (TLC) was significantly decreased by 20.9%. There was also a decrease in absolute counts of lymphocytes, monocytes, and basophils (Nazki et al. 1988). Similar results have also been reported in sheep ascending from Base camp, Dachi (1500 m), to highland pasture, Dagwan (4000 m) (Shah et al. 1991). In sheep at the base farm before migrating to high altitude, the values of hemoglobin (g%), TEC (10^6/mm^3), hematocrit (%), TLC (10^3/mm^3), lymphocytes (%), neutrophils (%), monocytes (%), eosinophils (%), and basophils (%) were 10.1 ± 1.35, 9.1 ± 0.65, 30.1 ± 2.68, 6.6 ± 0.42, 61.2 ± 1.98, 27.7 ± 1.36, 2.95 ± 0.23, 2.35 ± 0.23, and 3.05 ± 0.63, respectively. After staying at the alpine pasture, the animals showed significantly higher values of hemoglobin (11.9 ± 1.67 g%), TEC (9.1 ± 0.65 10^6/ mm^3), and hematocrit (30.1 ± 2.68%), while TLC (6.6 ± 0.42 10^3/ mm^3), lymphocytes (61.2 ± 1.98%), neutrophils (27.7 ± 1.36%), monocytes (2.95 ± 0.23%), eosinophils (2.35 ± 0.23%), and basophils (3.05 ± 0.63%) were significantly decreased.

A few studies on the effect of migration to alpine pastures and high altitude on small ruminants have been reported from Nepal, where sheep pastoralism is also an important livestock production system. Bhatt et al. (2018) investigated physiological and metabolic indices in grazing Baruwal sheep at two altitudes 2431msl and 3885 msl along the transhumance routes. Blood samples were collected from animals at lower and higher grazing altitudes after an adaptation period of 1 month. There was an increase in respiratory and heart rates, but the blood levels of glucose, albumin, and total protein were significantly decreased at higher altitude. It was suggested that effects of adverse environmental conditions may be managed by slowing the upward movement and availability of quality feed for enhancing production efficiency of transhumant sheep. In yet another recent study from Nepal, blood samples were collected from sheep after an adaptation period for 3 weeks at both lower (2431 m.s.l) and higher (3885 msl) pasture sites. Erythrocyte count, hemoglobin, and packed cell volume were significantly increased after stay at higher altitude, but the leukocyte count was found to be lowered (Barsila et al. 2020).

During our project work in Himachal Pradesh, almost identical alterations in hematological parameters have been documented in grazing crossbred sheep (Kumar et al. 2002a). The flocks of crossbred sheep (Gaddi with Rambouillet

(G × R) and Gaddi with Merino (G × M) maintained at the State Government Farm base camp, Jeory town (2086 m amsl), participated in the annual summer climbing to high-altitude pastures named Kara and Kanta (3400 m amsl) in Bhaba valley of Kinnaur district of Himachal Pradesh. Every summer, these sheep accompanied by shepherds take a long walk to the alpine pastures, stay there for around 4 months, and then migrate back to the base camp Jeory as winter sets in. To study the changes in hematological and biochemical parameters during their stay at the high-altitude pastures, blood samples were collected from the sheep at the base camp and after completion of 1 month stay at the alpine station. The blood samples were analyzed for various parameters. Stay at high-alpine meadows were marked by a significant rise in TEC, hemoglobin, and hematocrit values, which are obviously adaptable changes to the altitude and low-oxygen environment. There was a significant rise in plasma cholesterol but blood glucose did not show any fixed pattern (Table 5.2). Increase in plasma cholesterol in sheep after their stay at the high-alpine pastures indicates an attempt on the part of animals to cope up with the strenuous situation of climbing and grazing for long hours. Sejian et al. (2012) had shown that prolonged walking stress resulted in significant increase in blood cortisol in Malpura ewes. Since cholesterol is the precursor of cortisol, its rise in the grazing crossbred sheep is a reflection of stress on animals.

5.2.2.2 Behavioral Responses

There is hardly any published report on behavioral aspects of sheep at the alpine pastures. In Chap. 8, we have described a typical salt-craving behavior (salt hunger) of sheep suffering from sodium or salt deficiency at the alpine pastures. Other behavioral responses in grazing sheep at the Kara alpine pastures of Himachal Pradesh during our study were as follows (Kumar et al. 2002b).

(a) Leadership qualities: It was observed that the grazing sheep generally remained in groups of four or five, and one of them coming forward in the beginning appeared to assume a leadership role. The group kept changing with different grazing periods. During the resting period, they were observed to sit in groups. If the sun was bright, especially at the noon, they sat keeping their face away from direct sun.
(b) Geophagia: In a flock, approximately 10 to 15% sheep were found to exhibit geophagia.
(c) In general, the animals grazed for 9–10 h per day with intervening time spent in rest and rumination (3–4 h/day). Distance traveled during grazing varied from 4 to 7 km/day. Most preferred grass for grazing was Neeru grass (*Festuca alpinum*).

5.2.2.3 A Unique Strategy for Adaptation of Sheep to Extremely Cold Climate of Tibetan Plateau

Traditionally, the animals adapt to cold altitude by increasing their erythrocyte count, hemoglobin concentration, hematocrit, and efficiency of air-blood exchange in the lungs. But the question always remained how small ruminants like Tibetan sheep adapt to extreme cold climate in regions like Qinghai–Tibet Plateau of China, which is characterized by hypoxia and low temperature. The average temperature remains less than 8 °C and reaches as low as −2.5 °C during winter. To answer the question, Fan et al. (2021) adopted a new approach of performing sequencing technology on rumen microbial communities of sheep during different seasons. It was shown that rumen microbes develop a versatile and flexible role and modulate their functions according to season. During summer when the forage is of high nutritional quality, sheep showed increased forage degradation and fermentation through increased counts of important bacteria such as Bacteroidetes, *Ruminococcus_1*, *Saccharofermentans,* and *Ruminococcaceae_UCG-014* to improve growth performance. During winter, forage degradation and fermentation were low. There was an increase in Fermicutes and Fermicutes/Bacteroidetes ratio and rumen microbial

Table 5.2 Changes in hematological indices, plasma cholesterol, and blood glucose levels of crossbred sheep after a month long stay at the high-altitude pastures in Himachal Pradesh (Kumar et al. 2002a)

	Crossbred sheep			
	G × R		G × M	
Parameter	Base camp	Alpine pasture	Base camp	Alpine pasture
Hemoglobin (g/dl)	9.20 ± 0.18^{a}	12.18 ± 0.45^{c}	8.54 ± 0.14^{b}	12.05 ± 0.45^{c}
Total erythrocyte count (10^6/ul)	6.11 ± 0.29^{a}	6.58 ± 0.35^{ab}	8.58 ± 0.29^{c}	10.84 ± 0.77^{b}
Hematocrit (%)	27.54 ± 0.70^{a}	34.68 ± 1.13^{b}	28.73 ± 0.89^{a}	39.14 ± 1.85^{c}
Blood glucose (mg/dl)	51.6 ± 1.92^{b}	43.2 ± 2.49^{ab}	47.1 ± 2.04^{a}	53.6 ± 3.57^{c}
Plasma cholesterol (mg/dl)	75.7 ± 4.24^{a}	160.8 ± 10.30^{c}	96.7 ± 15.47^{ab}	166.0 ± 13.41^{b}

G × R = Gaddi × Rambouillet cross; G × M = Gaddi × Merino cross
Values bearing different superscript in the columns differ significantly ($P < 0.05$) between base camp and alpine pasture

diversity and harmony between microbes. The interactive activity between microorganisms and metabolism was significantly increased showing that microbiota could promote the decomposition of low-quality forage and meet the energy demands of the animal.

5.3 Genomics of Adaption to High Altitude and Hypoxia

The mechanism of adaptation to altitude is different in animals visiting the mountains temporarily and the ones which are native to the place and residing there for centuries. In contrast to the natives, migratory sheep and goats exposed to high altitude temporarily during summer grazing get adapted to the hypoxic environment mainly by modification of cardiovascular and pulmonary systems leading to an increase in red blood cell count, hemoglobin, hematocrit, and cardiac output and reduction in the metabolic rate (Zubieta-Calleja et al. 2007; Titaouine and Meziane 2015).

The native mammals such as yak, Tibetan sheep and goats, and Tibetan gazelle and mules adapt to very low partial pressure of oxygen by developing anatomical and physiological alterations that equip them for survival. These include large heart and lungs, high level of arterial saturation, modification of hemoglobin structure, high energy metabolism, larger lung volume, and lack of pulmonary vasoconstriction. Scientists have marveled at the mechanism of adaptation to such challenging conditions and have turned to the genetic basis of adaptation and the role played by the genes. Whole-genome sequencing has been carried out in different domestic animal species, and various candidate genes responsible for adaptation have been identified. Population genomic studies of high-altitude animals have generated several theories about the genes and pathways that may have contributed to hypoxic adaptation (Storz and Cheviron 2021).

Investigations of the genetic basis of high-altitude adaptation in ruminants like sheep, goat, and yak have revealed involvement of several genes (Friedrich and Wiener 2020). The whole-genome sequencing study of domestic yak (*Bos grunniens*) has offered important prospects for the identification of selection signatures for high-altitude adaptation. The yak is known to express hypoxia-Inducible factor-1 related genes *ADAM17*, *ARG2*, and *MMP3* responsible for altitude adaptation (Ayalew et al. 2021). In yak, the main features that contributed to the tolerance at high altitude were identified as lack of hypoxic pulmonary vasoconstriction (Dolt et al. 2007), a significantly shorter tongue than cattle modifying their grazing mechanism (Shao et al. 2010), and better nutrient utilization due to more effective microbial synthesis (Wang et al. 2011).

New studies have provided new insights and confirmed the role of *DSG3* gene in high-altitude hypoxia adaptation in the Tibetan cashmere goats (Kumar et al. 2018). Four other genes namely *LEPR*, *LDB1*, *EGFR*, and *FGF2* were found to have roles in the high-altitude adaptation of the highland Tibetan goats. Some common genes at the molecular level in cattle, sheep, and goat were identified as *EPAS1*, *JAF1*, *DKK2*, and *SPON1*, whereas *EGLN1* and *HIF3A* genes were reported as targets for the hypoxia response for Tibetan cattle (Friedrich and Wiener 2020; Wu et al. 2018, 2020).

Most of the recent research on genetic adaptation to altitude has been conducted in sheep and goats and reported mainly from China, Nepal, and India. Sheep native to high-altitude environment were found to show selection signatures for genes involved in hypoxia-induced factors (HIF), angiogenesis, vasodilation, and glycolysis pathways. Gene *FGF-7* was found to play an active role in the high-altitude adaptation of sheep at the Himalayas. This gene seems to potentially increase lung function by regulating its expression level in high-altitude sheep by altering its binding of specific transcription factors (Yang et al. 2016; Gorkhali et al. 2016). Investigations in Tibetan sheep (*Ovis aries*) and goats showed that they possessed capability to adapt to a wide variety of agro-ecological conditions on the Qinghai-Tibetan Plateau by involving several genes including *DSG3* and *FGF5* (Kumar et al. 2018; Guo et al. 2019, 2020). Gene *EPAS1* was also found to play a dynamic role in high-altitude adaptation in the Tibetan sheep and goats (Wu et al. 2020).

High-altitude sheep breed Changthangi of Indian Himalayas is adapted to cold climate and hypoxic conditions. Analysis of selection signatures has shown important insights into the adaptability of Changthangi. For detection of genomic regions under positive selection, Ahmad et al. (2021) analyzed population structure of Changthangi sheep and contrasted it with selected Indian and European commercial breeds of sheep. The principal component analysis revealed Changthangi sheep to cluster with Tibetan sheep. Genomic comparisons of Changthangi using cross population extended haplotype homozygosity showed multiple functional regions present on *Ovis aries* (Oar) chromosomes 2, 3, 6, and 18 to be under selection in Changthangi sheep. It was further reported that these regions were related with adaptation to climatic and hypoxic stressors, fleece characteristics, and functioning of immune and reproductive systems. *UCP* genes reported to be associated with adaptation to cold and hypoxic conditions were the main loci under positive selection in Changthangi sheep population. Indian researchers (Ahmad et al. 2021) have recently worked on high-altitude hypoxia-resistant sheep breed Changthangi. Bharti et al. (2018) from the Defense Research and Development Organization (DRDO) had reported success in developing a crossbred high-altitude-resistant goat by mixing genes of local adaptive Changthangi breed with Black Bengal and Sirohi goats from the plain areas.

Regarding goats, several genes, viz., *LEPR*, *LDB1*, *EGFR*, and *FGF2* were found related to the high-altitude adaptation of the Nagqu Tibetan goats (Jin et al. 2020). Population genomic analyses combined with genome-wide association studies conducted by Li et al. (2022) have shown contribution of *3′-phosphoadenosine 5′-phosphosulfate synthase 2* (*PAPSS2*) gene to high-altitude adaptability in Tibetan goats on the Qinghai-Tibetan Plateau. Moreover, this gene was implicated in hypoxia-related pathways in Tibetan goats.

5.4 Effect of Loading, Unloading, Road Transportation, and Long Walk on Physiological and Hematological Indices and Hormonal Profile of Sheep and Goats

As mentioned elsewhere, there is a general practice at several government sheep and goat farms in north-western Himalayan states to transport the animals in a truck up to the road head after which the flocks take a long walk to the alpine pastures. However, studies especially related to transport stress or walking stress on migratory animals in these hilly states have received only minimal attention. In some European and Western countries, sheep and goats are loaded in vehicles, carried to variable distances for marketing and other operations, and offloaded at the desired destinations. Transportation is also sometimes necessary to shift wild animals from zoo/sanctuary to other places for alternate accommodation or treatment. It has been suggested that transportation is a risky and stressful process for farm animals. The following paragraphs summarize results of some scientific investigations pertaining to the effect of stressful operations on migratory sheep and goats and other animals.

5.4.1 Effect of Loading, Unloading, and Road Transport by Vehicles

Transport of animals is an important exercise in livestock husbandry, but it serves as a stress factor affecting the animal health. Previous studies on some animal species have often highlighted that loading, unloading, and road journey by vehicles serve as a stressful situation for animals and affect their production potential. This was observed in sheep, goats, as well as in guanacos belonging to camelids family. It was shown that certain physiological, hematological, and biochemical parameters such as heart rate, hematocrit value, leukocyte number, especially neutrophil and lymphocyte, and blood levels of adrenocortical hormones can be helpful in evaluating the transport stress in animals.

Rounding up and loading of sheep onto a vehicle and driving them for a distance of 15 km were found to significantly increase the plasma cortisol and prolactin levels and a rise in heart rate. On the contrary, the hematocrit and body osmolality were shown to be decreased (Broom et al. 1996). Similarly, sheep transported for a period of 30 and 48 h had higher plasma protein concentration than the animals transported for 12 h. There was an increase in neutrophils and a decline in lymphocytes resulting in increased neutrophil: lymphocyte ratio (Fisher et al. 2010). The study revealed that healthy adult sheep transported under good conditions can tolerate transport durations of up to 18 h without undue compromise to their welfare. Fazio et al. (2011) investigated the effects of a short road journey of 125 km for 3 h on hematological variables and plasma cortisol level in 138 healthy sheep. Platelet counts (10^3/ul) and plateletocrit (%) in pregnant ewes increased from 238.5 ± 104.58 to 340.2 ± 152.83 ** and from 0.20 ± 0.08 to 0.30 ± 0.16**, respectively. The plasma cortisol level rose significantly after transportation compared to the basal values.

These hematological variables and plasma cortisol level were advocated to be effective markers in assessing short-term transport stress in sheep.

Rajion et al. (2001) studied the quantum of stress in goats that were transported by road for a distance of 46 km in an open truck. Immediately following the road transportation, there was a significant increase in neutrophil: lymphocyte ratio from pre-partition value of 0.74 ± 0.06 to 1.64 ± 0.27 and plasma glucose concentration from 5.00 ± 0.65 to 14.03 ± 1.25 mmol/L, but there was no change in plasma cholesterol level. In an Iranian study, Tajik et al. (2016) investigated the effects of transportation stress in goats by driving them in a closed truck for 3 h for about 180 km. Transportation caused a significant increase in serum cortisol concentration, but there was no change in the level of serum thyroid hormones.

In a recent study, Tozlu Çelik et al. (2021) measured the stress effect caused by transporting Karayaka sheep from sea level to an altitude of 1500 m in Turkey. The distance covered was 182 km in 5 h. Blood samples were collected before the transportation and after arrival at the higher altitude. Transportation caused a significant increase in the serum levels of triidothyronine, T_3 (2.02 to 2.34 ng/ml), and thyroxine,T_4 (from 7.46 to 9.57 µg/dl). Surprisingly, cortisol level was not affected. On the other hand, the oxidative stress parameter malondialchehyde was significantly decreased, and protein carbonyl values were not affected in sheep after transport. It was concluded that road transportation and altitude constituted a stress on sheep, and stress-reducing measures were recommended.

Among other studies abroad, Lendrawati et al. (2020) from Indonesia investigated 4, 8,and 12 h of road transport duration on body weight loss and hematological and biochemical responses of sheep. Their findings revealed that increased transport time significantly increased body weight loss and decreased hematocrit and serum levels of glucose and cortisol. In a study in Mongolia, Xiao et al. (2024) investigated the effects of transport and pre-transport feeding during summer on the well-being of transported sheep. They compared the effects of 1 h and 3 h journeys by two identical vehicles on the same route. It was reported that serum levels of triglycerides and free fatty acids after 3 h transport were significantly lower than after 1 h transport. The serum concentrations of T_4 and malondialdehyde were higher after the 3 h transport than the 1 h transport, whereas serum creatine kinase levels tended to be lower after the 3 h transport than after the 1 h transport.

Several studies in India and abroad have recently been conducted to study the impact of various kinds of stresses on the body systems of animals.

Among Indian studies, Mir and Ashutosh (2020) conducted a detailed study on the effects of transporting goats belonging to Krishi Vigyan Kendra (KVK) Leh in the territory of Ladakh during winter, hot dry, and hot humid seasons. Physiological, hematological, blood biochemical, hormonal, and blood minerals were studied. Of the physiological parameters, respiration rate, pulse rate, rectal temperature, and skin temperature were increased significantly due to transportation during all seasons. The maximum values of skin temperature were observed during the hot dry season compared to the hot humid and winter seasons. Among hematological indices, hemoglobin concentration, PCV, and white blood cell (WBC) concentration were significantly increased due to transportation during all the seasons. There was a significant increase in neutrophil percentage and significant decrease in lymphocyte percentage in all the transported goats during all seasons. During the winter season, neutrophil-to-lymphocyte ratio was significantly higher after transportation. Transportation and unloading caused an increase in plasma glucose level to the extent of 10, 53, and 86% during winter, hot dry, and hot humid seasons, respectively. There was a significant increase in plasma concentrations of cortisol and prolactin hormones of goats during all seasons. Sodium, potassium, and magnesium concentrations in plasma were significantly decreased, while phosphorous concentrations were significantly increased in goats during transport in all the seasons. Blood gas analysis revealed a significant decrease in pCO_2 concentrations in goats during all seasons, and the maximum decrease was observed during the hot humid season followed by hot dry and winter season. There was a significant increase in pO_2 value in transported goats during all seasons. Blood pH was not affected by transportation.

In another study from India conducted in the *Animal Physiology Division of National Dairy Research Institute (NDRI), Karnal,* Gupta et al. (2019) studied the physiological response to 8 h transportation stress at different flocking densities in hot humid and winter seasons in goats. The stress resulted in significant increase in respiration rate, rectal temperature, pulse rate, and skin temperature in both the seasons which declined to the basal values within 6–12 h post-transportation.

In the study of Ambore et al. (2019) at the College of Veterinary Science, Shirwal, Maharashtra, goats transported by road for 12 h over a distance of 350 km exhibited leukocytosis and significant increase in hematocrit values. The differential leukocyte count revealed significant increase in neutrophil percent in transported goats. Panchal et al. (2022) conducted a study in Mumbai Veterinary College to evaluate the effect of short journey transportation stress on the serum biochemical parameters in adult goats. Goats transported by road for about 200 km exhibited significant increase in serum levels of total proteins, albumin, globulin, total calcium, phosphorus, and glucose after transportation.

It is evident from most studies that loading, unloading, and prolonged road journey constituted a stressful situation for the animals to various degrees. The transport stress was shown to cause a sharp increase in heart rate, hematocrit value, neutrophil: lymphocyte ratio, and the plasma levels of cortisol and prolactin and thus served as good indicators of stress on the animals. The characteristic features of increase in plasma cortisol and neutrophil: lymphocyte ratios due to transport stress have also been reported in adult male camelids (Zapata et al. 2004).

5.4.2 Effect of Long Distance Walking on the Animal's Health

In majority of cases in the Himalayan states, sheep and goat flocks travel long distances on foot for several days with intermittent rest to reach their destination. The quantum of stress of long walks under such situations has not been investigated. However, it could be reasonably presumed that the prolonged walking stress would certainly be of much higher order than the transport stress in vehicles as discussed

Table 5.3 Effect of 14 km long walk by Malpura sheep in semi-arid climate on some hormones (Sejian et al. 2012)

Parameters	Control group	Walking stress group
Plasma cortisol (nmol/L)	16.1	28.1**
T3 (nmol/L)	2.27	1.82**
T4 (nmol/L)	150.4	135.7*

Significantly different (*$P < 0.05$, ** $P < 0.01$) from the control value

under Sect. 5.4.1. In the hilly terrain, the flocks are not only subjected to walking stress but also stress due to limited access to feed and water. Animals exposed to long walks with restricted feeding are known to mobilize body energy reserves as indicated by increase in non-esterified fatty acids in plasma. The increased energy demand during walking is believed to stimulate adrenal cortex which, in turn, increases secretion of cortisol hormone. It has been reported that cortisol is the principal stress-relieving hormone in sheep (Sejian and Srivastava 2010).

In a study in the Jammu region, Reothia et al. (2016) measured the serum levels of thyroid hormones (T_3 and T_4) and cortisol in migratory Rambouillet ewes climbing from the Jammu and Kashmir State Government Sheep Farm, Bilawer (844 m), to mid-station (Mandhar,1580 m) on way to the alpine pastures at Sarthal (2134 m). The animals covered a distance of 4–5 km per day with intermittent resting. The serum levels of both thyroid hormones (T_3 and T_4) were found to increase non-significantly during uphill migration and the highest concentrations were recorded immediately after reaching at the high altitude. The non-significant changes in thyroid hormone levels were also evident during downhill migration. Similarly, non-significant increase of serum cortisol level was recorded in both uphill and downhill migration indicating migration imposed stress on the animals.

Effect of long-distance walking in a semi-arid tropical environment prevailing in the state of Rajasthan has been investigated on endocrine responses in Malpura ewes by Sejian et al. (2012). The ewes were made to walk a distance of 14 km in ambient temperature and humidity of 32–44 °C and 12–26%, respectively. The values of T_3, T_4, and cortisol were monitored in animals of control and experimental groups (Table 5.3). The walking stress resulted in a significant increase in cortisol level, whereas the values of plasma T_3 and T_4 were significantly decreased. It was suggested that sheep had the capability to adapt to long-distance walking and that adrenal and thyroid gland hormones play a significant role in such adaptation.

It is plausible that similar alterations might occur in migratory sheep and goats walking a longer distance for several days. The level of walking stress is expected to be of higher magnitude in migratory animals because of lower ambient temperature and uphill journey in mountains.

References

Ahmad SF, Mehrotra A, Charles S, Ganai NA (2021) Analysis of selection signatures reveals important insights into the adaptability of high-altitude Indian sheep breed Changthangi. Gene 799:145809. https://doi.org/10.1016/j.gene.2021.145809

Ambore BN, Maini S, Patodkar VR, Borikar ST (2019) Effect of transportation on body weight and hematology in goats. Int J Curr Microbiol App Sci 8(07):947–952

Ayalew W, Chu M, Liang C, Wu X, Yan P (2021) Adaptation mechanisms of yak (*Bos grunniens*) to high-altitude environmental stress. Animals (Basel) 11(8):2344. https://doi.org/10.3390/ani11082344

Barsila SR, Kreuzer M, Devkota NR, Ding L, Marquardt S (2014) Adaptation to Himalayan high altitude pasture sites by yaks and different types of hybrids of yaks with cattle. Livest Sci 169:125–136

Barsila SR, Bhatt K, Devkota B, Devkota NR (2020) Haematological changes in transhumant Baruwal sheep (*Ovis aries*) grazing in the western Himalayan mountains in Nepal. Pastoralism 10:4. https://doi.org/10.1186/s13570-019-0156-6

Bharti VK, Kumar P, Biswas A, Kumar K, Gogoi D, Pawar DD, Deshmukh PB, Srivastava RB, Kumar B (2018) Development of region specific hybrid goat and their performance evaluation under high altitude condition. Def Life Sci J 3(2):165–171

Bhatt K, Devkota NR, Tiwari ICP, Barsila SR (2018) Changes in physiological and metabolic parameters of sheep (*Ovis aries*) during transhumance at western Himalayan pastures. J Agric For Univ 2:171–175

Broom DM, Goode JA, Hall SJ, Lloyd DM, Parrott RF (1996) Hormonal and physiological effects of a 15 hour road journey in sheep: comparison with the responses to loading, handling and penning in the absence of transport. Br Vet J 152(5):593–604

Dolt KS, Mishra MK, Karar J, Baig MA, Ahmed Z, Qadar Pasha MA (2007) cDNA cloning, gene organization and variant specific expression of HIF-1α in high altitude yak (*Bos grunniens*). Gene 386(1–2):73–80

Fan Q, Cui X, Wang Z, Chang S, Wanapat M, Yan T, Hou F (2021) Rumen microbiota of Tibetan sheep (*Ovis aries*) adaptation to extremely cold season on the Qinghai-Tibetan plateau. Front Vet Sci 8:673822. https://doi.org/10.3389/fvets.2021.673822

Fazio E, Medica P, Mignacca S, Cravana C, Ferlazzo A (2011) Haematological and cortisol changes after a 3 h road journey in sheep. J Anim Vet Adv 10(19):2487–2492

Fisher AD, Niemeyer DO, Lea JM, Lee C, Paull DR, Reed MT, Ferguson DM (2010) The effects of 12, 30, or 48 hours of road transport on the physiological and behavioral responses of sheep. J Anim Sci 88(6):2144–2152

Friedrich J, Wiener P (2020) Selection signatures for high-altitude adaptation in ruminants. Anim Genet 51(2):157–165

Gorkhali NA, Dong K, Yang M, Song S, Kader A, Shrestha BS, He X, Zhao Q, Pu Y, Li X, Kijas J, Guan W, Han J, Jiang L, Ma Y (2016) Genomic analysis identified a potential novel molecular mechanism for high-altitude adaptation in sheep at the Himalayas. Sci Rep 6:29963. https://doi.org/10.1038/srep29963

Guo J, Zhong J, Li L, Zhong T, Wang L, Song T, Zhang H (2019) Comparative genome analyses reveal the unique genetic composition and selection signals underlying the phenotypic characteristics of three Chinese domestic goat breeds. Genet Sel Evol 51(1):70. https://doi.org/10.1186/s12711-019-0512-4

Guo J, Zhong J, Liu GE, Yang L, Li L, Chen G, Song T, Zhang H (2020) Identification and population genetic analyses of copy number variations in six domestic goat breeds and bezoar ibexes using next-generation sequencing. BMC Genomics 21(1):840. https://doi.org/10.1186/s12864-020-07267-6

Gupta D, Ashutosh M, Kashyap G, Punetha M, Patel B, Ashutosh PI, Tejaswi V, Jalmeria N, Ahirwar M (2019) Physiological response to transportation stress at different flocking densities in hot humid and winter seasons in goats and assessment of pretreatment of vitamin C, electrolyte and jaggery. Indian J Anim Res 53(3):299–306

Hall JE, Vaz M, Kurpad A, Raj T (2013) Guyton and Hall text book of medical physiology. Elsevier, p 376. (Guyton Book) https://shop.elsevier.com/books/guyton-and-hall-textbook-of-medical-physiology-3rd-sae/vaz/978-81-312-5773-9

Jin M, Lu J, Fei X, Lu Z, Quan K, Liu Y, Chu M, Di R, Wei C, Wang H (2020) Selection signatures analysis reveals genes associated with high-altitude adaptation in Tibetan Goats from Nagqu. Tibet Animals (Basel) 10(9):1599. https://doi.org/10.3390/ani10091599

Johnston R (2015) Blog author. https://www.pftforum.com/blog/author/arpijay/

Katoch A, Sharma A, Negi M, Mandial RK, Wadhwa DR, Sharma D (2020) Haematobiochemical and mineral status in Chegu Pashmina goats of Himachal Pradesh. Indian J Vet Med 40(2):20–22

Koizumi T, Ruan Z, Sakai A, Ishizaki T, Matsumoto T, Saitou M, Matsuzaki T, Kubo K, Wang Z, Chen Q, Wang X (2004) Contribution of nitric oxide to adaptation of tibetan sheep to high altitude. Respir Physiol Neurobiol 140(2):189–196

Kumar R, Kumar N, Singh SP, Rastogi SK (1992) Hematological and physiological responses of yak on transportation to lower altitude in spring and winter. J Appl Anim Res 2(1):9–12

Kumar N, Kumar R, Rastogi SK, Singh SP (1994) Thyroid status of migratory and farm-managed Gaddi goats of Himachal Pradesh. Indian J Anim Sci 64(7):748–749

Kumar R, Rajvir S, Sharma KB, Singh SP (2002a) Hematologcal changes in exotic and crossbred sheep migrating to alpine pastures of north-western Himalayan region. SARAS J Livest Poult Prod 18(3–4):41–47

Kumar R, Sharma KB, Singh SP (2002b) Behavioral studies in migratory sheep at alpine pastures of north-western Himalayan region. A Preliminary Note- The Vet 26:1–2

Kumar C, Song S, Jiang L, He X, Zhao Q, Pu Y, Malhi KK, Kamboh AA, Ma Y (2018) Sequence characterization of *DSG3* gene to know its role in high-altitude hypoxia adaptation in the Chinese cashmere goat. Front Genet 9:553. https://doi.org/10.3389/fgene.2018.00553

Lendrawati L, Priyanto R, Jayanegara A, Manalu W, Desrial D (2020) Effect of different transportation period on body weight loss, hematological and biochemical stress responses of sheep. J Indones Trop Anim Agric 45(2):115–123

Li C, Wu Y, Chen B, Cai Y, Guo J, Leonard AS, Kalds P, Zhou S, Zhang J, Zhou P, Gan S, Jia T, Pu T, Suo L, Li Y, Zhang K, Li L, Purevdorj M, Wang X, Li M, Wang Y, Liu Y, Huang S, Sonstegard T, Wang MS, Kemp S, Pausch H, Chen Y, Han JL, Jiang Y, Wang X (2022) Markhor-derived introgression of a genomic region encompassing *PAPSS2* confers high-altitude adaptability in Tibetan goats. Mol Biol Evol 39(12):msac253. https://doi.org/10.1093/molbev/msac253

Luks AM (2017) What constitutes High altitude? Chapter 10 - "High-Altitude Travel". The Travel and Tropical Medicine Manual (Fifth ed). Pages 138–151. https://doi.org/10.1016/B978-0-323-37506-1.00010-6 Get rights and content – https://www.sciencedirect.com/science/article/abs/pii/B9780323375061000106

Milledge JS, West J (2007) High altitude medicine and physiology, 4th edn. CRC Press. https://www.amazon.com/High-Altitude-Medicine-Physiology-John/dp/0340913444

Mir NA, Ashutosh A (2020) Seasonal variation of stress response parameters due to induced transportation stress in goats supplemented with vitamin C and jaggery under tropical Indian climatic conditions. Biol Rhythm Res 51(6):847–857

Moneva P, Yanchev I, Dyavolova M, Gudev D (2016) Hematocrit as a potential marker of acclimatization capacity and stress sensitivity in sheep exposed to transport and high altitude. Bulg J Agric Sci 22(6):999–1005

Nazki AR, Matto PA, Khan M (1988) Haematological studies in Kashmir sheep. Effect of migration to highland pasture. Ind J Anim Sci 58(9):1060–1062

Panchal P, Dalvi S, Ambade R, Dighe D, Deshmukh R (2022) Serum biochemical changes during transportation stress in adult goats. Indian J Vet Sci Biotechnol 18(1):119–121

Paralikar SJ, Paralikar JH (2010) High-altitude medicine. Indian J Occup Environ Med 14(1):6–12

Rajion MA, Saat IM, Zulkifli I, Goh YM (2001) The effects of road transportation on some physiological stress measures in goats. Asian-Aust J Anim Sci 14(9):1250–1252

Rastogi SK, Singh SP (1990) Normal hemogram and blood analytes of mountain Gaddi goats. Indian J Anim Sci 60(11):1338–1339

Reothia A, Devi J, Upadhyaya R, Kour K, Koul A, Sarma K (2016) Studies of some hormonal parameters in Rambouillet sheep during seasonal migration in Jammu region. J Anim Res 6(6):1037–1041

Sejian V, Srivastava RS (2010) Pineal-adrenal-immune system relationship under thermal stress: effect on physiological, endocrine, and non-specific immune response in goats. J Physiol Biochem 66(4):339–349

Sejian V, Maurya VP, Naqvi SM (2012) Effect of walking stress on growth, physiological adaptability and endocrine responses in Malpura ewes in a semi-arid tropical environment. Int J Biometeorol 56(2):243–252

Shah ZA, Pandia D, Qureshi MQ (1991) Effects of high altitude on some haematological parameters in sheep in Kashmir valley. Indian Vet J 68:385–386. https://eurekamag.com/research/002/083/002083166.php

Shah F, Gupta M, Poonia JS (2010) Studies of different haematological parameters in male beetal goats at different physiological stages. Haryana Vet 49(12):38–39. https://www.luvas.edu.in/haryana-veterinarian/archive-2010.php?AM6

Shao B, Long R, Ding Y, Wang J, Ding L, Wang H (2010) Morphological adaptations of yak (*Bos grunniens*) tongue to the foraging environment of the Qinghai-Tibetan Plateau. J Anim Sci 88(8):2594–2603

Somvanshi R, Biswas JC, Sharma B, Koul GL (1987) Haematological studies on Indian pashmina goats. Res Vet Sci 42(1):124–126

Storz JF, Cheviron ZA (2021) Physiological genomics of adaptation to high-altitude hypoxia. Annu Rev Anim Biosci 9:149–171

Tajik J, Nazifi S, Eshtraki R (2016) The influence of transportation stress on serum cortisol, thyroid hormones, and some serum biochemical parameters in Iranian cashmere (Raini) goat. Vet Arh 86(6):795–804

Titaouine M, Meziane T (2015) The influence of altitude and landforms on some biochemical and hematological parameters in Ouled Djellal ewes from arid area of south East Algeria. Vet World 8(1):130–134

Tozlu Çelik H, Aslan FA, Us Altay D, Kahveci ME, Konanç K, Noyan T, Ayhan S (2021) Effects of transport and altitude on hormones and oxidative stress parameters in sheep. PLoS One 16(2):e0244911. https://doi.org/10.1371/journal.pone.0244911

Wang H, Long R, Liang JB, Guo X, Ding L, Shang Z (2011) Comparison of nitrogen metabolism in yak (*Bos grunniens*) and indigenous cattle (*Bos taurus*) on the Qinghai-Tibetan plateau. Asian-Australas J Anim Sci 24(6):766–773

Wu DD, Ding XD, Wang S, Wójcik JM, Zhang Y, Tokarska M, Li Y, Wang MS, Faruque O, Nielsen R, Zhang Q, Zhang YP (2018) Pervasive introgression facilitated domestication and adaptation in the *Bos* species complex. Nat Ecol Evol 2(7):1139–1145

Wu DD, Yang CP, Wang MS, Dong KZ, Yan DW, Hao ZQ, Fan SQ, Chu SZ, Shen QS, Jiang LP, Li Y, Zeng L, Liu HQ, Xie HB, Ma YF, Kong XY, Yang SL, Dong XX, Esmailizadeh A, Irwin DM, Xiao X, Li M, Dong Y, Wang W, Shi P, Li HP, Ma YH, Gou X, Chen YB, Zhang YP (2020) Convergent genomic signatures of high-altitude adaptation among domestic mammals. Natl Sci Rev 7(6):952–963

Xiao J, Han Z, Li X, Phillips CJC, Shi B (2024) Effects of transport duration and pre-transport fasting on blood biochemistry in Dorper × Mongolian sheep. Animals (Basel) 14(10):1482. https://doi.org/10.3390/ani14101482

Yang J, Li WR, Lv FH, He SG, Tian SL, Peng WF, Sun YW, Zhao YX, Tu XL, Zhang M, Xie XL, Wang YT, Li JQ, Liu YG, Shen ZQ, Wang F, Liu GJ, Lu HF, Kantanen J, Han JL, Li MH, Liu MJ (2016) Whole-genome sequencing of native sheep provides insights into rapid adaptations to extreme environments. Mol Biol Evol 33(10):2576–2592

Zapata B, Gimpel J, Bonacic C, González BA, Riveros JL, Ramírez AM, Bas F, Macdonald DW (2004) The effect of transport on cortisol, glucose, heart rate, leukocytes and body weight in captive-reared guanacos (*Lama guanicoe*). Anim Welf 13(4):439–444

Zubieta-Calleja GR, Paulev PE, Zubieta-Calleja L, Zubieta-Castillo G (2007) Altitude adaptation through hematocrit changes. J Physiol Pharmacol 58(Suppl 5 (Pt 2)):811–818

Part III

Impact of Climate Change on Transhumance and Herders' Perceptions to Changing Climate on the Migratory System

Effects of Changing Global Climate on Small Ruminant Production System

6

Abstract

Climate change and global warming have become household names in the society during the last few decades. The words like greenhouse gases have also become popular in public without understanding what the word stood for. Change in climate is emerging as the main environmental challenge to all living beings including domestic large and small ruminants on the Himalayas, threatening their well-being and existence. Global warming caused by the accumulation of greenhouse gases such as carbon dioxide, methane, and nitrous oxide in the atmosphere is now a reality. Many reports have indicated that hill and mountain environment was particularly sensitive to the effect of global warming. The impact of climate change and consequential heat stress conditions have been found to affect the growth and production, feed intake, milk yield, reproduction, and other biological functions of the productive animals. As per the prediction of the intergovernmental panel on climate change, by the year 2100, the global surface temperature may be increased by 3.7–4.8 °C. It has been advised that the impact of climate change on pastoral nomads and their animals is likely to be devastating. The centuries-old practice of nomadic transhumance on the mountains is under serious threat. In case of pastoral nomads, global warming and heat stress would not only affect the productive performance of migratory animals but also badly disturb their transhumant cycle and routine of upward and downward journeys on the hills. Scientists are predicting an increase in unpredictable weather events as a result of global warming. Various reports show that the nomads are already experiencing increased temperature resulting in drying up of water resources, reduced and untimely rainfall and other extreme events, and reduced quality and quantity of pastures replaced by weedy and unwanted plants.

During the past few years, we have come a long way to understand the climate and its role in warming the earth. This chapter endeavors to explain the basic

S. P. Singh, J. K. Malik, *Migratory Small Ruminant Farming System in the Himalayas-Peculiarities, Problems, Prospects, Climate Effects, Nutrition and Health*,
https://doi.org/10.1007/978-981-96-6691-1_6

concepts of climate, definition of climate and related terms, historical journey on the role of atmospheric gases in global warming, and the process of buildup of greenhouse gases (GHG) in the atmosphere leading to a rise in global temperature. We have elaborately covered the role of livestock in contributing to the GHG emission and global warming and also the source, sector, and country-wise emissions of GHG from the livestock. The special features of Indian livestock's contribution to the emission of GHG has been discussed separately.

The most important variable in the climate affecting the exposed animals is heat stress. This chapter addresses in detail the impact of heat stress on physiological and productive attributes of large and small ruminants. Thermal stress was found to reduce feed intake and cause a significant decline in milk yield and milk quality in both large and small ruminant species. Heat stress caused a reduction in milk protein, milk fat, casein, and fatty acids. Exposure of animals to hot environment was accompanied with increase in incidence of infertility, cases of anestrus and embryonic mortality in females, and impairment of reproduction in males causing reduced quantity and quality of sperm production. Heat stress affected immune system negatively, causing a reduction of cytokines. Global warming is having a devastating effect on health status of livestock. Rising temperature has been found to increase the development of certain pathogens and affecting non-vector borne diseases. Among small ruminants, sheep and goats are well adapted to the hot environment. An important observation reported in sheep was that ewes were most affected by heat stress from 5 days prior to estrus, during estrus, and then for ensuing 5 days suggesting vulnerability of this period to thermal stress. Finally, major strategies related to climate stress comprising of measures to reduce enteric methane emissions by reducing number of ruminant livestock, reducing the enteric methane emission, selecting heat-tolerant and disease-resistant breeds, and alleviation of heat stress physically have been discussed elaborately.

Keywords

Climate change · Global warming · Green house gas emission · Impact of climate change on ruminants · Heat stress mitigation

Abbreviations

CH_4	Methane
CO_2	Carbon dioxide
FAO	Food and Agriculture Organization
FAOSTAT	Food and Agriculture Organization Corporate Statistical Database
GAAC	Global average atmospheric carbon dioxide
GHG	Green House Gases

$GtCO_{2e}$	Gigatons carbon dioxide equivalent
GW	Global warming
GWP	Global Warming Potential
IEA	International Energy Agency
IPCC	Intergovernmental Panel on Climate Change
N_2O	Nitrous oxide
NASA	National Aeronautics and Space Administration
NOAA	National Oceanic Atmospheric Administration
UEPA	United States Environmental Protection Agency
UNCA	United Nations Climate Action
UNEP	United Nations Environmental Program
UNFCCC	United Nations Framework Convention on Climate Change
WMO	World Meteorological Organization
WRI	World Resources Institute

6.1 Climate, Weather, Meaning of Climate Change, and Related Issues

Climate

Climatologists define climate as the long-term averages of climatic elements such as ambient temperature, humidity, wind velocity, radiation, precipitation (rainfall), and atmospheric pressure at a place. The term weather, on the other hand, denotes short time to time variation in the values of these climatic elements.

Climate Change

With the awareness of the effects of changing world climate and global warming, a number of definitions have come up which properly explain the climate change. Sawyer and Narayan (2019) defined climate change as the long-term misbalance of customary weather conditions such as temperature, wind, and rainfall characteristics of a specific region, and it is likely to be one of the main challenges that mankind faces during the present century.

National Aeronautics and Space Administration (NASA 2018) defined climate change as "A broad range of global phenomena created prominently by burning of fossil fuel, which add heat trapping gases to earth's atmosphere." According to the United Nations Climate Action (UNCA), climate change refers to long-term shifts in temperature and weather patterns. These shifts may be natural such as through variations in the solar cycle. But since 1800, human activities have been the main driver of climate change, primarily due to burning of fossil fuel such as coal, oil, and gas and emitting gases like carbon dioxide (CO_2), methane (CH_4), and nitrous oxide (N_2O). The United Nations Framework Convention on Climate Change (UNFCCC 1992) defined climate change as "a change of climate which is attributed directly or indirectly to human activity that alters the composition of the global atmosphere and which is in addition to natural climate variability observed over comparable time periods."

6.1.1 Composition of Atmospheric Air We Inhale and Exhale

A healthy normal atmosphere contains mixture of gases such as nitrogen (N_2), oxygen (O_2), and carbon dioxide (CO_2) and water vapor. Oxygen is carried to our body cells during respiration to provide energy. Carbon dioxide generated during chemical reactions is breathed out. Nitrogen is an inert gas and hardly performs any significant function. It is just inhaled and exhaled during respiratory cycle. Table 6.1 presents the composition of inspired and expired air during respiratory cycle in human subjects.

About 99% of the atmosphere is made up of O_2 and N_2 which do not absorb the infrared radiation emitted by the earth and have nothing to do with warming effect. The remaining about 1% includes CO_2 and other gases produced during anthropogenic activities by human population and have an important role in warming of the earth.

6.1.2 What Atmospheric Gases Have to Do With Warming of the Planet Earth

As explained earlier, 90% of the atmosphere is made up of nitrogen and oxygen gases which don't absorb solar radiation and let it pass through them. Gases such as carbon dioxide added to the atmosphere during anthropogenic activities of humans and animals absorb and trap the heat in the atmosphere and work in the same manner as a greenhouse works; hence, the name greenhouse gas (GHG) has been assigned. As the hot sun shines, most of the solar radiation reaches the earth heating it up. The warm earth, in turn, emits infrared radiations which are picked up by the greenhouse gases such as CO_2, CH_4, and N_2O are added to the atmosphere by human and animal activities. The vibration caused by radiation to the molecules of the greenhouse gases heats up the atmosphere. The faster the molecules vibrate, the hotter they are. The speed of vibration decides the warming potential of the gas. Carbon dioxide and other gases thus act like a cap or blanket trapping some of the heat around the earth, which, otherwise, could have been radiated into space (Gates 2021). To keep things simple and to express the combined warming power of all the

Table 6.1 Normal composition of inspired and expired air during respiratory process in human subjects. (IEA web 2020)

Gas	Inspired air (%)	Expired air (%)
Nitrogen	78.0	78.0
Oxygen	21.0	16.10
Carbon dioxide	0.04	4.40
Other gases including water vapor	>1	

Table 6.2 The lifetime duration of GHG in atmosphere and their global warming potential. (Moore 2008; web: https://blogs.edf.org >ghg_lifetimes)

GHG	Lifetime in atmosphere (years)	Global warming potential (100 years)
CO_2	300 to 1000	1
CH_4	12	25
N_2O	114	298

greenhouse gases into a single unit, experts have suggested a measure called "carbon dioxide equivalent (CO_2e or CO_{2}_eq)." The greenhouse gases have a lifetime and stay in the space for variable periods and thus have variable warming potential (Table 6.2).

6.1.3 Awareness About Global Warming—A Journey into the History

As early as 1850, Arrhenius showed that an air rich in carbon dioxide retained more heat from solar radiation causing an increase in air temperature. It was also observed that if lots of carbon dioxide is released into the atmosphere, more heat would be retained. A scientist named Gust Ekholm (1901) used the term greenhouse effect for the first time in the year 1901. Scientists became aware of a possible role of greenhouse gases in the global warming way back in 1960s. Around 1960, Charles D Keeling showed that the concentration of CO_2 in the atmosphere was gradually increasing, staying there, and was not immediately absorbed in the ocean. A decade later, around 1972, John Sawyer published a scientific report linking global warming and greenhouse gas concentration. Alarmed by further reports and evidences, leading nations sought the role of the United Nations (UN) which finally led to the establishment of Intergovernmental Panel on Climate Change (IPCC) in the year 1988 by the World Meteorological Organization (WMO) and the United Nations Environment Programme (UNEP). The establishment of the IPCC was endorsed by the UN General Assembly in the year 1988. Since the creation of IPCC for assessing the science related to climate change, Working Groups have contributed to the assessment reports, special reports, and methodology reports. More recently, the Working Group II contributed to the IPCC sixth assessment report entitled "Climate Change 2022: Impacts, Adaptation and Vulnerability" (IPCC 2022).

6.2 Build Up of GHG in the Atmosphere and a Rise in Global Temperature

Atmospheric concentrations of carbon dioxide, methane, and nitrous oxide have been gradually increasing with time. During the pre-industrial times, around mid-eighteenth century, the atmospheric concentrations of greenhouse gases were in a balance, i.e., as much gases were produced, as much were used up. For example of

Table 6.3 Year-wise estimate of global average atmospheric carbon dioxide (GAAC) concentration (Lindsey 2023)

Year	GAAC (ppm)
2000	369.0
2019	409.9
2020	412.5
2021	419.0

carbon dioxide, as much CO_2 was produced it was absorbed by the plants and vegetation. But then there was a boom in industrial activity and the world started using fossil fuel in various operations especially in automobiles. Fossil fuel is totally a carbon source of plants compressed under the earth millions of years ago. After that, the greenhouse gases emissions in the atmosphere started increasing tremendously (Table 6.3). According to some estimates, CO_2 and CH_4 concentrations in the atmosphere have increased by 149% and 262%, respectively, by the year 2020 as compared to the pre-industrial level (Lindsey 2023).

Scientists from the Institute of Oceanography San Diego, California, have estimated that we are adding to the atmosphere roughly 40 billion metric tonnes of CO_2 per year. According to them "it is a mountain of carbon that world digs out of the earth, burn and release into the atmosphere as CO_2 year after year" (National Oceanic and Atmospheric Administration; NOAA 2021). There is a general feeling that global temperature rise due to rising greenhouse gases should be limited to no more than 1.5 °C to avoid devastating impact on climate. Reports from various quarters suggest that temperature of the earth is projected to rise by around 3.2–5.8 °C by the end of the century if concentrations of the GHG continued to increase unabated.

The International Center for Integrated Mountain Development (ICIMOD), Nepal, in a study of surface air temperature of 103 years reported that annual mean surface air temperature has significantly increased in Hindu Kush Himalayan Region (HKHR) at a rate about 0.10 °C per decade (Tiwari et al. 2020). Similar results were obtained by Rayamajhi and Manandhar (2020). Sawyer and Narayan (2019) from the Sydney University reported that the current climate models indicated a 0.28 °C temperature increase per decade for the next two decades and predicted the increase in global average surface temperature between 1.88 and 4.08 °C by 2100. According to NASA, the global average surface temperature increased from 0.6 to 0.9 °C between 1906 and 2005, and the rate of temperature increase has been doubled in last 50 years. A global climate model predicted the average Australian temperature relative to 1990 to increase by 0.1 to 1.3 °C by 2020, 0.3 to 3.4 °C by 2050, and up to 6.7 °C by 2080.

6.2.1 What Livestock Has to Do with Greenhouse Gas Emission

It is well known that the digestive system of animals such as cattle, buffaloes, sheep, and goats, called ruminants, possess a complex four-chambered stomach where millions of microbes naturally live and attack the plant material, feed, and fodder that the animals ingest. The rumen works like a fermentation vat breaking down the hard plant material into small molecules. The resulting breakdown products include useful nutrients as well as gases such as carbon dioxide and methane. The gases are emitted into the atmosphere during eructation and burping. It has been estimated that an adult cow or buffalo emits around 150 to 200 liters of gas per day. The value from sheep and goat touches 20 to 25 liters per day (Gates 2021). The number of ruminant animals in the world run into millions (Opio et al. 2013), and all of them contribute substantial quantity of greenhouse gases into the atmosphere promoting global warming. Of these gases, the main culprit is methane which has a warming power 25 times more than that of carbon dioxide (Moore 2008).

Although the role of livestock in contributing to increasing concentration of greenhouse gases in the atmosphere is now well established, some experts have questioned the very claim that agriculture and livestock husbandry are responsible for climate change anyway. Attention to the possible role or no role of the livestock in global warming came into light after the publication of 2006 report of Food and Agriculture Organization of the United Nations (FAO) "*Livestock Long Shadows*" which reported, perhaps erroneously, that livestock contributed greenhouse gases more than the entire transport sector (FAO 2006). Enraged by the controversy, Glatzle (2018) claimed that livestock-borne emissions had nothing to do with climate change, and the earth warming was most likely driven by natural factors. Another animal scientist Mitloehners (2020) has objected *to paint the livestock as villain* in climate crisis ignoring the role of fossil fuel which was the main culprit.

Many scientists have felt that the readings of greenhouse gas emissions from ruminants have been exaggerated while quantifying the emissions or perhaps deliberately. Sejian et al. (2016) have reported lack of consistency in published data in estimating the enteric methane emissions perhaps due to different approaches adopted by different investigators. Mitloehners (2020) has argued that if the feeds and fodders of the ruminant animals were suitably modified, the cows will consume more carbon dioxide than they would emit.

6.2.2 Greenhouse Gas Emissions from the Livestock—Culprit or Victim?

The emissions of GHG that affect climate come from every part of the world and every human activity like burning of fossil fuels, heating buildings using coal, clearing land and forests, industries, transport, agriculture including livestock, and landfills for garbage. It is reported that ten most rich countries of the world contribute largest emissions to the tune of 68%, whereas one hundred least GHG emitting countries generate just 3% of the total emissions. It is also a big dilemma that only seven G20 members namely China, the United States, the European Union, India, Indonesia, Brazil, the Russian Federation, and international transport contributed more than 55% of total global GHG emissions in 2020. The G20 member countries collectively contributed 75% of the total global GHG emissions. There is great variation in per capita emissions among the countries. In 2020, the world average per capita GHG emissions was 6.3 tCO_2e. While the United States continues to be far above this level at 14 tCO_2e, least developed countries on average emit 2.3 tCO_2e per capita yearly (UNEP 2020). Obviously, people and countries creating more problems to the climate have a greater responsibility to act fast and cut down their emissions.

The role of livestock, especially ruminants, in greenhouse gas emissions has often been talked about during past few decades. While ruminants play an important role in providing high-quality proteins, they are an important source of greenhouse gases. In a way, livestock could be considered as a culprit as well as a victim. Climate directly or indirectly affects livestock production, which, in turn, harms the climate quality by emitting lots of greenhouse gases in the atmosphere.

In the years 2019 and 2020, total greenhouse emission reached a new increase of 51 billion tonnes of CO_2e per year from 49 billion tonnes in 2005, showing that atmospheric concentration of CO_2 has been continuing to rise (Gates 2021). NASA reported that the year 2022 was the warmest year on record.

6.2.3 Livestock's Contribution to GHG

The global ruminant population was estimated at 3612 million in the year 2010 with cattle making up to 40%, sheep and goats 55%, and buffaloes remaining 5% (Opio et al. 2013). Livestock from developing countries including India and China have generally been blamed for excessive emission of GHG, but scientific studies have shown that China, not India, was on the top of the list in the matter of adding GHG emission to the atmosphere. Chhabra et al. (2013) have argued that although the Indian livestock contributed substantial methane, the per capita or per head value was quite low, 24.2 kg CH_4/animal/year. There are mainly two sources of GHG emissions from ruminant animals: (a) enteric fermentation during which lots of gases especially methane is produced as a result of microbial action on feed and fodder inside the rumen and (b) anaerobic fermentation of livestock manure producing methane and denitrification of manure generating nitrous oxide.

6.2.3.1 Sector-wise Emissions

Using the IPCC estimates for total global anthropogenic GHG emissions to the tune of 59 ± 6.6 $GtCO_2$-eq for the year 2019, the major sectors contributing to the GHG emissions include energy supply 34%(20 $GtCO_2$-eq), industry 24%(14 $GtCO_2$-eq), agriculture, and forestry and land use 22% (13 $GtCO_2$-eq). Two other important sectors comprise transport and buildings contributing 15% (8.7 $GtCO_2$-eq) and 6% (3.3 $GtCO_2$-eq), respectively. Emissions from global total agriculture and related land use reached 9.3 $GtCO_{2}$_eq in 2018. While contribution of crop and livestock activities within the farm gate generated 5.3 $GtCO_2$-eq, the land use and land use change activities generated approximately 4 $GtCO_2$-eq.

It has been estimated that the main sources of global livestock emissions are cattle which contribute to the extent of 65–77% GHG emission. Contribution of GHG emissions from livestock in low- and middle-income countries from ruminants and monogastric livestock such as pigs and poultry is estimated at 70% and

53%, respectively, and these are likely to increase as requirement for animal products increases in these countries. While absolute GHG emissions have shown increasing trend, GHG emissions intensities have decreased at the global level mainly as a result of improved meat and milk productivity of cattle breeds (FAOSTAT 2018; Davis et al. 2015; Mbow et al. 2019).

6.2.3.2 Source-wise GHG Emission

Emissions from different sectors in the form of methane, nitrous oxide, and carbon dioxide are presented below.

Methane (CH_4) Irrespective of the species, the largest source of GHG emission in the ruminants is methane with more than 85% from enteric fermentation and rest from manure (Opio et al. 2013). The livestock sector contributes 35% methane and 65% nitrous oxide from anthropogenic sources. Ruminant animals produce lots of methane during anaerobic microbial digestion/fermentation in their rumen which helps in the breakdown of carbohydrates into simple molecules. Poor-quality/fibrous rations cause higher CH_4 emissions (Gerber et al. 2013). Over the years, enteric methane emission has been rising gradually. Solid and liquid dung and manure also act as source of methane when undergoing anaerobic fermentation. Large amount of methane is also produced during storage of dung. Among the leading world countries whose livestock release most enteric methane are Latin America, Africa, China, India, and the United States (WRI (2011) (Table 6.4).

Nitrous Oxide (N_2O) Nitrous oxide is produced from manure by nitrification and denitrification, leaching, and volatilization. Nitrous oxide emissions from manure storage and processing and from application of manure on pastures and crops contribute up to three million tonnes of nitrogen (FAO 2016).

Carbon Dioxide (CO_2) Carbon dioxide is the primary GHG emitted through ruminant digestive process and through human activities (normal respiratory process)

Table 6.4 Sources of emissions of GHG (FAO 2016)

GHG	Main source	% of emission
CH_4	Enteric fermentation	3.9
	Manure management	4.3
	Feed/processing rice	0.4
N_2O	Applied and deposition manure	16.4
	Fertilizer and crop residues	7.7
	Manure management	5.2
CO_2	Feed	13.0
	Land use activities (soybean)	6.0
	Land use activities (pasture expansion)	3.2
	Direct energy use	1.5

and is naturally present in the atmosphere as part of earth's carbon cycle. Main human activities adding CO_2 in atmosphere include use of fossil fuel (coal, natural gas), transportation, industrial processes, and land use changes.

6.2.3.3 Country-wise GHG Emissions from Livestock

There is a wide variation in GHG emissions from livestock of different countries of the world. Assuming latest available figure of total GHG as 49 gigatonnes CO_2-eq in the year 2005 as 100%, the country-wise values varied from as high as 16.38% (China) to as little as 1.85% (Canada). China and the United States were among the most GHG emitting countries, whereas India is on the seventh position emitting just 4.25% of GHG emission. Almost 50% of the global greenhouse gas is emitted by first three countries, viz., China, the United States, and Europe (Bhatta et al. 2021) (Table 6.5).

6.2.3.4 Special Features of India's Contribution of GHG, Especially Methane and Nitrous Oxide

1. Higher enteric methane emission

India has around 13% of global population of cattle and 53% of global population of buffaloes (20th Livestock census, Govt. of India). These two domestic ruminant species are responsible for over 85% of total enteric methane emission of India. Daily enteric methane emission was significantly higher in cattle compared to buffalo. Estimates of enteric methane from Indian livestock varied from7.25 to as high as 18.5 teragram per year (Tg/year) with an average of 9–10 Tg/year, although the authors have admitted lack of consistency in the values (Sejian et al. 2016).

Table 6.5 Countries arranged in descending order with regard to the extent of GHG emissions (especially methane) by their livestock (WRI 2011)

	Emissions of GHG			Emission of enteric methane	
Serial number	Country	% of total GHG	Serial number	Country	Million tonnes CO_2-eq
1	China	16.38	1	Latin America	460
2	USA	15.72	2	Africa	280
3	Europe	12.10	3	China	259
4	Brazil	6.47	4	India	218
5	Russian Fed.	4.65	5	Rest of Asia	175
6	Indonesia	4.62	6	USA	130
7	India	4.25			
8	Japan	3.15			
9	Canada	1.85			

Table 6.6 Species-wise enteric methane emission[a]

	Emission		
	Reference		
Species	Sejian et al. (2021b)	Swamy and Bhattacharya (2006)	Patra (2014)
Cattle	46.2		
Buffalo	45.1		
Goat	5.06	4.2	5.19 kg × 10^9
Sheep	2.6	2.0	6.18 kg × 10^9
Yak	0.04		
Mithun	0.04		

[a] Values are expressed as per cent emission unless otherwise stated

2. Nitrous oxide emission from Indian livestock is little around 1.42 gigagram per year (Gg/year)(Chhabra et al. 2013). Swamy and Bhattacharya (2006) had reported total NO_2 emission to the extent of 0.99 Gg/year.
3. Presence of more than100 million buffaloes, besides cattle.
4. Large number of migratory small ruminants on Himalayas belonging to the nomads.

6.2.3.5 Species-wise Enteric Methane Emission

Among the domestic animals, cattle and buffalo are the main methane-emitting species followed by goat, sheep, yak, and mithun (Table 6.6). The average enteric methane emission from sheep and goats has been reported as 475 million tonnes of CO_2-eq (Opio et al. 2013; Gerber et al. 2013).

6.3 Impact of Climate Change on Livestock Production

Experience of past two decades and increasing incidences of environmental hazards including recent heat waves in Europe, Australia, and other countries have shown that the climate change was no longer a fallacy. It is real and has come to stay. Warming planet and its impact on humans, animals, crops, and other species is becoming an issue of concern for the scientists and world organizations. The FAO in its report in 2007 has stated that 20% of the animal breeds in the world were at risk and almost one breed per month is becoming extinct (FAO 2007 https://www.fao.org).

Global warming has emerged as the top most environmental challenge to mankind threatening the well-being of future generations. Impacts of global warming are greatest in the tropical, subtropical, and hilly regions of developing countries such as India which have large human and animal population, and demand for food/feed is expected to be very high (FAO 2006; Negi et al. 2021). Detailed studies by Krishnan et al. (2019) have shown that even if global temperature rise is kept to 1.5° centigrade, the warming in the Hindu Kush Himalaya region (HKH) would likely to be at least to 0.3° centigrade higher and in the north-western Himalayas to 0.7° higher than the plains. Besides direct impact of rising global surface temperature and extreme weather on human and animal body systems, climate has an indirect effect on food crops, pastures, grasslands, and water resources affecting food security in the future. Global human population is expected to increase from 7.2 billion to 9.6 billion by the year 2050. Evidently, the demand for agricultural products is also expected to go up significantly (Sejian et al. 2021a). Among the agriculture sector which is the third largest source of GHG, livestock accounts for 40% of the world's agriculture gross domestic product (GDP) (Sejian et al. 2013), provides 33% of proteins, and employs 1.3 billion people (Escarcha et al. 2018). This sector, therefore, is expected to play a vital role in meeting food security demand of animal protein in the form of milk, meat, egg, fiber, and other animal products.

The most serious effect of planet warming is in the form of direct effect comprising of heat stress which affects the productive traits of animals such as milk and meat production, reproduction, and health and diseases. Indirect effects include effects on quality and quantity of forage production and water demand (Cheng et al. 2022). The concept of how the thermal stress harms the body system has also changed. Effect of heat was traditionally thought to result from decreased feed intake. However, it has now been clearly established that thermal stress disturbs and disorganizes the steady state of free radicals and metabolic rate, finally causing cellular and oxidative damage to the tissues and Organs (Belhadj Slimen et al. 2016). Although livestock production is facing increased pressure from changing climate, the efforts toward meeting the challenges and developing suitable mitigation strategies have not been systematically drawn. In a recent analysis, Escarcha et al. (2018) have conducted a systematic literature review and screened the world literature on the subject of climate change and reported that only 23% of the available publications were from the developing countries, of which 62% were on large ruminants and 38% on small ruminants.

The purpose of this chapter is to present the current state of knowledge and include scholarly recent research findings and summarize the effects of thermal stress on large and small ruminants in the background of climate change.

6.3.1 Effects of Heat Stress on Attributes of Large Ruminants

Climate change and resulting heat stress have been known to affect animal welfare and animal production very severely. However, it has been documented that, compared to the large ruminants, small ruminant species are less prone to the thermal effects and are better adaptive to the adverse effects of climate change particularly in terms of reproductive performance, survival, production yield, and resistance to re-emerging diseases. Danmaigoro et al. (2024) have recently reported that this was mainly because of their morphological features against harsh climate effects. Furthermore, as compared to the small temperate ruminants, tropical breeds are relatively more adaptive to the adverse effects of climate change. It was also reported that among large ruminants, buffaloes exhibited greater sensitivity to the heat stress and a strong negative correlation of the Energy Corrected Milk (ECM) with the meteorological data (Piscopo et al. 2024).

6.3.1.1 Effects on Milk Yield and Milk and Meat Quality

Temperature humidity index (THI) is a convenient tool to assess heat stress on animals (Upadhyay et al. 2009). A THI value of 70 or less is considered comfortable for ruminants. In India and neighboring countries of the Indian subcontinent, the THI values exceed 80 during hot dry and humid seasons which are very stressful. The first sign of thermal stress in most dairy animals is reduced feed intake which declines further as the temperature rises. In *Bos taurus* milch animals, this happens when the temperature rises above 26 to 30 °C. Reduced feed intake itself was found to be responsible for around 35% decrease in milk production (Rhoads et al. 2009). Numerous studies have shown that thermal stress had a pronounced effect on milk yield and quality of milk in almost all species. Summer et al. (2018) observed that the daily milk yield was highly affected by climate change. The increment of temperature and humidity was found to cause a significant decrease in milk production. The effects on milk fat and milk lactose content were controversial, but both the milk protein and casein contents tend to decrease when the temperature–humidity index value was higher than 75. Ekine-Dzivenu et al. (2020) in a Tanzanian study reported decline in milk production by 4.16–14.42% under THI values varying from 71 to 86. Uddin and Kebreab (2020) reported that heat stress could reduce daily milk yield up to 25% along with significant reduction in feed intake. As is well known, the tropical cattle breeds are better adapted to the heat stress than the breeds from temperate regions. Recently, Johnson (2018) established that at identical level of heat stress, tropical cattle had comparatively less reduction in feed intake (25 vs 30%) and daily milk yield (20 vs 28%) than the temperate cattle.

Chauhan and Ghosh (2014) in the Indian studies have reported annual loss of milk yield due to thermal stress to the extent of 1.8 million tonnes of milk costing approximately Rs 2661.62 crores. Liu et al. (2019) have estimated annual economic losses to the dairy industry in the United States to the tune of 1.69 to 2.36 billion US dollars. Heat stress was accompanied with not only a decline in milk yield, feed intake, and body weight but also decreased meat yield and egg production. Pragna et al. (2017) have observed that heat stress increased somatic cell counts and can affect various components of milk such as fat per cent, solid-non-fat, protein, casein, and lactose contents.

Meat production and quality is another economic trait affected by thermal stress, and it has recently attracted increasing public awareness and concern. Heat was found to significantly reduce growth rate, feed conversion efficiency, and quality. Chronic heat stress was reported to result in dark, firm, and dry meat. Animals exposed to acute heat stress before slaughter were reported to exhibit stimulation of muscle glycogenolysis leading to pale, tender, and exudative meat characterized by low water-holding capacity (Bejaoui et al. 2023). The impact of heat stress on meat quality was found to vary between livestock species (Sejian et al. 2021b).

6.3.1.2 Effects on Physiological, Behavioral, and Metabolic Responses

The animals subjected to thermal stress show quick alterations in physiological, behavioral, and metabolic parameters. There is elevation of respiratory rate, pulse rate, and body temperature. Heat stress in beef cattle caused high mortality, increased respiration rate, decreased rumination and feed intake, and increased frequency of drinking water (Summer et al. 2018). Behavioral changes were marked by shade seeking, higher water intake, reduced feed intake, spending more time in standing, and increased urinary and defecating frequency. In buffaloes, wallowing was a frequent and unique characteristic (Sejian et al. 2021b). Metabolic responses include effects on proteins, fats, and carbohydrates and related enzymes. The very first response to heat stress is reduction in feed intake which is believed to happen due to secretion of two adipokines which work on hypothalamic axis. Heat stress depresses body protein deposition and reduces fat oxidation and lipolytic enzyme activity. Although protein synthesis is disturbed, it does not affect the synthesis of heat shock proteins (Belhadj Slimen et al. 2016). Carbohydrates are readily available and converted into a number of metabolic intermediates as source of energy. Heat stress induces oxidative stress and enhances reactive oxygen species (ROS) production. An important result of the stress is to increase antioxidant enzymes, viz., superoxide dismutase, catalase, and glutathione peroxidase (Maan et al. 2013).

6.3.1.3 Effect on Reproductive System

Heat stress affects various processes of reproductive system in both female and male animals. In females, heat stress was found to reduce estrus period and fertility and increase the incidences of anestrus and embryonic mortality (Cheng et al. 2022). The response was observed to vary from species to species. High ambient temperature and humidity were shown to be responsible for low reproductive performance and summer infertility in cattle and buffaloes. Buffalo is characterized by silent heat and diurnal pattern of estrus behavior. A temperature rise of more than 2 °C was found to cause negative effect on hypothalamo-hypophyseal axis leading to disbalance of the hormone secretions (Upadhyay et al. 2009). Heat stress was found to adversely affect follicle development and reduce pregnancy rate in cattle (Liu et al. 2019). Lozano et al. (2005) reported a reduction in pregnancy rate from 39.4 to 31.6% when temperature-humidity index on the day of service increased from less than 72 to greater than 78. Dairy cows exposed to thermal stress during gestation had delivered calves with lower birth weights compared to the calves born to control cows (Collier et al. 1982).

Hyperthermia was found to impair reproduction in males causing reduced quality and quantity of sperm production and decreased fertility. Sperm production and percentage of normal spermatozoa were decreased in *Bos indicus* bulls during hot summers in Africa (Uddin and Kebreab 2020). Serious defects in sperm morphology of head and tail were observed in buffalo bulls exposed to short-term heat stress by scrotal insulation for 7 days. Heat stress was found to significantly destabilize scrotal thermoregulation (Gonçalves et al. 2021).

6.3.1.4 Effect on Immune System

The immune system of farm animals is their main body defense that provides them protection from environmental stressors. Heat stress can negatively impact immune functions through cell-mediated as well as humoral immune responses. Consequently, hot weather conditions can cause livestock to be more susceptible to diseases and increase the incidence of certain diseases, resulting in an increased potential of morbidity and mortality (Chirico et al. 1997; Dahl et al. 2020; Cheng et al. 2022). Sejian et al. (2021a) have explained the mechanism of how cytokines, and small proteins that control immune system, are released and regulate immune system. Animals suffering from heat stress are susceptible to diseases which affect their performance. Immune suppression causes activation of hypothalamo-pituitary axis leading to release of glucocorticoids and catecholamines that inhibit the synthesis and release of cytokines. Cytokines have an important role in the immune defense. Heat stress was found to increase the proportion of B cells (CD4–CD21+) and tended to decrease the proportion of monocytes (CD11b + CD172a+) in Holstein cows (Joo et al. 2021). It has been suggested that immune markers are very valuable indicators for the detection of heat stress in dairy cattle (Lemal et al. 2023).

6.3.1.5 Effects on Livestock Health

Impact of climate on health and disease aspects of different species of animals is a topic of prime importance but only scattered literature is available. As the time passes and the earth is getting warmer, changing climate is having a devastating effect on health of livestock. The Office International des Epizooties (OIE) Scientific Commission warned about the quick spreading of some diseases, especially those that are vector-borne. The two most mentioned emerging and re-emerging cattle diseases are blue tongue and Rift Valley fever (OIE 2009). The global distribution of blue tongue virus changed drastically for which climate change has been blamed (Husen Harun et al. 2022). Climate is affecting animal health directly as well as indirectly through change of disease pattern and making disease outbreaks difficult to control (FAO 2000). Directly, climate-driven fluctuations and extreme environmental conditions such as heat strokes, droughts, floods, and uncertain weather affect body system's physiological and immune responses especially of migratory small ruminants. A simple example of how climate affects biological function is reported by Das (2017). As is well known, increased incidences of bovine clinical mastitis occur during hot summers. Reduction of thermal stress by providing proper shade management to the affected animal resulted in lower frequency of clinical mastitis in cows than those exposed to their natural environments. For infectious disease in an area, the timing and intensity of the disease outbreak is also linked to climate. In some regions, climate could generate new transmission modes. Alteration in rainfall pattern could also lead to expansion of vector population. Similarly, a combination of temperature and humidity may favor helminth infections at some places (Swarankar and Singh 2013). Indirect effects are linked to changing feed resources, scarcity of water, grain production, and range and abundance of vectors and pathogens. The stress caused by direct and indirect factors could affect animal health and disease burden from pathogens such as bacteria, parasites, and their vectors (Husen Harun et al. 2022).

6.3.1.5.1 Effects on Pathogens and Vectors

Rise in ambient temperature due to climate change has been found to increase the rate of development of certain pathogens that have one or more life cycle outside the host animal and thus shorter generation time resulting in higher pathogen population. On the contrary, some pathogens are very sensitive to high temperature and may perish. The third category of pathogens are those that are extremely sensitive to moisture and thus their survival would depend on the degree of precipitation (Das 2017).

Climate change affects the vectors including flies, ticks, and mosquitoes in a similar fashion. Temperature and moisture impose limits on their survival and propagation. Vector-borne diseases are especially prone to climate change (rainfall and temperature) which may affect both the distribution and abundance of the disease (Das 2017). High temperature is especially limiting because it causes excessive loss of moisture from their body surface. For this reason, cooler regions which were previously too cold for certain vectors may begin to entertain them to flourish due to a rise in temperature (Swarankar and Singh 2013). Pathogens and parasites sensitive to humidity may be affected by alterations in precipitation. Certain pathogens and vectors could also spread due to wind velocity.

6.3.1.5.2 Effects on Non-vector Diseases

An important feature of non-vector borne diseases is that the pathogen or parasite spends some period of life outside the body of host. In addition, climate affects non-vector borne diseases more frequently. The commonly known non-vector diseases include foot-and-mouth disease (FMD), peste des petits ruminants (PPR), blue tongue, helminthiasis including fascioliasis and haemonchosis. Among the parasitic infections, haemonchosis is very common in small and large ruminants and causes significant loss in terms of productivity. Survival of the eggs and larvae depend on suitable temperature and humidity till they are ingested by another animal. Scientists have worked out the correlation between the timing, distribution, and quantity of rainfall for *Haemonchus contortus* development and probed the threshold rates of precipitation and evaporation needed for the survival of *H. contortus* larvae to the infective stage (Swarankar and Singh 2013). Taking the example of some parasites, parasitologists and clinicians have developed the prediction models depicting how increase or decrease in the climatic element would affect the disease pattern. The predicted effects of ambient temperature on some parasites are summarized in Table 6.7 (Swarankar and Singh 2013).

The impacts of climate change on livestock have been recently reviewed in depth by Escarcha et al. (2018). A summarized view of his findings is presented in Table 6.8.

Table 6.7 Predicted effects of ambient temperature on some parasites

Change in ambient temperature	Parasite			
	Nematodes	Lung worms	Flukes	Ticks
Increase	Rapid larval development, increased generation interval	Decreased survival of larvae and intermediate host	Increased mollusk activity	Hot dry conditions reduce activity and survival
Decrease	Slow larval development	Slower larval development	Cold winters decrease survivability of snail	Reduce risk of establishment

Table 6.8 Climate change impacts on livestock

(a) Productive traits:
Decrease in milk yield, production efficiency in cattle, production of meat and wool, and live weight losses in beef cattle and sheep
(b) Reproductive performance:
Reduction of reproductive parameters such as lower conception rates, animal fertility, general fitness, birth rates, and an increase in age at first calving in beef cattle as well as increase in still births, embryo mortality, and abortions
(c) Animal diseases:
Change in disease pattern and distribution of infectious diseases, increased persistence and survival of pathogens and parasites, higher prevalence of endemic diseases such as helminths in ruminants, increase in incidence of mastitis in dairy animals
(d) Forages, pastures, and feedstuffs:
Decline in productivity of pasture/rangelands/forages, worsening of feeds quality such as herbage and forage digestibility, decrease in quantity and availability of forage/grains feedstuffs
(e) Heat stress effects:
Changes in the physiological, behavioral, and metabolic functions of livestock, reduced forage intake, nutrient absorption, and feed conversion efficiency, negative impact on reproductive and breeding performance of animals, changes in milk lipid profile

6.3.2 Effects of Heat Stress on Attributes of Small Ruminants

Small ruminants such as sheep and goats, which make up approximately 56% of the world ruminant population have a great relevance in the world as they significantly contribute toward global food security and provide livelihood to a large number of graziers and pastoral herders (FAO 2016). United Nations has projected that by 2050 global human population is likely to touch 10 billion, and feeding this large population would be a challenge. Ruminant animals, especially small ruminants, are expected to serve as the main source of good-quality protein and fiber in future. Generally, sheep and goats are reared on grazing lands or under a mixed system, and more than 50% of the world's small ruminants are distributed in arid, semi-arid, and mountainous regions. On the mountains, for instance, it is estimated that about 30 to 40 million nomads herding the animals on the mountains of China, Central Asia, Russia, and Eastern Europe would be an important source to meet the demand for food (Uddin and Kebreab 2020). Similarly, an estimated population of approximately ten million livestock herders with large population of small ruminants is known to inhabit Himalayan grazing lands including Bhutan (Namgay et al. 2013). They are part of the mountains of Hindu Kush Himalayan region serving a population of 240 million people, mostly poor and marginal farmers (Wester et al. 2019).

Although lesser attention has been paid to the small ruminants, many studies report that sheep and goats were better adapted to hot environment. It was reported that the values of THI above which ewes start to suffer from thermal stress was higher compared to the large ruminants, although it varied from breed to breed (Chauhan and Ghosh 2014). Small ruminants are considered most easily reared animals possessing rare qualities of heat tolerance and ability to withstand feed and water scarcity. Among the multiple climatic stresses, heat stress and extremes of climate change are the principal climatic attributes affecting sheep and goat performance in their habitat. The specific details pertaining to the two main small ruminant species, sheep and goats, have been discussed in the following sections.

Sheep is recognized as one of the most resistant species to climatic extremes and is less prone to heat stress compared to the large ruminants (Caroprese et al. 2021). Comparative study of three Indian arid zone sheep breeds namely Chokla, Magra, and Marwari showed that Magra had highest adaptability to heat stress followed by Marwari and Chokla (Singh et al. 2016). Respiratory mechanism is the principal mode utilized by animals to maintain their body temperature and tolerate heat stress. The main mechanisms involved in the tolerance of heat stress in desert goats were shown to be shallow rapid panting and decreased metabolic rate (Naga et al. 2021). Sheep has also been described possessing other attributes like higher sweating rate, higher respiratory rate, and lower basal metabolic rate and water conservation capacity. However, Harle et al. (2007) reported poor reproductive performance by sheep when exposed to high ambient temperature accompanied by high humidity which resulted in increased incidences of embryonic mortality and reduced fetal growth.

Goats have been recognized as the first ruminant animal to be domesticated around 11,000 years ago. Goats are renowned among ruminant species to possess a number of special characteristics which no other species can match (Kaliber et al. 2016). For poor and landless marginal farmers, goat is considered as an ideal climate resilient animal model for possessing better thermo tolerance, higher resistance to disease, rearing easiness, ability to thrive during feed and water scarcity, utilize low-quality fodder and grasses, have efficient browsing behavior(Koluman 2023), emit less methane, and capacity to thrive and perform better under harsh environment (Sejian et al. 2016). Cheng et al. (2022) from the Texas University have reported that goats were far less susceptible to heat stress than other ruminants and their voluntary feed intake declined when the temperature was more than 10 °C above their thermal comfort zone. In another study, goats (a mix of 75% temperate breed with 25% local breed) comfortably thrived in hot environment at the average daily temperature of 34.2 °C and relative humidity of 68.2% and showed a capacity of significant water restriction for more than 30 days (Kaliber et al. 2016) .

In the Indian studies, Salwm Black goat breed from south India was identified as the most heat tolerant having better thermal resilience quality compared to other south Indian breeds (Amitha et al. 2019; Rashamol et al. 2019; Archana et al. 2018). In addition, goat meat is lean and delicious having low cholesterol. Goat milk is famous being easily digestible and having no religious taboo (Sejian et al. 2021c). From the point of view of pastoralists on the hills, goat is more resilient animal possessing ability to resist multiple stresses and has the ability to undertake long walks in difficult terrains.

Effect on Productive Attributes

Besides alterations in physiological, biochemical, and other responses, thermal stress was found to hamper productive attributes such as milk and growth and quality of milk and meat in both sheep and goats. Several studies have shown the adverse impact of heat stress on milk yield and quality of milk in ewes and goats. Todaro et al. (2014) reported a decrease in milk yield and reduction in total protein, fat, casein, oleic acid, linoleic acid, and linolenic acid, and an increase in saturated fatty acids in milk of sheep exposed to hot summer season. In another study, Sevi and Caroprese (2012) observed that physiological and production performance of lactating sheep were found to be affected at temperature higher than the upper critical temperature. Working with Sarda ewes, Peana et al. (2007) recorded a 15% reduction in milk yield at environmental temperature over 21 to 24 °C. Sevi et al. (2001) observed that when the ambient temperature exceeded 35 °C, milk production declined by 20% in Comisana sheep breed. Furthermore, Comisana ewes were found to have higher proportion of short chain and saturated fatty acids in milk when they were exposed to solar radiation (Sevi et al. 2002).

Similar to sheep, thermal stress was found to have a negative effect on milk yield and its composition in goats. Salama et al. (2014) reported a decrease in milk yield of dairy goats with an increase of THI value. An increase of one unit of THI resulted in a decrease of 1% milk yield. Hamzaoui et al. (2013) observed a reduction of protein and casein levels by 12.5% and 11.5%, respectively in milk of heat stress-exposed Murciano-Granadina dairy goats. Exposure of dairy goats to considerable heat stress at 34 °C air temperature and 25% relative humidity for 5 weeks resulted in reduction in milk yield and solids and fat levels in milk more in Alpine goats compared to the Nubian goats (Brown et al. 1988). A number of other studies in goats have also established the impact of thermal stress on milk (Salama et al. 2014) and meat production (Archana et al. 2018).

6.3.2.1 Effect on Endocrine System

After prolonged exposure to heat stress, compensatory physiological, endocrine, and behavioral mechanisms are activated for the survival of the animal. Heat exposure negatively affects oxidative stress in sheep along with stimulation of the receptors in hypothalamus to initiate a series of physiological changes including a rise in respiration rate, rectal temperature, and heart rate. The stress induces several neuroendocrine responses which includes activation of hormonal axis and release of tropic hormones. Hormones from pituitary gland act on thyroid and adrenal glands causing secretion of respective hormones such as triiodothyronine (T_3), thyroxine (T_4), glucocorticoids, and cortisol (Mcmanus et al. 2020). Prolactin from pituitary gland is the principal hormone and reportedly has a thermoregulatory role in small ruminants including stimulation of sweat glands. Prolactin secretion also reportedly played a role in improving the ability of small ruminants to adapt to warm environment (Choy et al. 1997). There is increased secretion of cortisol which stimulates gluconeogenesis to provide extra energy to fight thermal stress and help to cope up with heat stress. During heat stress, thyroid hormones exhibit reduced functioning resulting in a decrease in the secretion of T_3 and T_4.

6.3.2.2 Effect on Immune System

In most domesticated livestock, thermal stress is reported to depress the immune system through alterations of cytokines secretions, making the animals more vulnerable to diseases. In two different studies on Polwarth and Merino sheep inside a climatic chamber, thermal stress was found to decrease white blood cells. da Silva et al. (1992) reported a decrease in blood neutrophils and eosinophils in Polwarth, whereas Wojtas et al. (2014) observed a reduction in total white blood cell counts in Merino sheep.

Adaptive responses to thermal stress are regulated at molecular level and involve increased expression of heat shock proteins (HSPs) (Banerjee et al. 2013). HSPs are biological markers playing a role in antigen processing. The HSPs work as escort to prevent protein and cell damage. Various stressors cause either a suppressive or a stimulatory effect on HSPs. Previous studies have shown higher expression of *HSP70*, *HSP60*, *HSP90*, and *HSP27* genes in small ruminants during heat stress (Mohanarao et al. 2014; Hooper et al. 2018; Joy et al. 2020). The newest idea is that when animal is exposed to stress, the endogenous cells secrete danger-associated molecular patterns (DAMPs). Among the hormones that play a role in heat stress are prolactin and cortisol. Prolactin can reduce cytokines secretion, and cortisol secretion affects expression of genes related to innate immune response (Caroprese et al. 2021).

6.3.2.3 Impact on Reproduction

Rising ambient temperature year by year has been badly affecting livestock performance, especially in tropical and subtropical regions which are home to prized breeds of sheep and goats. Almost all studies have shown that heat stress affects reproductive functioning of sheep negatively. Working in pastoral environment, Harle et al. (2007) showed that high temperature accompanied by high humidity was a major factor for poor reproductive performance of sheep with increased level of embryonic mortality and reduced fetal growth. Thermo-neutral zone for sheep has been identified between 12 and 32 °C, which would mean that ambient temperature beyond 32 °C will be uncomfortable for most sheep breeds (Al-Dawood 2017). After prolonged controlled (hot room) studies, Van Wettere et al. (2021) have recently documented an important observation concluding that the reproductive performance of ewes was most affected by heat stress from 5 days prior to estrus, during estrus, and then for ensuing 5 days suggesting vulnerability of this period to thermal exposure. Thermal stress was also found to impair estrous behavior, follicular development, ovulation rate, and fertility. In these experiments, exposure of pregnant ewes for prolonged period (> 50 days) was found to reduce placental and fetal growth and reduction in birth weight of lambs.

As a consequence of heat stress in males, the biological phenomena such as sexual activity, endocrine secretions, testicular functions, spermatogenesis, and physical and chemical characteristics of semen are affected. High temperature also interfered with oxidative metabolism of glucose in spermatic cells and accumulation of reactive oxygen species which was reflected in sperm primary defects (Krishnan et al. 2017). The stress was also found to seriously affect male fertility and impair sperm production badly affecting semen quality. It increased the proportion of morphologically abnormal spermatozoa and changed seminal characteristics such as ejaculate volume, semen pH, sperm concentration, and sperm motility.

6.4 Mitigation Strategies Related to Climate Change

The story of climate change starts with the releasing of excess greenhouse gases such as carbon dioxide, methane, and nitrous oxide into the earth's atmosphere and setting the cycle of warming the earth. The resulting heat stress from warm climate has brought unparalleled misery to the humans, animals, plants, crops, oceans, and earth's ecosystem. During preindustrial times, the earth's carbon cycle was roughly in balance, i.e., CO_2 emitted was equal to CO_2 absorbed by plants and oceans. GHG emissions have increased dramatically since 1850 due to industrial revolution and other human activities. So, in a way, the key to the mitigation lies in fixing the very cause responsible for warming of earth, i.e., removing the greenhouse gases. Researchers are trying to implement all sorts of ideas to cut down on GHG emissions, even wondering whether the gases could be sucked up (carbon capture) from atmosphere by special devices (Gates 2021). The central idea of all the efforts is to cut down the emissions of GHG, if not to zero, but to a significant low level. This has to be done by all sectors emitting GHG including industries, manufacturers, agriculture, and animal husbandry.

We have attempted to summarize below some of the suggested options by experts to cut down or reduce the GHG and also directly provide relief to the animals from heat stress.

6.4.1 Strategies to Reduce Enteric Methane Emissions

Experts have suggested a number of methods and approaches to cut down the enteric methane emission from ruminants. The two main approaches include (a) by reducing the number of ruminant animals at the global level especially large ruminants which are the main source of enteric methane and (b) by dietary management and nutritional interventions.

6.4.1.1 Measures to Reduce the Number of Ruminant Livestock

It is well known that ruminant animals such as cattle, buffaloes, sheep, and goats generate large volume of greenhouse gases, chiefly methane and carbon dioxide, during enteric fermentation which are released into the atmosphere. Estimated daily output of gas from an adult cow is around 150 to 200 liters. With the global ruminant population of approximately 3700 million (Opio et al. 2013), the volume of GHG emitted into the atmosphere daily would be staggering. Out of the total global emission, 14.5% is attributed to agriculture sector. As mentioned earlier, a few scientists are of the opinion that gas emission from ruminants has been exaggerated.

There have been suggestions from various quarters that reducing number of ruminant animals at the global level could be one of the strategies to alleviate the problem of global warming for distant future. It has been observed that healthier animals were more productive and generated lower emissions. It was, therefore, advantageous to keep fewer but healthier animals (FAO 2020). As a measure of coping strategy to tackle climate change, Ali et al. (2022) from Pakistan have reported that after getting convinced by this idea, the herders in the area have been reducing number of livestock. In many parts of the world, reduction in animal number has been suggested as an effective strategy to significantly reduce the carbon footprint (Capper et al. 2009). In the Netherlands, with rapid rise in milk production per cow from 6270 kg in the year 1990 to 8350 kg in 2008, there has been a decrease in methane production from 17.6 to 15.4 g/kg fat- and protein-corrected milk (FPCM) (Bannink et al. 2011). This could be an example to reduce the number of ruminant animals for developing countries such as India where the ruminants' population is already quite large. But to reduce number of animals is not an easy task in the Indian conditions as there is a ban on slaughter of cattle progenies in a number of Indian States.

6.4.1.2 Measures to Reduce the Enteric Methane Emission from Ruminants

These methods comprise mainly of dietary management, adoption of improved pastures, genetically better ruminant breeds emitting lower GHG, lowering grazing intensity, feeding high-quality fodder as compared to animals consuming low-quality fodder, or by grazing on less matured pastures Sejian et al. 2011, Eckard et al. 2010, Gerber et al. 2013)).

Some of the strategies relating to nutritional interventions to reduce enteric methane emission from ruminants include feeding of different dietary combinations (Beauchemin et al. 2008), special supplements, feed additives, diets to improve feed digestibility, removal of protozoa from the reticulo-rumen, application of genetic technologies, etc. A brief account of suggested interventions is presented in Table 6.9.

Some of the other novel methods suggested by Astuti et al. (2024)to curtail GHG emission from livestock include use of *p*lant secondary metabolites (tannins, saponins, and phenolic monomers), use of chemical inhibitors (chloral hydrate, nitroethane, etc.), application of genetic technologies as a nutritional tool, immunization as a biological approach, and use of rumen microbial metagenomics as an emerging tool.

Table 6.9 Nutritional intervention strategies for reduction of enteric methane emission from ruminants

Measure	Reference
Improving feed digestibility by replacing structural carbohydrates with non-structural carbohydrates in the diet	Hristov et al. (2015)
Feeding diets composed of plant-based food like serials and pulses	Clark and Tilman (2017)
Supplementation with feed additives such as 3-nitrooxypropanol and red algae	Roque et al. (2019)
Supplementation with concentrate feed which works as a sink of hydrogen in the rumen	Wattiaux et al. (2019) and Chauhan et al. (2023)
Feeding nitrates/sulfates which can reduce methane emission and methanogens	Bhatta et al. (2021)
Inclusion of inophores in diets which causes inhibition of microbes, but there is a ban on their use in some countries	Bhatta et al. (2021)
Use of biohydrogenation which restricts the hydrogen supply to methanogens resulting in mitigation of enteric methane	Bhatta et al. (2021)
Providing concentrate within the feed is negatively correlated with methanogenesis in rumen. Propionate rather than acetate is preferable	Astuti et al. (2024)

6.4.2 Selection of Heat-Tolerant and Disease-Resistant Breeds

An important parameter for domestic animals to survive climate change is their breed. Breeds that have inbuilt characteristics of heat tolerance and disease resistance need to be identified and promoted (Singh et al. 2021). In case of hilly terrain or for pastoral farming, we need breeds that are able to walk long distances and bear migration stress (Escarcha et al. 2018). The cattle, sheep, and goat breeds belonging to tropical countries are known to possess these characteristics.

6.4.3 Alleviation of Heat Stress

There should be provision of proper shelter with shaded accommodation with trees all around. The sheds should be oriented in a manner that there is enough space per head for the animals to avoid radiating heat load on the animal. Trees would not only protect animals from sunlight but also act as heat sink by removing the heat from the air. Shades provided with trees have been found as the most effective cooling devices. Besides shelter, there must be easy availability of clean drinking water in the premises. Singh et al. (2021) have reported that a simpler and potentially more cost-effective method of improving animal production during thermal stress is the modification of the diet and/or the provision of nutritional supplements during stressful situations. In case of migratory hill animals, heat stress could be avoided by movement/migration of the herds to higher altitudes and choosing proper migration time depending on weather conditions.

References

Al-Dawood A (2017) Towards heat stress management in small ruminants – a review. Ann Anim Sci 17(1):59–88

Ali I, Raza G, Hussain A, Zafar M, Akram W, Begum F (2022) Herders' perspectives on pastoral systems in the face of changing climate in Karakoram mountains of Northern Pakistan. Res Sq. https://doi.org/10.21203/rs.3.rs-1484818/v1

Amitha JP, Krishnan G, Bagath M, Sejian V, Bhatta R (2019) Heat stress impact on the expression patterns of different reproduction related genes in Malabari goats. Theriogenology 131:169–176

Archana PR, Sejian V, Ruban W, Bagath M, Krishnan G, Aleena J, Manjunathareddy GB, Beena V, Bhatta R (2018) Comparative assessment of heat stress induced changes in carcass traits, plasma leptin profile and skeletal muscle myostatin and HSP70 gene expression patterns between indigenous Osmanabadi and Salem Black goat breeds. Meat Sci 141:66–80

Arrhenius (1850) Website: https://youmatter.world/en/definition/climate-change-meaning-definition-causes-and-consequences/

Astuti PK, Ayoob A, Strausz P, Vakayil B, Kumar SH, Kusza S (2024) Climate change and dairy farming sustainability; a causal loop paradox and its mitigation scenario. Heliyon 10(3):e25200. https://doi.org/10.1016/j.heliyon.2024.e25200

Banerjee D, Upadhyay RC, Singh SV, Hooda OK, De S (2013) HSP-gene expression profile and their relationship with physiological and hormonal variables in heat and cold adapted goats. In: Sahoo A, Kumar D, Naqvi SMK (eds) Climate resilient small ruminant production. National initiative on climate resilient agriculture (NICRA), CSWRI, Avikanagar, pp 84–92

Bannink A, van Schijndel MW, Dijkstra J (2011) A model of enteric fermentation in dairy cows to estimate methane emission for the Dutch National Inventory Report using the IPCC Tier 3 approach. Anim Feed Sci Technol 166–167:603–618

Beauchemin KA, Kreuzer M, O'mara F, McAllister TA (2008) Nutritional management for enteric methane abatement. A review. Aust J Exp Agric 48:21–27

Bejaoui B, Sdiri C, Ben Souf I, Belhadj Slimen I, Ben Larbi M, Koumba,S, Martin P, M'Hamdi N (2023) Physicochemical properties, antioxidant markers, and meat quality as affected by heat stress: a review. Molecules 28(8):3332. https://doi.org/10.3390/molecules28083332

Belhadj Slimen I, Najar T, Ghram A, Abdrrabba M (2016) Heat stress effects on livestock: Molecular, cellular and metabolic aspects, a review. J Anim Physiol Anim Nutr (Berl) 100(3):401–412

Bhatta R, Malik PK, Sejian V (2021) Enteric methane mitigation in livestock: an indian perspective. In: Sejian V, Chauhan SS, Devaraj C, Malik PK, Bhatta R (eds) Climate change and livestock production: recent advances and future perspectives. Springer, Singapore, pp 197–211

Brown DL, Morrison SR, Bradford GE (1988) Effects of ambient temperature on milk production of Nubian and Alpine goats. J Dairy Sci 71(9):2486–2490

Capper JL, Cady RA, Bauman DE (2009) The environmental impact of dairy production: 1944 compared with 2007. J Anim Sci 87(6):2160–2167

Caroprese M, Ciliberti MG, Albenzio M, Sevi A (2021) Heat stress associated changes in the immune system related responses in sheep. In: Sejian V, Chauhan SS, Devaraj C, Malik PK, Bhatta R (eds) Climate change and livestock production: Recent advances and future perspectives. Springer, Singapore, pp 49–58

Chauhan DS, Ghosh N (2014) Impact of climate change on livestock production: a review. J Anim Res 4(2):223–239

Chauhan SS, Zhang M, Osei-Amponsah R, Clarke I, Sejian V, Warner R, Dunshea FR (2023) Impact of heat stress on ruminant livestock production and meat quality, and strategies for amelioration. Anim Front 13(5):60–68

Cheng M, McCarl B, Fei C (2022) Climate change and livestock production: a literature review. Atmos 13:140. https://doi.org/10.3390/atmos13010140

Chhabra A, Manjunath KR, Panigrahy S, Parihar JS (2013) Greenhouse gas emissions from Indian livestock. Clim Chang 117:329–344

Chirico J, Jonsson P, Kjellberg S, Thomas G (1997) Summer mastitis experimentally induced by *Hydrotaea irritans* exposed to bacteria. Med Vet Entomol 11(2):187–192

Choy VJ, Nixon AJ, Pearson AJ (1997) Distribution of prolactin receptor immunoreactivity in ovine skin and changes during the wool follicle growth cycle. J Endocrinol 155(2):265–276

Clark M, Tilman D (2017) Comparative analysis of environmental impacts of agricultural production systems, agricultural input efficiency, and food choice. Environ Res Lett 12:064016. https://doi.org/10.1088/1748-9326/aa6cd5

Collier RJ, Beede DK, Thatcher WW, Israel LA, Wilcox CJ (1982) Influences of environment and its modification on dairy animal health and production. J Dairy Sci 65(11):2213–2227

da Silva RG, da Costa MJ, Sobrinho AG (1992) Influence of hot environments on some blood variables of sheep. Int J Biometeorol 36(4):223–225

Dahl GE, Tao S, Laporta J (2020) Heat stress impacts immune status in cows across the life cycle. Front Vet Sci 7:116. https://doi.org/10.3389/fvets.2020.00116

Danmaigoro A, Muhammad MA, Abubakar K, Magiri RB, Bakare AG, Iji PA (2024) Morphological and physiological features in small ruminants: an adaptation strategy for survival under changing climatic conditions. Int J Biometeorol 68(8):1497–1505

Das S (2017) Impact of climate change on livestock, various adaptive and mitigative measures for sustainable livestock production. Appro Poult Dairy Vet Sci 1(4):64–70

Davis KF, Yu K, Herrero M, Havlik P, Carr JA, D'Odorico P (2015) Historical trade-offs of livestock's environmental impacts. Environ Res Lett 10(12):125013. https://doi.org/10.1088/1748-9326/10/12/125013

Eckard RJ, Grainger C, De Klein CA (2010) Options for the abatement of methane and nitrous oxide from ruminant production: a review. Livest Sci 130:47–56

Ekine-Dzivenu CC, Mrode R, Oyieng E, Komwihangilo D, Lyatuu E, Msuta G, Ojango JMK, Okeyo AM (2020) Evaluating the impact of heat stress as measured by temperature-humidity index (THI) on test-day milk yield of small holder dairy cattle in a sub-Sahara African climate. Livest Sci 242:104314. https://doi.org/10.1016/j.livsci.2020.104314

Escarcha JF, Lassa JA, Zander KK (2018) Livestock under climate change: a systematic review of impacts and adaptation. Climate 6(3):54. https://doi.org/10.3390/cli6030054

FAOSTAT (2018) FAOSTAT. Food and Agriculture Organization Corporate Statistical Database. www.fao.org/faostat/en/#home

Food and Agriculture Organization of the United Nations [FAO] (2006) Livestock's long shadow-environmental issues and options. http://www.fao.org.pdf

Food and Agriculture Organization of the United Nations [FAO] (2007). https://www.fao.org

Food and Agriculture Organization of the United Nations [FAO] (2016). www.fao.org/climate-change@FAO,2016

Food and Agriculture organization of the United nations 2000©FAO (2020) CA8946EN/1/05.20 Animal health and climate change, https://www.fao.org/3/ca8946en/CA8946EN.pdf

Gates B (2021) How to avoid a climate disaster: the solutions we have and the breakthroughs we need, 1st edn. Random House Large Print, New York

Gerber PJ, Steinfeld H, Henderson B, Mottet A, Opio C, Dijkman J, Falcucci A, Tempio G (2013) Tackling climate change through livestock – a global assessment of emissions and mitigation opportunities. Food and Agriculture Organization of the United Nations (FAO), Rome. https://www.fao.org/4/i3437e/i3437e.pdf

Glatzle A (2018) Domestic livestock and its alleged role in climate change https://doi.org/10.5772/intechopen.80389

Gonçalves AA, Garcia AR, Rolim Filho ST, Silva JARD, Melo DN, Guimarães TC, Tavares HR, Silva TVG, Souza EB, Santos SDSD, Ohashi OM (2021) Scrotal thermoregulation and sequential sperm abnormalities in buffalo bulls (*Bubalus bubalis*) under short-term heat stress. J Therm Biol 96:102842. https://doi.org/10.1016/j.jtherbio.2021.102842

Gust Ekholm (1901) Website: https://youmatter.world/en/definition/climate-change-meaning-definition-causes-and-consequences/

Hamzaoui S, Salama AAK, Albane IIE, Such X, Caja G (2013) Physiological responses and lactational performance of late-lactation dairy goats under heat stress conditions. J Dairy Sci 96(10):6355–6395

Harle KJ, Howden SM, Hunt LP, Dunlop M (2007) The potential impact of climate change on the Australian wool industry by 2030. Agric Syst 93(1–3):61–89

Hooper HB, Dos Santos SP, de Oliveira SA, Merighe GKF, Negrão JA (2018) Acute heat stress induces changes in physiological and cellular responses in Saanen goats. Int J Biometeorol 62(12):2257–2265

Hristov AN, Oh J, Giallongo F, Frederick TW, Harper MT, Weeks HL, Branco AF, Moate PJ, Deighton MH, Williams SR, Kindermann M, Duval S (2015) An inhibitor persistently decreased enteric methane emission from dairy cows with no negative effect on milk production. Proc Natl Acad Sci USA 112(34):10663–10668

Husen Harun MS, Bekere H, Hussen DH, Yusuf MA (2022) Review on impact of climate change on livestock health and productivity. BAOJ Nutr 1(1):1001. https://doi.org/10.2139/ssrn.3517784

IEA web (2020) Composition of inhaled and exhaled air. https://byjus.com/biology/composition-gases-breathe

Intergovernmental Panel on Climate Change. IPCC (1988) History of the IPCC. https://www.ipcc.ch/about/history

IPCC Climate Report (2022) Summary. https://www.ipcc.ch/report/ar6/wg2/

John Sawyer (1972) Climate: Sawyer predicted rate of warming in 1972 https://www.nature.com/articles/448992c

Johnson JS (2018) Heat Stress: impact on livestock well-being and productivity and mitigation strategies to alleviate the negative effects. Anim Prod Sci 58(8):1404–1413

Joo SS, Lee SJ, Park DS, Kim DH, Gu BH, Park YJ, Rim CY, Kim M, Kim ET (2021) Changes in blood metabolites and immune cells in Holstein and Jersey dairy cows by heat stress. Animals (Basel) 11(4):974. https://doi.org/10.3390/ani11040974

Joy A, Dunshea FR, Leury BJ, Clarke IJ, DiGiacomo K, Chauhan SS (2020) Resilience of small ruminants to climate change and increased environmental temperature: a review. Animals (Basel) 10(5):867. https://doi.org/10.3390/ani10050867

Kaliber M, Koluman N, Silanikove N (2016) Physiological and behavioral basis for the successful adaptation of goats to severe water restriction under hot environmental conditions. Animal 10(1):82–88

Charles D Keeling (1960) Website: https://youmatter.world/en/definition/climate-change-meaning-definition-causes-and-consequences/

Koluman N (2023) Goats and their role in climate change. Small Rumin Res 228:107094. https://doi.org/10.1016/j.smallrumres.2023.107094

Krishnan G, Bagath M, Pragna P, Vidya MK, Aleena J, Archana PR, Sejian V, Bhatta R (2017) Mitigation of the heat stress impact in livestock reproduction. In: Payan-Carreira R (ed) Theriogenology. IntechOpen, London, pp 63–86

Krishnan R, Shresh AB, Ren G, Rajbhandari R, Saeed S, Sanjay J, Syed MA, Vellore R, Xu Y, You Q, Ren X (2019) Unravelling climate change in the Hindu Kush Himalaya: rapid warming in the mountains and increasing extremes. In: Wester P, Mishra A, Mukherji A, Shreshtha AB (eds) The Hindu Kush Himalaya Assessment- mountains climate change, sustainability and people. Springer Nature Switzerland AG, Cham., pp 57–96

Lemal P, May K, König S, Schroyen M, Gengler N (2023) Invited review: From heat stress to disease-Immune response and candidate genes involved in cattle thermotolerance. J Dairy Sci 106(7):4471–4488

Lindsey R (2023) Climate change: atmospheric carbon dioxide. https://www.climate.gov/news-features/understanding-climate/climate-change-atmospheric-carbon-dioxide

Liu J, Li L, Chen X, Lu Y, Wang D (2019) Effects of heat stress on body temperature, milk production, and reproduction in dairy cows: a novel idea for monitoring and evaluation of heat stress -A review. Asian Australas J Anim Sci 32(9):1332–1339

Lozano DRR, Vásquez PCG, González PE (2005) Effect of heat stress and its interaction with other management and productive variables on pregnancy rate in dairy cows in Aguascalientes. Mexico Vet Mex 36(3):245–260

Maan R, Kataria N, Pilania PK, Sharma A, Sankhala LN, Kataria AK (2013) Fluctuations of serum glutathione reductase activities due to changes in extreme ambient temperatures in Marwari sheep from arid tracts. ELBA Bioflux 5(1):9–13

Mbow C, Rosenzweig C, Barioni LG, Benton TG, Herrero M, Krishnapillai M, Liwenga E, Pradhan P, Rivera-Ferre MG, Sapkota T, Tubiello FN, Xu Y (2019) Food security. In: Shukla RP, Skea J, Calvo Buendia E, Masson-Delmotte V, Pörtner H-O, Roberts DC, Zhai P, Slade R, Connors S, van Diemen R, Ferrat M, Haughey E, Luz S, Neogi S, Pathak M, Petzold J, Portugal Pereira J, Vyas P, Huntley E, Kissick K, Belkacemi M, Malley J (eds) Climate change and land: an IPCC special report on climate change, desertification, land degradation, sustainable land management, food security, and greenhouse gas fluxes in terrestrial ecosystems. Cambridge University Press, Cambridge, UK/New York, pp 437–550. https://doi.org/10.1017/9781009157988.007

McManus CM, Faria DA, Lucci CM, Louvandini H, Pereira SA, Paiva SR (2020) Heat stress effects on sheep: Are hair sheep more heat resistant? Theriogenology 155:157–167

Mitloehners F(2020) The scientist setting the record straight on cows and climate change. https://www.darigold.com/the-scientist-setting-the-record-straight-on-(cows-and-climate-change/

Mohanarao GJ, Mukherjee A, Banerjee D, Gohain M, Dass G, Brahma B, Datta TK, Upadhyay RC, De S (2014) *HSP70* family genes and *HSP27* expression in response to heat and cold stress *in vitro* in peripheral blood mononuclear cells of goat (*Capra hircus*). Small Rumin Res 116(2–3):94–99

Moore L (2008) Greenhouse gases: how long will they last? https://blogs.edf.org/climate411/2008/02/26/ghg_lifetimes/

Naga AM, Khalek TA, Osman MA, Elbeltagy AR, Abdel-Aal ES, Abou-Ammo FF, El-Shafie M (2021) Physiological and genetic adaptation of desert sheep and goats to heat stress in the arid areas of Egypt. Small Rumin Res 203:106499. https://doi.org/10.1016/j.smallrumres.2021.106499

Namgay K, Millar J, Black R, Samdup T (2013) Transhumant agro-pastoralism in Bhutan: Exploring contemporary practices and socio-cultural traditions. Pastoralism 3:13. https://doi.org/10.1186/2041-7136-3-13

NASA (2022) Says fifth warmest year on record, warming trend continues warmest. https://climate.nasa.gov/news/3246/nasa-says-2022-fifth-warmest-year-on-record-warming-trend-continues/

NASA web (2018). http://youmatter.world/en/definition/climate-change-meaning-definition-causes-and-consequences/

Negi VS, Thakur S, Dhyani R, Bhatt ID, Rawal RS (2021) Climate change observations of indigenous communities in the Indian Himalaya. Weather Clim Soc 13(2):245–257

NOAA (2021) Carbon dioxide peaks near 420 ppm at Mauna Loa observatory. IOC web: https://research.noaa.gov/article/ArtMID/587/Article/D/2764/coronavirus-response-barely-slows-rising-carbon-dioxide

OIE (2009) currently known as World Organisation for Animal Health (WOAH) coordinate and support animal diseases.en.wikipedia.org/wiki/World_Organisation_for_Animal_Health

Opio C, Gerber P, Mottet A, Falcucci A, Tempio G, MacLeod M, Vellinga T, Henderson B, Steinfeld H (2013) Greenhouse gas emissions from ruminant supply chains – a global life cycle assessment. Food and Agriculture Organization of the United Nations (FAO), Rome

Patra AK (2014) Trends and projected estimates of GHG emissions from Indian livestock in comparisons with GHG emissions from world and developing countries. Asian Australas J Anim Sci 27(4):592–599

Peana I, Fois G, Cannas A (2007) Effects of heat stress and diet on milk production and feed and energy intake of Sarda ewes. Ital J Anim Sci 6(sup1):577–579

Piscopo N, Matera R, Cotticelli A, Trapanese L, Tamburis O, Cimmino R, Salzano A (2024) Investigation of climate effects on the physiological parameters of dairy livestock (cow vs. buffalo). Sensors (Basel) 24(4):1164. https://doi.org/10.3390/s24041164

Pragna P, Archna PR, Aleena J, Sejian V, Krishnan G, Bagath M, Manimaran A, Beena V, Kurien EK, Varma G, Bhatta R (2017) Review article: Heat stress and dairy cow: impact on both milk yield and composition. Int J Dairy Sci 12:1–11

Rashamol VP, Sejian V, Bagath M, Krishnan G, Beena V, Bhatta R (2019) Effect of heat stress on the quantitative expression patterns of different cytokine genes in Malabari goats. Int J Biometeorol 63(8):1005–1013

Rayamajhi N, Manandhar B (2020) Impact of climate change and adaptation measures on transhumance herding system in Gatlang, Rasuwa. Air Soil Water Res 13:1–10. https://doi.org/10.1177/1178622120951173

Rhoads ML, Rhoads RP, VanBaale MJ, Collier RJ, Sanders SR, Weber WJ, Crooker BA, Baumgard LH (2009) Effects of heat stress and plane of nutrition on lactating Holstein cows: I. Production, metabolism, and aspects of circulating somatotropin. J Dairy Sci 92(5):1986–1997

Roque BM, Salwen JK, Kinley R, Kebreab E (2019) Inclusion of *Asparagopsis armata* in lactating dairy cows' diet reduces enteric methane emission by over 50 percent. J Clean Prod 234:132–138

Salama AAK, Caja G, Hamzaoui S, Badaoui B, Castro-Costa A, Façanha DAE, Guilhermino MM, Bozzi R (2014) Different levels of response to heat stress in dairy goats. Small Rumin Res 121(1):73–79

Sawyer G, Narayan EJ (2019) A review on the influence of climate change on sheep reproduction. In: Narayan E (ed) Comparative endocrinology of animals. IntechOpen, London, pp 1–21

Sejian V, Lakritz J, Ezeji T, Lal R (2011) Forage and Flax seed impact on enteric methane emission in dairy cows. Res J Vet Sci 4:1–8

Sejian V, Ravindra JP, Prasad CS (2013) Livestock production under ensuing climate change scenario: resilience versus performance. In: Sahoo A, Kumar D, Naqvi SMK (eds) Climate resilient small ruminant production. National initiative on climate resilient agriculture (NICRA), CSWRI, Avikanagar, pp 8–19

Sejian V, Bhatta R, Malik PK, Madiajagan B, Al-Hosni YAS, Sullivan M, Gaughan JB (2016) Livestock as sources of greenhouse gases and its significance to climate change. In: Moya BL, Pous J (eds) Greenhouse gases. IntechOpen, Croatia, pp 243–259

Sejian V, Chauhan SS, Devaraj C, Malik PK, Bhatta R (eds) (2021a) Climate change and livestock production: recent advances and future perspectives. Springer. https://doi.org/10.1007/978-981-16-9836-1

Sejian V, Silpa MV, Reshma Nair MR, Devaraj C, Krishnan G, Bagath M, Chauhan SS, Suganthi RU, Fonseca VFC, König S, Gaughan JB, Dunshea FR, Bhatta R (2021b) Heat stress and goat welfare: adaptation and production considerations. Animals (Basel) 11(4):1021. https://doi.org/10.3390/ani11041021

Sejian V, Silpa MV, Reshma Nair MR, Devaraj C, Devapriya A, Ramachandran N, Awachat VB, Shashank CG, Fonsêca VFC, Bhatta R (2021c) Goat as the ideal future climate resilient animal model. In: Sejian V, Chauhan SS, Devaraj C, Malik PK, Bhatta R (eds) Climate change and livestock production: recent advances and future perspectives. Springer, Singapore, pp 279–292

Sevi A, Caroprese M (2012) Impact of heat stress on milk production, immunity and udder health in sheep. A critical Review. Small Rumin Res 107(1):1–7

Sevi A, Annicchiarico G, Albenzio M, Taibi L, Muscio A, Dell'Aquila S (2001) Effects of solar radiation and feeding time on behavior, immune response and production of lactating ewes under high ambient temperature. J Dairy Sci 84:629–640

Sevi A, Albenzio M, Annicchiarico G, Caroprese M, Marino R, Taibi L (2002) Effects of ventilation regimen on the welfare and performance of lactating ewes in summer. J Anim Sci 80(9):2349–2361

Singh KM, Singh S, Ganguly I, Ganguly A, Nachiappan RK, Chopra A, Narula HK (2016) Evaluation of Indian sheep breeds of arid zone under heat stress condition. Small Rumin Res 141:113–117

Singh G, Samad HA, Karthiga K, Priyanka K, Sarma L, Chouhan VS, Maurya VP (2021) Comparative assessment of thermo-tolerance of crossbred and indigenous cattle breeds. In:

Sejian V, Chauhan SS, Devaraj C, Malik PK, Bhatta R (eds) Climate change and livestock production: recent advances and future perspectives. Springer, Singapore, pp 73–81

Summer A, Lora I, Formaggioni P, Gottardo F (2018) Impact of heat stress on milk and meat production. Anim Front 9(1):39–46

Swamy M, Bhattacharya S (2006) Budgeting anthropogenic greenhouse gas emission from Indian livestock using country-specific emission coefficients. Curr Sci 91(10):1340–1353

Swarankar CP, Singh D (2013) Effect of climate change on livestock diseases. In: Sahoo A, Kumar D, Naqvi SMK (eds) Climate resilient small ruminant production. National initiative on climate resilient agriculture(NICRA), CSWRI, Avikanagar, pp 53–68

Tiwari KR, Sitaula BK, Bajracharya RM, Raut N, Bhusal P, Sengel M (2020) Vulnerability of pastoralism: a case study from the high mountains of Nepal. Sustainability 12(7):2737. https://doi.org/10.3390/su12072737

Todaro M, Bonanno A, Scatassa ML (2014) The quality of Valle del Belice sheep's milk and cheese produced in the hot summer season in Sicily. Dairy Sci Technol 94:225–239

Uddin ME, Kebreab E (2020) Review: Impact of food and climate change on pastoral industries. Front Sustain Food Syst 4:543403. https://doi.org/10.3389/fsufs.2020.543403

UNEP Copenhagen Climate Centre (2020) Emission gap report. http://www.unep.org/emissions-gap-report-2020

UNFCCC (1992) United Nations Framework Convention on Climate Change. https://unfccc.int/resource/docs/convkp/conveng.pdf

Upadhyay RC, Ashutosh KA, Gupta SK, Gupta SV, Singh SV, Rani N (2009) Inventory of methane emission from livestock in India. In: Aggarwal PK (ed) Global climate change and Indian agriculture, Case studies from the ICAR network project. ICAR, New Delhi, pp 117–122

van Wettere WHEJ, Kind KL, Gatford KL, Swinbourne AM, Leu ST, Hayman PT, Kelly JM, Weaver AC, Kleemann DO, Walker SK (2021) Review of the impact of heat stress on reproductive performance of sheep. J Anim Sci Biotechnol 12(1):26. https://doi.org/10.1186/s40104-020-00537-z

Wattiaux MA, Uddin ME, Letelier P, Jackson RD, Larson RA (2019) Invited review: emission and mitigation of greenhouse gases from dairy farms: the cow, the manure, and the field. Appl Anim Sci 35(2):238–254

Wester P, Mishra A, Mukherji A, Shrestha AB (eds) (2019) The Hindu Kush Himalaya assessment- mountains, climate change, sustainability and people. ICIMOD, Springer Nature, Cham

Wojtas K, Cwynar P, Kołacz R (2014) Effect of thermal stress on physiological and blood parameters in merino sheep. Bull Vet Inst Pulawy 58(2):283–288

WRI (2011) World resources institute annual report 2011. https://www.wri.org/research/wri-annual-report-2011-2012

7 Himalayan Pastoralists' Perceptions to Changing Climate and How It Is Affecting the System and Their Livelihoods

Abstract

Climate change or global warming is no more a fantasy but a reality affecting all living beings on planet earth. Recent reports have suggested that the effects of climate change, especially on livestock and animal husbandry activities, were more visible in the Indian Himalayas, with a warming rate higher than the global average. Direct and indirect effects of climate change on the productive aspects of livestock have been studied and reported previously, but how the global warming has affected the life and livelihood of pastoralists and their animals and how they perceive the effects of changing climate have received little attention. This chapter provides an overview of feedback and perceptions of migratory pastoralists on the effects of climate change on grazing ruminants in different Himalayan countries, namely India, Nepal, Bhutan, China, Pakistan, and Afghanistan. In most cases, the feedback was obtained by direct interviews, through questionnaires or participatory appraisal from large number of randomly selected respondents. Besides the scientific and peer reviewed papers, the feedback or perceptions of herders were also obtained from the stories appearing in the local print and electronic media. A brief account of country-wise perceptions of the pastoralists to global warming is presented.

In India, active migratory pastoralism is practiced in union territories of Jammu and Kashmir and Ladakh and north-western states of Himachal Pradesh and Uttarakhand. Most studies from these states involving the Bakarwals, Gaddis, and Bhutias revealed that majority of pastoralists perceived rising of ambient temperature, drying up of water resources due to warming, incidence of unpredictable weather conditions, cloud bursting, significant decline in snowfall, and thinning of glaciers. Global warming is responsible for shifting of alpine rangelands to higher altitudes and disturbing the time schedule of upward and downward journeys. For instance, early onset of summer has forced the shep-

S. P. Singh, J. K. Malik, *Migratory Small Ruminant Farming System in the Himalayas-Peculiarities, Problems, Prospects, Climate Effects, Nutrition and Health*,
https://doi.org/10.1007/978-981-96-6691-1_7

herds to early movement of flocks. In all areas, unpredictable weather conditions, especially erratic rains, were pointed by the respondents as major setback. The commencement of their upward journey from the foot hills and downward journey from alpine grasslands always depended on weather conditions which is a big rider on their travel plan. In Nepal, several studies on local pastoralists' perception of climate change have been reported from different districts of Nepal namely Mustang, Dolpe, Lamjung, Gorkha and also from national parks and conservation areas. Eighty percent of herders perceived an increase in air temperature, erratic and decreasing trend in precipitation, and appearance of new pasture species. In central Nepal, although precipitation did not show a clear pattern, herders observed an increased frequency of drought. On questioning, many herders felt that the warming effects of climate change were threatening the practice of transhumance. In Bhutan, yak farming is practiced in alpine region, and result of a study on how local yak herders perceive the effects of changing and warming climate is available. The study was undertaken in five villages. It was found that most participants were aware of global warming. They believed that climate change was responsible for increased rainfall, decreased water availability, more frequent droughts and flash floods, and ascent of snow lines. Many herders revealed that new vegetation was becoming established and rhododendrons were encroaching the grasslands leading to decline in forage availability.

There are not many reports from China on herders' perception to the problem of climate change. Some studies have been undertaken in Tibetan Autonomous Region (TAR). From Daiqian village of Zhuaxixiulong township belonging to Tianzhu county, majority of pastoralists were found aware of climate change issues and observed grassland degradation and poor health and productivity of local livestock. In a recent report, agro-pastoralists perception to climate change was studied in Qilan mountains in north-west of China. A total of 565 pastoralists in four counties were selected to asses the effect of climate change. Majority of the respondents were found aware of climate change effects. Most of them perceived a decrease in rainfall and some felt that rainfall was getting unpredictable. From Pakistan, results of two major studies on herders' response to climate change are available, one in Haramosh valley of Gigit Baltistan located at 3027 m amsl in karakoram and the other from three agro-ecological zones of Punjab province. In Gilgit study, 150 households were selected and given questionnaires to collect information. Results showed that majority of respondents perceived increase in temperature and 25 percent felt a decrease in rainfall. A total of 97% of herders indicated outbreak of new diseases, growth of invasive weeds, decreased water level in springs, landslides, floods, and avalanches. In a Punjab study, individual perception and vulnerabilities to climate change were studied on 405 small livestock herders from 30 villages. Majority felt that rainfall pattern had changed and became unpredictable. Livestock herders also felt emergence of new diseases.

Although Afghanistan has been described as a pastoral country, it has been difficult to obtain scientific and relevant data from it because of its isolation and volatile political history of wars and conflicts. There are not many studies available on pastoralists perception of changing climate. A few studies showed that the pastoralists were aware of rising ambient temperature. A recent report from Red Cross Red Crescent Climate Centre has stated that the climate warming effect, rising temperature, and decreased precipitation would cause loss of rangelands through degradation, and heat stroke would affect livestock reproduction and reduce the availability of animal feed leading to livestock starvation.

Keywords

Pastoralists' perception of global warming · Climate change and transhumance · Drying up of water resources · thinning of glaciers · Shifting of range lands

Abbreviations

DFID	Department for International Development
UNDP	United Nations Development Program

In the earlier Chap. 6, we had presented the effects of climate change, especially the rising ambient temperature and humidity (heat stress) on productive and reproductive functions and health and disease aspects of ruminants. The work relating to the effects of changing climate on hilly migratory flocks during ascending and descending journeys or during their stay at alpine pastures has received only minimal attention. Changing climate, especially unpredictable or uncertain weather conditions such as rain and snowfall, landslides, and avalanches, have been playing havoc on the pastoral life of shepherds as well as productive functions of accompanying pastoral animals, but these aspects do not seem to have been studied scientifically.

Many reports have indicated that hill and mountain environments are particularly sensitive to the effects of global warming and that the impacts of climate change and economic globalization were more drastically and intensely felt on animal husbandry practice on higher hills (Kreutzmann 2010a). Recent reports have confirmed that the consequences of climate change were most prominent in the high mountainous regions (Dhimal et al. 2021). Working on the effects of climate change on indigenous communities in the Indian Himalayan town of Almora, Negi et al. (2020) reported that Himalayas had some of the most visible signs of climate change with a warming rate greater than the global average. Many other investigators have also observed that climate in the hills changes rapidly through elevations and distances (Shrestha et al. 2012; Negi et al. 2020). Furthermore, the regions identified as most vulnerable have mountain people who are more susceptible to food insecurity and depend mostly on livestock (Romeo et al. 2020).

This chapter provides an overview of recent knowledge related to the perceptions and opinions of the nomadic herders who are directly involved in annual seasonal migration and face the impacts of changing climate on livelihood of migratory herders, on the health of Himalayan rangelands and on the future of this profession as a whole. We believe that the information is highly relevant and is based on systematic review of the studies undertaken on perceptions of Himalayan pastoralists on impacts of climate change on their livelihoods and pastoral animals in six major countries of Hindu Kush Himalayan (HKH) region namely India, China, Nepal, Bhutan, Pakistan, and Afghanistan.

The problems of pastoralists and herding system in the Himalayas are altogether different from the animals reared in plains as they follow an annual seasonal migration cycle moving the animals to the alpine rangelands with the start of the summer, staying there for four to five months, and returning to the base camps in the foothills with the onset of winter. Their movements up and down the hills involve long distance walking, scarcity of feed and water, non-availability of temporary shelters, conflicts with local farmers, protection from diseases, adverse weather conditions, and predator wild animals, and hosts of many other problems. As the mountain areas are reported to be more vulnerable to climate change, pastoralists and accompanying animals are subjected to various kinds of climatic stresses and weather extremes.

7.1 Feedback and Perceptions of Migratory Pastoralists on the Effects of Climate in Different Himalayan Countries

Most published reports relating to the feedback and perceptions from pastoralists and how climate change was affecting their lives and profession are based on information obtained by structured questionnaires or participatory rural appraisal or personal interviews with randomly selected respondents. The country-wise perceptions of pastoralists actively involved in the seasonal migratory practice are presented below.

7.1.1 India

India is the seventh largest country in the world and second largest country in Asia with a total land area of 3,287,263 km^2. It shares its borders with countries such as Myanmar, Bangladesh, China, Bhutan, Pakistan, and Nepal. It has vast resources of livestock and poultry (http://www.mapsofindia.com). As per the 20th livestock census (2019), Government of India, the bovine population in India was 302.7 million and population of small ruminants (sheep and goats) 223.1 million. Nomadic pastoralism involving small ruminants is commonly practiced in the Union territories of Jammu and Kashmir and Ladakh and states of Himachal Pradesh and Uttarakhand by local tribes, viz., Bakerwals, Gaddis, and Bhotias, respectively. Agriculture plays a vital role in the Indian economy, and over 60% of the total work force is engaged in agriculture and allied sectors.

The shepherds' perception of changing climate and how it is affecting their profession and livelihood in different north-western state of India are presented below.

7.1.1.1 Union Territories of Jammu and Kashmir (JK) and Ladakh

The union territories are located in most north-western part of India having a geographical area of 101,387 km^2 and representing 41.6% of Indian Himalayan region (ENVIS Bulletin web). Of the 14 local tribes in the union territories, two local tribes namely Bakarwals and Gujjars are most prominent. They are economically and socially marginalized groups. Sheep is the major animal species reared in the state followed by goats. Total number of both the species is estimated at 4,977,700 (GOI (2019) 20th Livestock census). The shepherds follow an annual seasonal migratory cycle leaving the foothills with animals for alpine pastures around mid-April, staying there for 4 to 5 months, and departing from the alpine grasslands for base camps around mid-October. The commencement of their upward and downward journeys always depends on weather conditions, which is a major concern in their travel plan. Generally, the migration is vertical but it could be horizontal from one pasture to another at the same elevation depending upon the availability of grasses. Livestock rearing through pastoral migration forms the major economic activity of Bakarwals and Gujjars.

The following quote from a shepherd of Jammu and Kashmir, Mukhtaar Chechi (Tufail 2014), provides a glimpse or perception of how seriously the herders take the warming effects of changing climate on their profession and how hard they are feeling about it. *The quote is loud and clear to describe how unpredictable the climate has become which is playing havoc on the livelihood of pastoralists of north-western Indian states.*

> We are all singing the same song and trying to tell the world, look we are seeing these things (climate extremes) and you are not doing anything. If the animals die, we die. If the pastures get dried, we will be no more. If the rivers die, we are gone.

The authors in their subsequent publications (Tufail et al. 2024) have studied in detail how changing climate has significantly affected the Gujjar and Bakerwal communities, particularly their seasonal movement patterns. The challenge of the climate stress directly impacted their activities through fluctuating weather patterns and some governmental policies.

A few other cases of perceived effects of climate change on herders and their animals are presented in shepherd's own language.

Wani et al. (2017) studied the problems and climate change perceptions of Bakerwal pastoralists. A sample of 120 respondents following six major routes of migration was selected randomly, out of which 39 cases were identified for their climate change perceptions. Data were collected using pre-tested structured interviews. It was found that the respondents perceived incidences of torrential rains, cloud bursting, and hail storms. When asked about their perception of climate change for the past two decades, majority replied that they were experiencing increase in ambient temperature (71.8%) and decrease in rainfall and snowfall by 64.0% and 56.4%, respectively. Unpredictability in weather, especially erratic rains, was pointed out by all respondents. In another study from Jammu and Kashmir, Tufail (2014) surveyed 400 households for their perceptions through questionnaires while they were grazing their animals both in summer and in winter pastures. Most respondents felt that the air temperature has been progressively increasing for the past few decades, drying up the water resources and causing repeated droughts. The pattern of rainfall and snowfall has become very unpredictable. During the past few years because of change in climate and warm weather during March, Bakerwals have started crossing the mountain passes one month before the scheduled time, i.e., in April. The author has described a deadly case of loss of large number of flocks and humans due to abrupt change in weather and their being trapped in the snow for days along with the flocks. According to him, this happened during the year 2009 when Gujjars and Bakerwals suffered heavy mortality in the flocks. Due to unusual rise in daily temperature in March, the herders started their migration to upper reaches one month before their routine time schedule of April. On reaching the heights, the season suddenly changed leading to heavy snowfall due to which shepherds and flocks were trapped in the mountainous range of Himalaya. This tragedy took the lives of 50 shepherds and thousands of animals. Tribal leaders have been blaming the climate change for their miseries.

Nissa (2019) has described yet another case of uncertain trapping of flocks due to unpredictability of weather. Normally the shepherds start moving to upper reaches of Himalayas during April. But during some years, they were unable to move to upward destinations because of continued cold and snow even during the month of May. On the contrary, there have been cases of early onset of summer and early movement of flocks to the higher elevations. However, when the pastoralists crossed over to higher reaches, the weather changed abruptly and flocks remained stranded at one place for weeks.

In a most recent field study on 235 nomadic herders of Rajouri district of Jammu and Kashmir, Ahmed and co-investigators (Ahmed et al. 2022) have described rising ambient temperature, unexpected rainfall, and uncertain weather conditions as three most important factors that were found to disturb the livelihood of herders. Continued increase in environmental temperature and reduced rain fall were responsible for drying up of water resources, reduced growth of pastures, and grazing areas causing drought-like situation. Increase in temperature and reduced availability of grasses have been forcing the herders to start seasonal migration for uphills 15 to 20 days earlier than usual time of April 15 every year. During the past few years, it is also being observed that the early migration of flocks was met with uncertain weather conditions like heavy rains, thunderstorms, and hailstorms on the way leading to unexpected halts and trapping of shepherds and their flocks in the storms. Haji Rashid, one of the respondents, claimed that during April 2020, hailstorms and bad weather killed approximately 120 sheep and goats in the Jai valley of Baderwah region.

In general, these studies tend to show that nomadic herders in the region were well aware and were experiencing the impacts of climate change in their day-to-day life.

7.1.1.2 The State of Himachal Pradesh (HP)

Himachal Pradesh, created in the year 1971, is a relatively young and small state over the Western Himalaya with a geographical area of 55,673 km^2 representing 10.4% of the geographical area of Indian Himalayas. The state is largely mountainous comprising hilly terrain (ENVIS Bulletin web). The state has 9 pastoral tribes, but the famous one involved in pastoral migration on a large scale is known as Gaddi, numbering around 0.2 million (Jitendra 2018). The total population of sheep and goats in the state is estimated at 1,899,700 (GOI (2019) 20th Livestock census).

Migratory pastoralism is quite relevant in the state as 90% sheep and 70% goats are reared under this system (Sankhyan et al. 2016). The Gaddis of Himachal Pradesh have been famously termed as "Grass Chasers." In every summer around April when snow starts melting in the mountains, Gaddis head upwards to alpine pastures along with their animals forming long queues on the highway roads. They stay there for 4 to 5 months and head back to the foothills with the onset of winter season. Due to their constant migratory nature, poetically, the Gaddis have been called migrants *who spend half the year travelling up and the other half travelling down* (Jitendra 2018).

The changing climate is proving a curse to the Gaddis' profession as the grazing lands/pastures have shrunk and the weather has turned erratic and unpredictable due to the effect of global warming. There is a general feeling that changing climate is playing havoc on the profession of agriculture and animal husbandry sector in the state, but research efforts to collect scientific climate data seem inadequate. Most available data relate to the media stories based on conversation with pastoralists or newsmen observations.

In a scientific study based on the interviews of rural and urban lifelong residents of Mandi district of Himachal Pradesh and local meteorological data, Padir and Bergeron (2018) reported continuing rise in air temperature in the area and adverse weather events such as hurricanes, tropical storms, prolonged period of drought, and erratic precipitation pattern. It was also reported that the alpine rangelands were gradually shifting to higher altitudes due to an increase in temperature (Asian Development Bank 2010). In another study on awareness or perception about climate change, Kimani and Bhardwaj (2015) collected data from 275 farm households of two mid-hill districts of Kullu and Solan. Results indicated that 88.4% of people perceived rise in ambient temperature and erratic change in pattern of rainfall. Perceived observations matched with the data from meteorological station. Rana et al. (2013) studied the people's and farmers' perception using multistage stratified sampling technique and data from people across different agro-climatic zones of Himachal Pradesh. The interviewed farmers were of the opinion that ambient temperature had been staying above normal for the past 10–15 years in their region. An observation of concern was drying up of surface waters including river water, springs, lakes, and wetlands. Farmers in Fatehpur Valley observed uneven distribution of rainfall and delay in onset of winter. In Theog, a high-altitude region of the state, interviewers noted a reduction in snowfall.

Besides the scientific publications on climate change perceptions, the local print and electronic media have also been telling the climatic change stories and problems of shepherds in their own ways and styles. Media do serve as a worldwide platform and has been successful in making the general public aware about the crucial role the changing climate is playing. Jitendra (2018) of "Down to Earth" has written about the decline in rainfall and other climatic effects in the area. In one of his reports, he has described the deficit in precipitation and reported that Kangra district of Himachal Pradesh has not received its average annual rainfall of 2022 mm since the year 2004. Due to rise in air temperature and continuous shortage of grasses at lower elevations, nomads have been heading to higher altitude to find enough grass for animals. The paper has also cautioned about a decline in snowfall. Choudhary Srishu of "Mint" (June 2019), writing from Kangra Valley of Himachal Pradesh, reported that air temperature in north-western Himalayas has risen by 1.6 °C during the last century, causing a rise in temperature humidity index (THI) and thermal stress on animals and humans. Erratic rainfall and shrinkage of grasslands have also been reported. Choubey (2018) of Deutsche Welle (DW) published the following observations about the effects of climate change in Kinnaur district of Himachal Pradesh: "*The Himalayan Gaddi community of pastoralists has long lived from the land. But with rising temperatures pushing them higher up into the mountains in search of grazing pastures, their way of life is under threat.*" Earth Journalism Network (2016) has similarly reported a rise in air temperature, unpredictable rainfall, decline in quality of pastures, earth erosion, and landslides in the Pin valley area of Lahaul and Spiti district. The district of Lahaul and Spiti has one of the largest concentrations of glaciers covering over about 2000 km^2. The paper reports significant decline in snowfall and thinning of most glaciers due to global warming. From the majority of aforementioned local media reports, it is obvious that the tradition of nomadic migration is seriously under threat due to climate change.

7.1.1.3 The State of Uttarakhand (UK)

This small state (10.02%) of the Indian Himalayan region with an area of 53,483 km^2 was created in November 2000 as the 27th state of Indian union. The state has mostly hilly terrain rich in natural resources with forest cover to the extent of 25.8% of the geographical area (ENVIS Bulletin web). There are five tribes in the state, out of which Bhotiyas is the major tribe involved in nomadic transhumance. The total number of sheep and goats has been estimated at 1,656,400 (GOI (2019), 20th Livestock census). There is ample evidence of global warming and changing climate affecting the day-to-day pastoral life of Bhotiyas who are gradually becoming aware of the incoming difficulties.

A number of scientific studies conducted in recent years have also documented the perceptions of indigenous communities to the impacts of climate change. Negi et al. (2020) conducted a field based study in the Indian part of Kailash Sacred Landscape (KSL) of Uttarakhand to obtain their feedback to the challenges of climate change. Questionnaires were distributed to the inhabitants of 14 villages and simultaneously meteorological data were recorded. Ninety seven per cent of the respondents of all groups agreed that climate was changing. Significant increase in maximum annual temperature of the past 3 decades (1980–2015) was recorded. Instrumental climate data and readings validated the observations. People of mid-altitude perceived erratic and unpredictable rainfall, drying up of water resources like springs and streams, drought incidences, and increase of pests and diseases. Respondents from high-altitude villages observed an increase in rainfall and cloudbursts, decreased snowfall, and faster melting of glaciers. Shepherds were keen observers and reported changes in composition of vegetation. In yet another study undertaken in 54 villages of Chamoli district of Uttarakhand, people's perceptions to climate change were recorded through questionnaires (Negi et al. 2017). Eighty per cent of the respondents were quite aware of the significant rise of air temperature, erratic rainfall pattern, which normally occurred in June every year but shifted to July, water scarcity, and incidences of dry spells. Inhabitants of higher altitude especially noted water shortage and drying up of water springs. A study was also conducted in villages located at higher altitude (5070 m amsl) in Joshimath division of Chamoli district of Uttarakhand bordering Tibet (Rautela and Karki 2015). Bhotiyas tribe constituted majority of the population in this area. In response to questionnaires, 86% of respondents perceived significant increase in air temperature and precipitation compared to data recorded in the year 1970. Ninety two per cent of the respondents perceived clear change in pattern of rainfall, lack of water resources, increase in incidences of dry spell, increased frequencies of landslides, earth shaking, reduced snowfall, and fast receding glaciers. Increase in pests and emergence of invasive weed species were also observed.

7.1.2 Nepal

Nepal is a landlocked country with a total land area of 147,516 km^2 bordered by India and China. Of the total land area, 86% is mountainous and 14% is flat. Approximately 2200 km of Himalayan range lies along the northern border of Nepal (Ichikawa 2012). Animal husbandry contributes about 11% of the national gross products. Main livestock species are goat (11.22 million), cattle (6.43 million), and sheep (0.8 million). Transhumant pastoralism is common and well developed in Nepal. Sherpa, Rai, and Limbus are the main tribal communities involved in transhumance pastoral system. Having tough terrain, eight climate zones and elevations ranging from 100 m amsl to 8000 m amsl, Nepal is an ideal example to record perceptions of agro-pastoralists. The pastoral system in the high mountains of Nepal involves free grazing of livestock on the rangelands. The pastoralists move to higher altitude pastures in summer and return to lower heights with the start of winters.

Studies to obtain the feedback on consequences of climate change on transhumance were undertaken in different districts of the country. Tewari et al. (2020) conducted a survey of 186 households and interviewed 38 key participants from three National Parks and conservation areas of Makalu, Annapurna, and Manaslew to assess the pastoralists' vulnerability to climate variability. Field-level workers and herders perceived an increase in air temperature steadily from 1901 through 2014 at a rate of 0.1 °C per decade. More than 80% of the respondents experienced decrease and erratic rainfall, water shortage, delay in arrival of monsoon by 15 to 20 days, and reduced snowfall. Warming trend was more pronounced at higher elevations. The local people's perception and scientific data correlated well with the warming climate. Almost similar results were obtained in another study in which perceptions of agro-pastoralists to climate change were obtained from the high- and trans-Himalayan districts of Dolpa and Mustang (Aryal et al. 2017). The perceptions corresponded with increasing trend of temperature, increasing as well as decreasing trends of precipitation and decreasing snowfall. Pastoralists also noticed appearance of new forage and pasture species in the rangelands. Ghermu village of Lamjung district of Nepal is home to indigenous and ethnic Gurung community and has very high vulnerability to climate change, particularly to landslides and outburst floods. A study was conducted in this village by Gentle and Thwaites (2016) to explore how transhumant pastoralism was being sustained in the mountains of Nepal and what the projected scenario was for transhumance? Results revealed that continuation of transhumance was threatened by impacts of climate variability on mountain ecosystem including conflicts between herders and nonherders. Perceived impacts included increase in surface temperature, erratic change in precipitation, decreased snowfall, scarcity of water, and frequent droughts. Out of nine, two water sources had completely dried up. Herders reported that farmlands were covered with invasive weeds such as *Eupatorium*. They also perceived increasing incidences of livestock diseases.

In a recent report from Nepal, Neupane et al. (2022) explored the perceptions from transhumant herders in two wards, Lho and Sirdibas of Chumnurbi Rural Municipality of Gorkha district, representing the western mountains of Nepal. These wards were selected based on the existence of transhumance livestock system and the availability of large number of herders and livestock. The survey was conducted using questionnaires and interviews to obtain feedback from herders along with the meteorological data. Results showed that over the past 32 years, the air temperature was gradually increasing. The average monthly rainfall trend from 1985 to 2017 showed a decrease by 0.26 mm and there was a decline in snow cover. A decrease was noticed in pasture resources and crop yield. Increased incidences of drought were also perceived by the herders. Other studies conducted by Rayamajhi and Manandhar (2020) at other locations in Nepal were generally along the lines presented above.

In a few more studies, Dahal et al. (2023) investigated the herders' perception of climate change in the central part of Nepal. The observations of farmers and data from nearby meteorological station were identical and revealed clear warming trend of ambient temperature. Although precipitation did not show a clear pattern, farmers observed an increased frequency of droughts. In yet another study, Darjee et al. (2022) gathered the herders' perception of climate variability through interviews and focus group discussion with the local residents of three ecological regions, namely, mountain, mid hills, and lowland areas of Nepal. Climatic measurements from meteorological stations of the regions were obtained for a period of thirty years. Results showed that the temperature increased in all the three regions, but on the contrary, the rainfall decreased. Approximately, 88% of the interviewers perceived rise in temperature, whereas 74% noticed a decline in rainfall.

7.1.3 Bhutan

Bhutan is also a landlocked country located in the eastern Himalayas bordering India in the east, west, and south and Tibet Autonomous Region of China in the north. The country is mostly mountainous with land area of approximately 38,000 km^2 and estimated population of 730,000.

Bhutan is an agrarian country with 69% of population living in rural areas and engaged in agriculture. Twenty two per cent of the rural household income comes from livestock rearing (Namgay et al. 2013). There are three main ethnic groups in Bhutan, namely, Bhutia, Nepalese, and Sherchop. Yak and cattle are the most common livestock species, besides mithun, horses, sheep, and goats. Herders keep large herds of *Bos indicus* type cattle, locally known as Thrabam or Siri, and of mithun x Thrabam crossbreds. Namgay and associates (2013) have described the transhumant agro-pastoralism in Bhutan. Migration to alpine pastures takes place in April/May and return journey in September/October, which may take four days to over a month. During this process, herders and their animals face all kinds of problems and cover ten to hundreds of kilometers by walk. Transhumant system is lately reported to be on a declining trend and herders are turning sedentary for different reasons. More recently, Namgay et al. (2021) have reported in detail the current status and future of transhumance livestock system in Bhutan.

Studies pertaining to the impacts of climate variability and perceptions of climate change by herders are not many (Chaudhary and Bawa 2011) in Bhutan. An interesting case of climate warming through the eyes of yak herders has been reported by Wangchuk and Wangdi(2018) and Chaudhary and Bawa(2011). Yak farming in Bhutan is practiced in the alpine region in the northern part of the country. The study in question was conducted in five villages located in rugged terrain of a remote area. A questionnaire survey was undertaken to understand how a mountain ecosystem is perceived by local yak herders to the changing and warming climate. A total of 100 elderly herders were selected using a two-stage sampling. It was found that most participants were aware of global warming and perceived that increasing air temperature was responsible for longer vegetation growing periods. Climate change was responsible for increased rainfall and flash floods, and ascent of snow lines. Many herders revealed that new vegetation was becoming established in barren lands. Rhododendrons were encroaching the grasslands leading to decline in forage availability. Few of them perceived a decrease in milk production and increased livestock predation. These problems affected the livelihood of herders who largely rely on yak farming.

7.1.4 China

The People's Republic of China is a vast country with total area of approximately 9,706,961 km^2 and a population exceeding 1.4 billion. It shares border with 14 countries and spanning five geographical time zones. The Tibetan plateau also known as Qinghai-Tibet plateau is called nature's precious gift serving as water tower and supporting 9.8 million pastoral population (Wikipedia web). The plateau is vulnerable to climate change threatening the grassland resources, watersheds, and livestock in the area.

The earliest evidence of agro-pastoralism and animal husbandry in China has been discussed vividly by Owlett (2016) of Stanford University. Li and associates (2019) studied the changing agro-pastoral livelihood under collective and private land use in China. China's agro-pastoral area is mainly distributed in the Yile Valley in west Yinjiag and eastern Tibetan plateau. Quite a few interesting but philosophical studies have been conducted to scrutinize the role of climate in global distribution of sheep and goats in China and how the climate change has affected the global nomadic migration (Zhang et al. 2021; Pei and Zhang 2014). As for the effect of climate change, the IPCC special report has documented that agriculture and food production would be increasingly affected by climate, especially the crop yield, and a decline in net grain production to the extent of 3% to 12% was expected in Asia by the end of twenty-first century.

The herders' perception on the problem related to changing climate doesn't seem to have been studied systematically in China as only very few papers are available on pastoralists and adaptation. In a recent report, Xie et al. (2022) studied the agro-pastoralists' perception to climate change and adaptation measures in the Qilan mountains in north west of China. Based on the participatory rural appraisal and qualitative and quantitative methods, 565 agro-pastoralists in four counties were selected to assess the perception of climate change in the area. Majority of the respondents were aware regarding climate change. In total, 83% of the respondents perceived the change in temperature, 77.08% in precipitation, and 72.74% in snowfall. Most of them perceived a decrease in rainfall and some felt that rainfall was getting unpredictable. Regarding temperature, majority felt a rising summer temperature and a warmer winter while some of the respondents expressed just opposite opinion. Five other sites presented similar trends between observed and perceived data. More than 50% of respondents felt that climate change had a medium impact on pasture quality. Some of the agro-pastoralists were enquired if they had adopted measures to respond to the impact of climate change and the answer was affirmative. The suggested adapted measures included adopting new technologies, new breeding methods, and building proper infrastructure (fences, greenhouses), land transfer, etc.

Some climate change perception studies have also been undertaken in Tibetan Autonomous Region (TAR). In a recent feedback from the pastoralists in Daiqian village of Zhuaxixiulong Township belonging to the Tianzhu county, Sharif et al. (2019) reported that majority of the pastoralists were aware of the climate change issues in the plateau and observed changes in climatic conditions leading to grassland degradation and poor health and productivity of the livestock. Rangelands are expected to be more affected by global warming (Kreutzmann 2010b) and, therefore, a lot of attention is being paid to the status of famous rangelands of Tibet and to improve the livelihood of nomads. Rangelands in Tibet cover vast areas in fragile ecological environment. The Tibet Autonomous Region (TAR) government have taken series of measures to tackle climate change and global warming and dedicated a whole research team to study the impacts of climate change.

7.1.5 Pakistan

Islamic Republic of Pakistan is a south Asian country with a total land area of around 796,096 km^2. The country is bounded by Iran to the west, Afghanistan to the north west, China to the northeast, and India to the east and southeast. Of the total area, about three-fifths consist of rough mountain terrain and plateau and the remaining two-fifths are plains. The climate is characterized by extreme variation in temperature and varies from very cold on northern mountains to very hot in the plains during summer. Livestock plays an important role in the country's economy. Majority of the livestock in northern areas are reared under transhumant system involving high-altitude grazing.

Very few studies seem to have been undertaken on the effect of climate change on transhumant pastoralism, especially on herders' perceptions and vulnerability to climate change. In a recent study, Ali et al. (2022) investigated the perceptions of pastoralists to the changing climate in Haramosh valley of Gilgit Baltistan located at 3027 m to 3352 m amsl in Karakoram. A total of 135, households were selected and given questionnaires and also interviewed to collect information. Meteorological data were also simultaneously recorded. Results showed that herders perceived effects of climate change and its negative impacts such as increase in temperature and unpredictable rainfall. Meteorological data also supported increasing trend of temperature (0.042 °C per year) and precipitation (1.42 mm per year) during the period from1984 to 2013. Majority of the subjects perceived increase in temperature. In total, 25% felt a decrease in rainfall, while 3% perceived an increase. Ninety seven per cent of herders indicated outbreaks of new diseases, growth of invasive weeds (55%), decreased water in springs, landslides, floods, and avalanches.

Faisal and associates (2021) conducted a study to explore individual perception and vulnerabilities to climate change on 405 small livestock herders from 30 villages in three agro-ecological zones, viz., Dera Ghazikhan (DGK), Rahim Yar Khan (RYK), and Faislabad Khan (FSK) of Punjab province. Majority of respondents observed an increase in high- and low-temperature variation. They also felt that rainfall pattern had changed and had become unreliable. Livestock herders also felt emergence of new diseases in animals. Perceptions about temperature change were consistent with the meteorological data. In another climate study from Pakistan, Ali et al. (2021) reported that climate change was expected to impact existence of large number of animals including several threatened mammals. They described how in Pamir-Karakoram mountains climate change had become a threat to two wild ungulates, Himalayan ibex (*Capra ibex sibirica*) and blue sheep (*Pseudois nayaur*), due to several human- and nature-induced factors. The authors investigated the populations of these two economically important wild ungulate species. Results indicated that the original area was a suitable habitat for the survival of Himalayan ibex and blue sheep, but the change in annual mean precipitation and climate over the years badly disturbed their habitat leading to their gradual loss.

7.1.6 Afghanistan

Afghanistan, a landlocked multiethnic country occupying 650,000 km^2 of land and having population around 30 million, is bordered by Pakistan on the east and south, Iran on the west, and Asian states of Turkmenistan, Uzbekistan, and Tajikistan in the north. Around 45% of the local land area is classified as rangeland. Geographically, 80% of the country is mountainous or desert. An outstanding feature of Afghanistan is its formidable mountain range, the Hindu Kush, that divides the country into three distinct geographical regions. Most of Afghanistan lies between 600 and 3000 m amsl with varying climate of extreme cold or hot (Ali and Shaoliang 2013).

Because of certain local constraints, it has been rather difficult to obtain scientific and relevant data from Afghanistan. For this reason Afghanistan has been called as a least surveyed country. Essentially, Afghanistan has been described as a pastoral country. Between 70% and 85% of the total land area is used as rangelands for livestock grazing, although rangelands have also often been the site of ethnic and communal conflicts. Livestock raising using rangelands is an essential component of local farming system and a livelihood strategy for over 80% of Afghan households. It is estimated that there are around 1.5 million nomadic pastoralists, popularly called Kuchis (ethnic Pashtuns, Takiks, Uzbeks, and Hazares), who depend entirely on migratory livestock or mobile livestock (Ali and Shaoliang 2013). According to a latest survey, Afghanistan has 8.7 million sheep, 7.28 million goats, 3.72 million cattle, and1.9 million other animals such as horse, donkey, and camel (Ali and Shaoliang 2013). The rangelands in the central highlands of the country form the major pastures for most livestock and are crucial for country's farming systems which comprise sedentary and migratory livestock. Over 35% of the export earning of Afghanistan comes from pastoralists.

Effects of Climate Change

Afghanistan is reported to be highly vulnerable to climate change and has been ranked 24th in global climate risk index of countries most affected by extreme weather events. The temperature has increased by 0.13 °C per decade since 1960 and precipitation has decreased by 2% per decade (Kapoor et al. 2021). Incidence of rains has decreased by 4–8 times per month. Projections have shown that by 2090 air temperature may increase by 2 to 6.2 °C. In alpine regions, summers have become longer and winter shorter. With the melting of glaciers in Hindu Kush Mountains, the risk of flash floods, flooding, and landslides are expected to increase (Price 2019). Disaster preparedness is poor and the country lacks early warning systems. In general, there is lack of data on almost every aspect of climate change effect on livestock and rangeland management. Some reports from the international organizations such as the Department for International Development (DFID), United Nations Development Programme (UNDP), and Red Cross relate to the effects of climate change on human health and diseases but none on animal husbandry aspects.

Few studies are available on pastoralists' perception of changing climate in Afghanistan. In a short study, Ali and Shaoliang (2013) reported that the pastoralists were aware of climate change, especially rising temperature and erratic precipitation. In another but a better organized study conducted in Herate province of Afghanistan, Iqbal et al. (2018) explored farmers' perception of drought prevalence and its socioeconomic impacts and resolution mechanism. Information was collected from 147 farming households through questionnaires. Results showed that drought had serious economic impact on agricultural and livestock production including reduction in crop yield and livestock productivity and welfare resulting in reduced livelihood options and weakening of their financial position. Farmers observed that lack of irrigation water was a major source of conflict. A recent report from Red Cross Red Crescent Climate Centre entitled "Climate change impacts on health and livelihoods: Afghanistan assessment" has summarized the effects of climate variability on the livestock and animal husbandry (Kapoor et al. 2021). According to the report, climate warming effect, rising temperature, and decreased precipitation would cause loss of rangelands through degradation and desertification and grazing potential. Heat stroke would affect livestock reproduction and reduce the availability of animal feed leading to livestock starvation. The report cites the incidence of severe droughts experienced during 1999–2004 that resulted in a loss of more than 50% of the pastureland and approximately three million livestock across Afghanistan, leading to humanitarian aid for almost one million people (World Bank 2018).

References

Ahmed S, Iqbal Z, Antahal PC (2022) Impact of climate change on nomadic herders' livelihoods: a case study of Bakarwal tribe in the western Himalayas in India. https://europepmc.org/article/PPR/PPR572633

Ali A, Shaoliang Y(2013) Highland rangelands of Afghanistan: significance, management issues and strategies. In: Ning W, Rawat GS, Joshi S, Ismail M, Sharma E (eds) High-altitude rangelands and their interfaces in the Hindu Kush Himalayas. International Centre for integrated Mountain Development (ICIMOD), Kathmandu, pp 15–24

Ali H, Din JU, Bosso L, Hameed S, Kabir M, Younas M, Nawaz MI (2021) Expanding or shrinking? Range shifts in wild ungulates under climate change in Pamir-Karakoram mountains, Pakistan. PLoS One 16(12):e0260031. https://doi.org/10.1371/journal.pone.0260031

Ali I, Raza G, Hussain A, Zafar M, Akram W, Begum F (2022) Herders' perspectives on pastoral systems in the face of changing climate in Karakoram Mountains of Northern Pakistan, pp 1–22. https://europepmc.org/article/PPR/PPR474599

Aryal S, Ghimire SK, Dhakal YR, Gaire NP, Bhandari S (2017) Perceptions of agro-pastoralists towards the change in temperature and precipitation in the trans-Himalayan regions of Nepal. Banko Janakari 27(1):21–30. https://doi.org/10.3126/banko.v27i1.18546

Asian Development Bank, Manila (2010) Climate Change Adaptation in Himachal Pradesh: sustainable strategies for water resources. https://www.adb.org/sites/default/files/publication/27999/cca-himachal-pradesh.pdf

Chaudhary P, Bawa KS (2011) Local perceptions of climate change validated by scientific evidence in the Himalayas. Biol Lett 7:767–770. https://doi.org/10.1098/rsbl.2011.0269

Choubey J (2018) Himalayan village life under threat. Global Ideas, Deutsche Welle (DW), December 5. www.dw.com. Climate-Change-Forces-Himalayan

Choudhary Srishu (2019) Pastoralists of Himachal Pradesh an unusual casualty of global warming. Mint, 26 June 2019. https://www.livemint.com/news/india/carpet-weavers-of-himachal-pradesh-an-unusual-casualty-of-global-warming-1561488151014.html

Dahal KR, Dahal P, Adhikari RK, Naukkarinen V, Panday D, Bista N, Helenius J, Marambe B (2023) Climate change impacts and adaptation in a hill farming system of the Himalayan region: climatic trends, farmers' perceptions and practices. Climate 11(1):11. https://doi.org/10.3390/cli11010011

Darjee KB, Neupane PR, Köhl M (2022) Do local perceptions of climate variability and changes correspond to observed climate changes? A comparative study from Nepal as one of the most climate-vulnerable countries. Weather Clim Soc 14(1):205–222

Dhimal M, Bhandari D, Dhimal ML, Kafle N, Pyakurel P, Mahotra N, Akhtar S, Ismail T, Dhiman RC, Groneberg DA, Shrestha UB, Müller R (2021) Impact of climate change on health and well-being of people in Hindu Kush Himalayan Region: a narrative review. Front Physiol 12:651189. https://doi.org/10.3389/fphys.2021.651189

Earth Journalism Network (2016) How climate change is affecting an old pastoral tradition in India's Himalayas. 17 October 2016. https://earthjournalism.net/stories/how-climate-change-is-affecting-an-old-pastoral-tradition-in-indias-himalayas

Faisal M, Abbas A, Cai Y, Ali A, Shahzad MA, Akhtar S, Haseeb Raza M, Ajmal MA, Xia C, Sattar SA, Batool Z (2021) Perceptions, vulnerability and adaptation strategies for mitigating climate change effects among small livestock herders in Punjab, Pakistan. Int J Environ Res Public Health 18(20):10771. https://doi.org/10.3390/ijerph182010771

Gentle P, Thwaites R (2016) Transhumant pastoralism in the context of socioeconomic and climate change in the mountains of Nepal. Mt Res Dev 36(2):173–182

GOI (Government of India) (2019) 20th livestock census: all India report. Department of Animal Husbandry and Dairying, Ministry of Fisheries, Animal Husbandry & Dairying, Govt of India. Available: https://dahd.nic.in/sites/default/filess/20thLivestockCensus2019AllIndiaReport.pdf

Ichikawa K (2012) Socio-ecological production landscapes in Asia. United Nations University Institute for the Advanced Study of Sustainability (UNU-IAS), Tokyo, Japan. https://www.semanticscholar.org/paper/Socio-ecological-Production-Landscapes-in-Asia-Ichikawa/fbe765a439bc45435fdce0b4a800d932dd02d8b3

Iqbal MW, Donjadee S, Kwanyuen B, Liu S (2018) Farmers' perceptions of and adaptations to drought in Herat Province, Afghanistan. J Mt Sci 15(8):1741–1756

Jitendra (2018) Climate change in India: mountains play hocus-focus in Himachal Pradesh Down to Earth, Tuesday 23 October 2018. https://www.downtoearth.org.in/climate-change/climate-change-in-india-mountains-play-hocus-focus-in-himachal-pradesh-61874

Kapoor A, Alcayna T, de Boer T, Gleason K, Bhandari B, Heinrich D (2021) Climate change impacts on health and livelihoods: Afghanistan assessment report published by Red Cross Red Crescent Climate Centre, Afghanistan. https://www.climatecentre.org/wp-content/uploads/Climate-change-impacts-on-health-and-livelihoods-AFGHANISTAN-assessment_April-2021_.pdf

Kimani NC, Bhardwaj SK (2015) Assessment of people's perceptions and adaptations to climate change and variability in mid-hills of Himachal Pradesh, India. Int J Curr Microbiol App Sci 4(8):47–60

Kreutzmann H (2010a) Pastoral practices on the move-recent transformation in mountain pastoralism on the Tibetan Plateau. In: Kreutzmann H, Yong Y, Richter J (eds) Regional workshop on Pastoralism and rangeland management on the Tibetan Plateau in the context of climate and global change, 21–25 October, Lhasa, Tibet, China. Deutsche Gesellschaft für Internationale Zusammenarbeit (GIZ) GmbH, Bonn, p 200

Kreutzmann H (2010b) Chapter 9: Summary of workshop and closing remarks. In: Kreutzmann H, Yong Y, Richter J (eds) Regional workshop on Pastoralism and rangeland management on the Tibetan Plateau in the context of climate and global change, 21–25 October, Lhasa, Tibet, China. Deutsche Gesellschaft für Internationale Zusammenarbeit (GIZ) GmbH, Bonn, p 200

Li D, Hruska T, Talinbayi S, Li W (2019) Changing agro-pastoral livelihoods under collective and private land use in Xinjiang. China Sustain 11(1):166. https://doi.org/10.3390/su11010166

Namgay K, Millar J, Black R, Samdup T (2013) Transhumant agro-pastoralism in Bhutan: exploring contemporary practices and socio-cultural traditions. Pastoralism 3:13. https://doi.org/10.1186/2041-7136-3-13

Namgay K, Millar JE, Black RS (2021) The future of transhumants' sustainable resource use in Bhutan: pressures and policies. Front Sustain Food Syst 5:618351. https://doi.org/10.3389/fsufs.2021.618351

Negi VS, Maikhuri RK, Pharswan D, Thakur S, Dhyani PP (2017) Climate change impact in the Western Himalaya: people's perception and adaptive strategies. J Mt Sci 14(2):403–416

Negi VS, Thakur S, Dhyani R, Bhatt ID, Rawal RS (2020) Climate change observations of indigenous communities in the Indian Himalaya. Weather Clim Soc 13(2):245–257. https://www.jstor.org/stable/27089894

Neupane M, Joshi R, Bhandari D, Awasthi N (2022) Perception of transhumant herders on climate change and their adaptation strategies. Our Nat 20(1):1–13

Nissa Z (2019) Is climate change affecting the migrations of Kashmir's pastoralists? https://www.downtoearth.org.in/climate-change/is-climate-change-affecting-the-migrations-of-kashmirs-pastoralists--64789

Owlett TE (2016) Finding greener pastures: the local development of agro- pastoralism in the Ordos region, North China. JIPA 40:42–53

Padir AL, Bergeron TP (2018) Perceptions of a changing climate, exposure, and vulnerability in Himachal Pradesh. Worcester Polytechnic Institute, Mandi. https://digital.wpi.edu/show/6969z1338

Pei Q, Zhang DD (2014) Long-term relationship between climate change and nomadic migration in historical China. Ecol Soc 19(2):68. https://doi.org/10.5751/ES-06528-190268

Price R (2019) Climate change as a driver of conflict in Afghanistan and other fragile and conflict affected states. The Institute of Development Studies and Partner Organisations. https://opendocs.ids.ac.uk/opendocs/handle/20.500.12413/14424

Rana RS, Bhagat RM, Kalia V, Lal H, Sen V (2013) Indigenous perceptions of climate change *vis-a-vis* mountain agricultural activities in Himachal Pradesh, India. Indian J Tradit Knowl 12(4):596–604

Rautela P, Karki B (2015) Impact of climate change on life and livelihood of indigenous people of higher Himalaya in Uttarakhand. India Am J Environ Prot 3(4):112–124

Rayamajhi N, Manandhar B (2020) Impact of climate change and adaptation measures on transhumance herding system in Gatlang, Rasuwa. Air Soil Water Res 13:1–10

Romeo R, Grita F, Parisi F, Russo L (2020) Vulnerability of mountain peoples to food insecurity: updated data and analysis of drivers. FAO and UNCCD, Rome. https://doi.org/10.4060/cb2409en

Sankhyan V, Dogra PK, Thakur YP (2016) Attributes of migratory goats and sheep farming and impact of some improved management strategies en-route migration in adopted flocks of Western- Himalayan region of India. Indian J Anim Sci 86(9):1079–1084

Sharif J, Rafiq MK, Rafiq MT, Aziz R, Qayyum A, Saleem AR, Nisa WU, Jenks MA, Li Y (2019) Climate change perceptions and adaptive actions by pastoral community on the Tibetan Plateau, China. Appl Ecol Environ Res 17(4):7987–8009

Shrestha UB, Gautam S, Bawa KS (2012) Widespread climate change in the Himalayas and associated changes in local ecosystems. PLoS One 7(5):e36741. https://doi.org/10.1371/journal.pone.0036741

Tiwari KR, Sitaula BK, Bajracharya RM, Raut N, Bhusal P, Sengel M (2020) Vulnerability of pastoralism: a case study from the high mountains of Nepal. Sustainability 12(7):2737. https://doi.org/10.3390/su12072737

Tufail M (2014) Impact of climate change on the seasonal movement of the Gujjar and Bakarwals: community perceptions. IOSR J Environ Sci Toxicol Food Technol 8(1):41–46

Tufail M, Kamble PN, Khumukcham J, Kumar K, Kumar A, Kachwaya DS, Sangma F, Tiwana AS (2024) Climate variability and transhumant adaptation among Gujjar and Bakarwal communities, India. Ecol Environ Conserv 30(1):222–231. https://doi.org/10.53550/eec.2024.v30i01.043

Wangchuk K, Wangdi J (2018) Signs of climate warming through the eyes of yak herders in northern Bhutan. Mt Res Dev 38(1):45–52

Wani SA, Meena BS, Ganaie AH, Mir NA (2017) Problem identification and climate change perception of Bakerwal pastoralists of Jammu & Kashmir. J Anim Res 7(1):99–105

Wikipedia web: http://en.wikipedia.org/wiki/Geography_of_china

World Bank (2018) Strengthening Hydromet and early warning services in Afghanistan: a road map. World Bank, Washington, DC. https://documents.worldbank.org/en/publication/documents-reports/documentdetail/976021545165642530/strengthening-hydromet-and-early-warning-services-in-afghanistan-a-road-map

Xie S, Ding W, Ye W, Deng Z (2022) Agro-pastoralists' perception of climate change and adaptation in the Qilian Mountains of northwest China. Sci Rep 12(1):12689. https://doi.org/10.1038/s41598-022-17040-2

Zhang Y, Wang G, Zhang Y, Zhao S, Han C (2021) Climate change is likely to alter sheep and goat distributions in mainland China. Front Environ Sci 9:748734. https://doi.org/10.3389/fenvs.2021.748734

Part IV

Minerals and Nutritional Status of Alpine Grasses and Tree Leaves of Indian Himalayas

8 Principal Minerals Limiting Production of Migratory Small Ruminants: Salt Hunger in Grazing Sheep

Abstract

Although undernutrition is recognized as a common factor for suboptimal productive performance in grazing ruminants in tropical countries, studies have revealed that minerals play specific and multiple roles in catalyzing and regulating body functions and body growth. Deficiencies and imbalances in mineral supply are also a major cause for reduced productive performance in migratory sheep and goats. The vast hilly terrain of Himalayan states has variable soils with wide range of pH which tend to interfere or promote uptake of certain mineral elements leading to their deficiencies or excess. Enormous investigative work is required to be conducted to systematically unravel the mineral status of animals in the Himalayan region. In our studies and literature search carried out in the northern Indian states of Himachal Pradesh, Uttarakhand, and union territories (UTs) Jammu and Kashmir and Ladakh, four minerals, namely, copper, zinc, phosphorus, and sodium were found to limit the migratory animal production. In some pockets of Jammu and Kashmir, calcium and iron, and in Uttarakhand, manganese were found to limit production. The animals examined in the middle hill areas of Himachal Pradesh exhibited borderline serum levels of copper and zinc. Sodium deficiency was widely prevalent in many areas, but there was no practice of regular mineral supplementation by the livestock owners. An important issue of significance which emerged from the literature search and our own work is the observation of low sodium content in forages, grasses, and tree leaves in Himalayan region. Studies carried out in hilly areas of other Himalayan countries such as China and Pakistan have also identified sodium as one of the prominent minerals limiting small ruminant production. Sodium deficiency of varying degree along with phosphorus has been reported in northern China and Tibetan plateau. In Pakistan, widespread deficiencies of copper, phosphorus, and sodium have been reported in herbages and tree leaves. Literature review shows that

S. P. Singh, J. K. Malik, *Migratory Small Ruminant Farming System in the Himalayas-Peculiarities, Problems, Prospects, Climate Effects, Nutrition and Health*,
https://doi.org/10.1007/978-981-96-6691-1_8

attention to sodium(salt) deficiency was originally drawn with the observation of salt craving by sodium-deficient cows. During our investigative work on migratory ruminants in Himachal Pradesh, we had observed an interesting phenomenon of intense salt hunger in the sheep flock after their prolonged stay at the high-altitude pastures. We feel that sodium deficiency appears to be an emerging prevalence in Himalayan region which needs to be thoroughly investigated systematically. This has been discussed in depth.

Keywords

Undernutrition · Mineral deficiencies and imbalances · Low sodium content in forages and herbages · Sodium deficiency · Salt hunger in grazing sheep

Abbreviations

amsl	above mean sea level
m	meter
Na	Sodium
Na:K	Sodium to potassium ratio

Minerals are essential for biological processes and fulfill the needs of growth and production in all animals. Presently, 19 mineral elements are known to be essential (Suttle 2010). Under natural conditions, soil is the main store house for plant mineral elements, and the treatment and status of the soil are known to affect their mineral contents. Similarly, the uptake of minerals from soil to the plant is not always a simple process. The bioavailability of minerals from the plant to the animal body is similarly governed by several factors.

From the point of wool production, mineral elements such as zinc and copper have been found to play an important role in sheep (Szigeti et al. 2020). As regards other minerals, sodium has been recognized as an essential element for water balance and acid base homeostasis. Sodium status in the animal body showing craving for salt can be assessed by various methods (Singh and Rani 1999). An interesting phenomenon observed in our study was the development of salt hunger in sheep after their prolonged stay at the high altitude pastures. It was generally observed that sheep were literally running behind the animal attendants who carried the salt bags on their shoulders in the pastures. Further details are presented in Sect. 8.3.2.

8.1 Minerals Limiting Ruminant Production

The small ruminant migratory rearing system has been going on in the Hindu Kush Himalayan countries for centuries but, unfortunately, systematic mapping or delineated studies on alpine pastures revealing specific minerals limiting animal

Fig. 8.1 Gaddi sheep grazing on pasture near the Sheep Farm, Tal District, Hamirpur, HP (msl 1180 m) (ICAR Project work/courtesy Director AH, HP)

production have not been carried out along the scientific lines. The reports on mineral status are available only from one or the other locations of Himalayan region. A few planned studies have been undertaken in the state of Himachal Pradesh under the Network project sponsored by the National Institute of Animal Nutrition and Physiology (NIANP), Bangalore, India, and under the ICAR Project at HP Agriculture University, Palampur, and some relevant information relating to the mineral status of large and small ruminants has been documented (Fig. 8.1). Lately, a number of authentic reports have been published from Sher-e-Kashmir University of Agricultural Sciences and Technology, Jammu and Kashmir, and its Mountain Livestock Research Institute, Manasbal, Highland Pastoral system and Extension Station, Nyome, Ladakh, and Indian Veterinary Research Institute, Izatnagar. In view of these handicaps, we have tried to gather information from pertinent sources in India and neighboring countries including Pakistan, China, and Nepal, where the problems of migratory small ruminants are almost identical to the problems faced by graziers in north-western Himalayan Indian states.

8.1.1 Indian Himalayan States

8.1.1.1 Himachal Pradesh and Uttarakhand

In the states of Himachal Pradesh and Uttarakhand, principal minerals limiting animal production have been identified as copper, zinc, calcium, sodium, and, in some pockets, phosphorus and manganese.

In three independent studies in Uttarakhand, one in Tarai region(foothills) and others in Haridwar, Pithorogarh, and Nainital districts, it was observed that most large ruminants had serum copper, zinc, and manganese levels below the required critical concentration for ruminants (Khan et al. 1979; Rajora and Pachauri 1993; Tiwari et al. 2012). Rajora and Pachauri (1993) studied copper, iron, manganese, and zinc contents in soil, fodder, and blood serum of buffaloes and cattle at three livestock farms to establish soil-plant and animal relationship. Analysis of fodder and soil samples revealed copper deficiency in 50% and zinc deficiency in 33% of samples, respectively. This was reflected in lower levels of serum copper and zinc in animals. Tiwari et al. (2012) investigated the mineral status of cattle and buffaloes in Nainital district of Uttarakhand. It was found that most samples of feed, soil, and blood serum from animals in the area had calcium, phosphorus, zinc, manganese, and copper level below the critical levels required for ruminants. Dietary supplementation of deficient minerals to anestrous and repeater animals for forty five days significantly decreased the incidences of the problem.

More systematic and well planned studies on the evaluation of mineral status of small and large ruminants have been carried out in the state of Himachal Pradesh. The state has wide variation of terrain from plain hilly region to alpine pastures at elevation range from 3000 to 3500 m asl. Kumar and Singh (2007) compared the mineral status of two groups of Gaddi x Merino crossbred sheep belonging to the government farms located in two separate hill zones. Zone I was sub-mountainous and subtropical with elevation below 650 m, whereas zone II was mid-hill subhumid zone with elevation ranging from 651 to 1800 m. Samples of plasma, wool, urine, and fodder species were collected and analyzed for macro- and microminerals. The main minerals limiting production were identified as sodium and copper. Sheep were found to have fairly good status for calcium, phosphorus, and magnesium. Some sheep in both the zones showed borderline serum copper levels. Sodium deficiency was found to be widespread. Almost all samples of fodder, pasture, and weeds collected from the two zones had sodium level (<0.06) far below the required critical levels (0.70 g/kgDM, Freer et al. 2007). Surprisingly, wool sodium level was considerably lower (33.3 to 44.6 mg/100 g) than the reported normal values for wool sodium (60 to 174 mg/100 g) in some European sheep breeds (Minson 1990; Patkowska-Sokoła et al. 2009) (Fig. 8.2).

In yet another study from Himachal Pradesh, Kumar et al. (2008) investigated the mineral status of the Amb and Gagret village livestock belonging to Una district representing agro-climatic zone I of Himachal Pradesh. Blood samples of local and crossbred animals were collected randomly along with samples of feed and fodder and analyzed for various macro- and microminerals. Results indicated that calcium, zinc, and copper were the main minerals limiting production. Most animals in Amb as well in Gagret had plasma Ca (8.17–8.68 mg/dl) and Zn levels (0.42–0.61 ppm) below the critical concentration recommended for ruminants. Marginally lower values of plasma copper were also recorded in the livestock of both the blocks. Some animals in Amb block were found to have significantly lower plasma sodium level. Analysis of feed and fodder samples showed that all minerals were within the normal range except copper, which was found to be lower in the animals of both blocks.

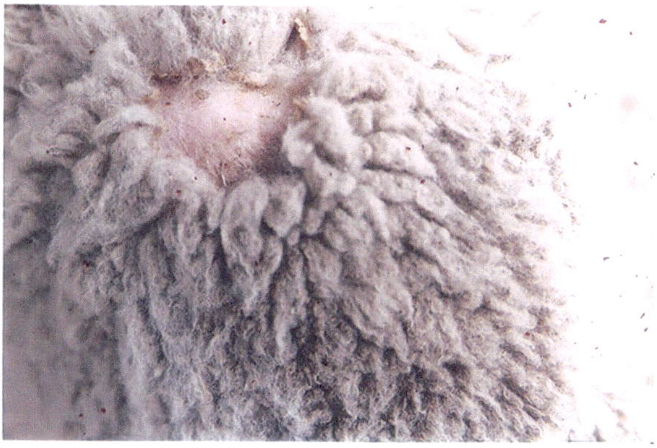

Fig. 8.2 Wool loss in migrating gaddi sheep (ICAR Project work/ courtesy Director AH, HP)

In an elaborate study in the mid hill zone of Himachal Pradesh, Verma et al. (2008) determined the mineral status of the crossbred dairy cows and tested the effectiveness of mineral supplementation in ameliorating the specific mineral deficiencies. The study was conducted on 100 cows of village Panchrukhi belonging to Kangra district. After initial assessment of mineral status of cows, the animals found deficient in one or more mineral elements were supplemented with appropriate mineral mixture for 30 days. Blood samples were collected on day 0 (pre-supplementation) and day 30 (post-supplementation) and analyzed for sodium, potassium, chloride, calcium, phosphorus, magnesium (macro minerals) and copper, zinc, and iron (microanimals). Results showed that the mean plasma concentrations of Ca (mg/dl), Na (mEq/l), Cl (mEq/l), Cu (ug/ml), and Zn (ug/ml) were significantly increased from the day 0 values of 8.85 ± 0.26, 138.8 ± 1.33, 94.54 ± 1.88, 0.59 ± 0.03, and 0.81 ± 0.04, respectively, to day 30 values of 9.56 ± 0.18, 144.8 ± 1.01, 101.60 ± 1.35, 0.70 ± 0.04, and 0.90 ± 0.04, respectively. Concentration of other minerals except iron also rose, but the values were not statistically significant. The authors noted a considerable decline in number of anestrous animals as a result of return of cyclic activity which may have been due to mineral supplementation and improvement in mineral status of the animals. The study concluded that strategic mineral supplementation was a right step in preventing mineral deficiencies.

In higher reaches of Himalayas in Himachal Pradesh, Kinnaur, at the elevation of around 2960 m and Lahaul Spiti, at the elevation over 3500 m, copper was the most recognized production-limiting mineral in dairy animals. On the other hand, iron was recorded above the animal requirements (Kumar et al. 2008; Sharma and Kumar 2007).

Minerals limiting production in the small and large ruminant species in mid and high hill regions of Himachal Pradesh and Uttarakhand are summarized in Table 8.1.

8.1.1.2 Union Territories Jammu and Kashmir and Ladakh

The UTs Jammu and Kashmir and Ladakh have a tough and highly variable terrain from foot hills to the alpine areas. There have been only few studies on the status of mineral elements in ruminant animals and feed and fodder. Yatoo et al. (2013)

Table 8.1 Minerals limiting productive potential of small and large ruminants in Himachal Pradesh and Uttarakhand

Study area	Species	Minerals limiting production	Reference
Foot hills—Tarai region, Uttarakhand (215 masl)	Large ruminants	Sodium	Rajora and Pachauri (1993)
		Calcium	
		Phosphorus	
		Copper	
		Zinc	
Kashipur, Haridwar, Pithoragarh, Uttarakhand (324 masl)	Large ruminants	Copper	Khan et al. (1979)
		Zinc	
Nainital district of Uttarakhand	Cattle and buffaloes	Calcium	Tiwari et al. (2012)
		Phosphorus	
		Zinc	
		Copper	
		Manganese	
Panchrukhi (Kangra), mid hill region, Himachal Pradesh (1320 masl)	Large and small ruminants	Sodium	Verma et al. (2008)
		Calcium	
		Copper	
		Zinc	
Amb and Gagret region of Shivalik hill Zone I, Himachal Pradesh	Large ruminants	Calcium	Kumar et al. (2008)
		Sodium	
		Zinc	
		Copper	
Kinnaur, Lahaul Spiti (3000–3500 masl)	Sheep	Copper	Kumar et al. (2008) and Sharma and Kumar (2007)

evaluated the prevalence of trace minerals in crossbred sheep and goats in Kashmir Valley. Fifty three crossbred sheep and thirty five crossbred goats grazing on the same pasture were randomly selected from three districts Budgam, Pulwama, and Srinagar. Blood/serum samples collected from these animals were analyzed for copper, zinc, and iron contents. The average prevalence of copper and zinc deficiency was recorded to the extent of 34 to 46% in sheep and goats. Surprisingly, the authors reported cases of iron deficiency in both sheep and goats to the extent of 18 to 20 per cent. This appears to be in contradiction with majority of published reports by other workers on nutritional profile of soil and fodder in the Himalayan countries as they have documented iron content in excess of ruminant requirements (Khan et al. 2012; Habib et al. 2013; Khan et al. 2006). Changthangi is one of the indigenous sheep breeds reared by the nomadic people Changpas of Ladakh. Yatoo et al. (2017) investigated the serum mineral status of sixty Changthangi ewes selected from farmer's herd. The animals were divided into two groups having 30 pregnant and 30 non-pregnant ewes, and serum levels of macro- (Ca and P) and microminerals (Cu, Zn, Fe, Co) were determined. It was reported that concentration of most minerals recorded was below the critical mineral concentration required for ruminants. Also significantly lower levels of serum calcium, phosphorus, copper, iron, and zinc were recorded in pregnant ewes during winter compared to summer possibly due to poor-quality feeds available during cold months.

Sheikh et al. (2019) assessed the nutritional profile of feeds and fodders fed to the cattle of Anantnag district of Jammu and Kashmir. A total of 369 samples were collected from six Tehsils and analyzed for macrominerals. Results showed that calcium, phosphorus, and magnesium contents were higher than the respective critical values, but sodium level was recorded below the requirement needed for the ruminants (0.7 to 1.0%; Freer et al. 2007). Sodium concentrations in mixed grass hay, rice bran, and wheat bran were very low (0.02, 0.03, and 0.04% respectively). In a more recent study, Ashraf et al. (2021) evaluated the mineral status of crossbred Jersey dairy cows in Pulwama district of Jammu and Kashmir. Macrominerals (Ca, P, Mg, Na, and K) were measured by the standard methods using commercial kits, while microminerals (Cu, Zn, Fe, Co, and Mn) were estimated using atomic absorption spectroscopy. Significantly ($P < 0.05$) lower blood levels of Ca, P, Mg, and Na were recorded in high milk yielders (above 10 kg/day) compared to the low milk yielders (0–5 kg/day) and dry animals. The values of plasma microminerals (Cu, Zn, Fe) were also lower in high milk yielders. Based on the results, it was recommended that there was need for supplementation of specific mineral mixtures for different stages of lactation.

8.1.2 Neighboring Himalayan Countries

The investigations carried out on comprehensive assessment of the mineral element status of livestock in the Himalayan belt in neighboring countries such as China and Pakistan have also identified a number of macro and trace minerals that may limit livestock productivity. Studies related to the principal minerals limiting production of migratory sheep and goats in Nepal still seem to be scarce.

8.1.2.1 China

Poor mineral nutrition had been recognized as a major impediment in limiting production in the hilly terrains of China as early as 1996 (Masters and White 1996). A massive joint Project sponsored by Chinese Ministry of Agriculture and Commonwealth Scientific and Industrial Organization (CSIRO), Australia, was launched in 1995 to identify the minerals which limit small ruminant production in sheep farms of northern China. Samples of blood, saliva, grasses, and feces of animals were collected for mineral estimation from several sheep farms across four provinces of northern China. Five mineral elements, namely, sodium, copper, zinc, selenium, and sulfur were identified as limiting the productivity of animals (Masters et al. 1996). Plasma copper concentration was measured below the critical level of 4.7 umol/L in sheep at many farms. This was confirmed by deficient copper concentration in liver (<80 umol/kg). Among other minerals, calcium and magnesium were within or above the recommended level, but the risk of phosphorus deficiency was widespread. Although the forages contained sufficient phosphorus during summer, during winter season, the values of phosphorus (0.9 to 3.0 g/kg) were found to be below the recommended level. Shen et al. (2019) investigated and worked on a mysterious disease locally called "Ruanguzheng Disorder" affecting Guizhou semi-fine wool sheep in Yunnan-Guizhou Plateau of China. The study revealed that the disease was caused by phosphorus deficiency. Phosphorus contents in forage from the affected areas were found markedly lower than those estimated in forage in the unaffected areas. Phosphorus contents of wool, blood, and bone from affected sheep were also significantly lower than the levels in healthy animals.

A distinctive feature observed in northern parts of China and Tibetan plateau was finding of sodium deficiency which was of a major concern. It was detected in many farms using salivary Na:K ratio as a criterion. The salivary Na: K ratio was reported to be far below the critical value (12 to 15) reflecting sodium deficiency. Later research conclusively proved that deficiency of sodium was a major problem in the area as sodium contents in most herbages were far below the requirement for ruminants (Xin et al. 2011). Fan et al. (2020) working on relationship of both macro- and micromineral elements in sheep grazing on Qilan mountains, China, concluded that winter herbage sodium levels in 24 farms in the area were below the recommended levels for macrominerals. Supplemental minerals were, therefore, recommended to avoid deficiencies. In yet another study carried out in the Qinghai province of China, Fan et al. (2019) investigated the mineral status of yaks and the relationship between macrominerals in soil, forage, and blood. Concentrations of sodium, phosphorus, and magnesium were found to be below the requirement levels for yaks. The herbage sodium concentration was in exceptionally low range of the requirement. Table 8.2 summarizes the results of the three major studies on sodium status in forages in China.

Table 8.2 Sodium contents (g/kg) in herbage/forages collected from different locations of Qinghai-Tibet Plateau and Qilian mountains of Tibet/China

Location (source of herbage)	Number of samples analyzed	Na level range in samples	Number of samples having Na below ruminant's requirement[a]	Reference
Qinghai-Tibet Plateau	Forage samples from each of the four county (20)	0.04–1.0 (g/kgDM)	All samples (low sodium unable to meet requirement)	Xin et al. (2011)
Qinghai-Tibet Plateau	Herbage samples from four yak farms (320)	71.81–119.2 (gkg^{-1})	All samples (herbage Na concentration exceptionally low)	Fan et al. (2019)
Qilian mountains	Herbage samples from typical alpine meadow (240)	45.17–173.67 (g/kg)	All samples	Fan et al. (2020)

[a]Recommended levels of Na for sheep (0.7–0.9 g/kg DM) and cattle (0.8–1.2 g/kg DM) (Freer et al. 2007)

8.1.2.2 Pakistan

A number of studies on mineral status of animals and feeds and fodder have been undertaken with regard to ruminant requirement. Many studies indicate deficient supply of sodium, copper, and phosphorus. Khan et al. (2006) investigated the mineral contents of soil and pastures in Leiah district of Punjab. The pasture forages had adequate levels of macrominerals calcium and magnesium but were deficient in copper and zinc considering the ruminant's requirement. Habib et al. (2013) assessed the mineral profile of tree leaves from eight local species utilized for feeding to sheep and goats. Leaves samples were collected from different locations in Kohat district of Khyber Pakhtunistan province of Pakistan. Calcium contents and trace minerals like Zn, Mn, and iron were either adequate or above the optimum level requirement for ruminants. On the contrary, phosphorus was in the lower range in all the samples analyzed resulting in wide Ca: P ratio. Average copper contents were recorded 3.32 ppm as against required dietary level of 4 to 10 ppm. The authors also analyzed the leaves for sodium content and found that sodium levels in all the samples were extremely low (0.02%) against the optimum requirement of 0.7 to 1.0% for ruminants. Another investigator from Pakistan, Khan et al. (2012), has more elaborately documented mineral profile of forages and specifically emphasized on widespread deficiency of sodium. Twenty six samples of forages and grass species from natural pastures were collected from Punjab farms and analyzed for five macro (Na, K, Ca, Mg, and Fe) and three microminerals (Cu, Zn, and Mn). Their results showed that forages had sufficient levels of K, Ca, Mg, Mn, and Zn, but the concentrations of sodium and copper were below the optimum level recommended for ruminants. With the exception of *Albizza lebbek* grass, most of the forage samples were found sodium deficient, having values below the critical level. The authors claimed that the minerals most widely present in inadequate amounts in the world were Na and Cu. Table 8.3 summarizes sodium levels of the forages and grass species of Pakistan.

It is obvious from Table 8.3 that out of total 26 samples analyzed from Punjab, Pakistan, 14 (53.8%) were deficient in sodium with some having sodium level as low as 0.06% (Khan et al. 2012). In all the 24 samples of tree leaves collected from Khyber Pakhtunistan, Pakistan, sodium levels measured were found to be far lower than the recommended level for sheep (Habib et al. 2013).

8.1.3 Sodium Deficiency—An Emerging Prevalence in Himalayan Regions and Poor Status of Sodium in Himalayan Herbage and Grazing Animals

All living systems require inorganic elements and minerals for normal life processes. Minerals cannot be synthesized by the body systems and soil is their store house. The plants derive the minerals from soil, and the animals, in turn, from the plants which they consume as feed and fodder. There is a well established relationship between soil, plants, and animals, and several factors regulate the transfer of minerals from one source to the other. By 1981, 22 mineral elements were believed

Table 8.3 Sodium contents (% DM basis) in different herbages and grasses collected from Punjab and Khyber Pakhtunistan, Pakistan

Location	Type of sample	Number of samples analyzed	Na level (g/kgDM)	Number of samples having Na below ruminant's requirement[a]	Reference
Punjab, Pakistan	Dried leaves of grasses I	11	0.06–2.13	5	Khan et al. (2012)
	Dried leaves II	4	0.43–0.76	3	
	Forage residues	6	0.30–0.60	6	
	Albizia lebbek grass	5	1.25–1.80	None	
Khyber Pakhtunistan	Tree leaves of 8 trees	24	0.02–0.08	24	Habib et al. (2013)

[a]Recommended levels of Na for sheep (0.7–0.9 g/kg DM) and cattle (0.8–1.2 g/kg DM) (Freer et al. 2007)

to be essential for animal life: seven major or macronutrients and 15 trace or micronutrients (Underwood, 1981). Numerous studies have shown that mineral deficiencies can result in many disorders including severe nutritional deficiencies, disturbance in biological processes, reproductive abnormalities, and decreased immune functions (Suttle 2010). Extensive research over the years has established that optimum quantity of a mineral is required for a specific function, and its presence in the tissues and fluids in the body forms the basis of diagnosis whether there is deficiency or excess of a mineral. By measuring the concentrations of minerals in the herbage, feeds, fodders, and animal tissues, fluids, and blood, the mineral status could be assessed.

Traditionally, among the macrominerals, nutritionist's main emphasis has been on elements like calcium, phosphorus, and magnesium but hardly on sodium. The reason is obvious as most of the disease conditions in the domestic animals are related to deficiency or excess of these minerals. Attention to sodium (salt) was drawn with the extensive experimental work of Denton (1982), in which Merino sheep were depleted of sodium. Denton's experiments produced huge data to unravel regulatory functions of sodium and discovered salt hunger and relation of sodium to blood pressure. Attention to sodium (salt) was also drawn with the observations of craving for salt by sodium-deficient cows. It was later reported by Olson et al. (1989) that salt was an essential mineral element for milk production, and prolonged sodium deficiency may lead to deceased milk production. It was also shown that sodium was essentially required for wool growth, and diets high in sodium chloride can improve wool growth efficiency in sheep (Thomas et al. 2007).

Regulatory function of sodium is associated with potassium in that they have an inverse relationship. As potassium level goes up, sodium level goes down. Exposure of animals to excess potassium was found to cause an immediate reduction in rumen sodium concentration (Singh 1989) and an immediate large increase in urinary sodium excretion. It was reported that animal diets excessively high in potassium were likely to cause sodium deficiency. When the animals were exposed to high potassium forages, they displayed salt deprivation symptoms (Dougherty et al. 1995). In modern agricultural practices, all kinds of fertilizers are being used on the crops changing the mineral composition of soil and herbages.

Poor Status of Sodium in Himalayan Herbage and Grazing Animals
Detailed and systematic analysis of sodium in alpine herbages, feeds, and fodders in different Hindu Kush Himalayan regions has not been perhaps undertaken, but available reports emanating from a few countries don't appear to give a rosy picture. As discussed under Sect. 8.1.2.1, several investigators from China have reported wide ranging deficiencies of sodium from northern part and Tibetan plateau areas (Fan et al. 2019, 2020; Xin et al. 2011). Cases of sodium deficiency have also been reported from Pakistan (Khan et al. 2012; Habib et al. 2013). Reports of sodium status of alpine herbage, feeds, and fodders from other Himalayan countries are scanty. Although not many reports of detailed experimentation to determine herbage sodium status are available from India, several investigators have documented sodium as a deficient mineral in feeds, fodders, and grasses. In a study conducted in Kamang district of Arunachal Pradesh, 117 samples of feed and fodders were subjected to detailed mineral analysis. Sodium level in all the samples ranged from 0.006 to 0.040%, which is far below the recommended level of sodium (0.7 to 1.0%) for ruminants (Chatterjee et al. 2011).

Literature search reveals that there could be some valid and possible reasons for lower sodium content in the alpine soil and vegetation of Himalayas. In a recent study, Gowda and Pal (2020) have reported the status of different minerals in agro-climatic zones of India and has depicted the Hilly and mountain zone (zone 6th) prone to sodium deficiency along with Ca, P, Zn, and Cu deficiency. Denton (1982) has stated that the contents of sodium in plants in mountains and alpine regions, which are far away from the sea coast, are very low. Long time ago, Australian Scientist, Hutton (1958) had put forward two views on presumably low sodium content in mountain herbage. According to him, sodium contents in rain water decrease with distance from sea, and the rain may be virtually distilled water (devoid of any salt) on the mountains which are hundreds of miles away from the sea. Secondly, in high mountain areas, melting of snow for many months leaches out the soluble sodium deep into the soil and it may not be available to the plants. In view of the above discussion, we feel there is dire need to undertake a thorough and systematic analysis of feeds, fodders, herbages, grasses, tree leaves, and livestock to assess the precise status of sodium in every region of Hindu Kush Himalayan countries.

8.2 Interaction Among Minerals Affecting Bioavailability

Some mineral elements have been found to interfere with the absorption of other minerals and affect their bioavailability. Molybdenum, sulfur, iron, and potassium are the principal minerals in this group which interfere in the systemic availability of copper, magnesium, and sodium. The susceptibility of grazing animals to copper, magnesium, and sodium deficiencies due to interfering mineral groups has been documented in ruminants. The mineral interactions are important from the point of view of migratory small ruminants when they happen to graze pastures having higher molybdenum, sulfur, and iron which reduce the bioavailability of copper resulting in its deficiency. In fact, copper supplementation in ruminants has been given special attention because of the narrow margin between deficiency and toxicity (Thorndyke et al. 2021). Antagonistic impact of molybdenum alone or in combination with elevated dietary sulfur on copper absorption is well known. Increasing concentration of molybdenum in forage was accompanied by significant decrease in copper availability. It was also shown that consumption of high iron in pasture or soil could reduce the copper status of sheep (Suttle et al. 1984). Recently, an intense case of copper deficiency in Tibetan sheep resulting from their grazing on sulfur-fertilized alpine pastures of Qinghai-Tibet Plateau has been reported by Zhai et al. (2021). The local herdsmen in the area have a practice of fertilizing pastures with ammonium sulfate [$(NH_4)_2SO_4$)] to improve the nitrogen content of the herbage. In this case as a result of antagonistic effect of sulfur on copper, the sulfur element combined with molybdenum to form a thiomolybdate complex in rumen, rendering copper nutrient unavailable for absorption. Consequently, the copper content in liver and blood dropped significantly resulting in copper deficiency in the animals.

Dietary potassium intake has been found to influence the status of at least two other minerals in the body, namely magnesium and sodium. High intake of potassium was found to decrease the absorption of magnesium in rumen resulting in increased incidence of grass tetany in sheep (Stewart 2022). Cows consuming high potassium and low sodium herbage were observed to excrete excessive magnesium through the urine (Martens et al. 2018).

When animals were exposed to high potassium forages, they displayed signs of salt deprivation. Dobson (1963) reported that sheep fed on grasses which had been fertilized with potassium fertilizers led to sodium deficiency in animals. Later it was shown that potassium loading of male buffalo calves interfered with reabsorption of sodium in renal tubules causing increased excretion of this element through the urine (Singh 1989). In a story from Kentucky University, some dairy farmers had reported that feeding hay and silage from a particular field was reducing the milk production in dairy cows (Dougherty et al. 1995). It turned out that the crop had been fertilized by potassium fertilizers and heavy potassium intake caused sodium deficiency in affected dairy cows. Warner and Stacy (1972) studied the inter-relationship between sodium and potassium in sheep and reported that, in general, the fate of ingested sodium in ruminants was intimately linked with the fate of potassium. This reasoning was corroborated by the observations in sheep wherein an increase of potassium intake resulted in a decrease of fecal sodium excretion.

8.3 Salt Hunger in Herbivores and Special Place of Sodium in Nutrition of Ruminants

Sodium has been recognized as one of the essential elements in the nutrition of mammals, deficiency of which results in decreased production, lower fertility, and lower body fluid osmotic pressure and membrane permeability (Aitken 1976). De Waal et al. (1989) reported depression in wool production in sheep consuming diet not supplemented with salt (NaCl). Denton (1982) has reported that it were the herbivores, rather than the carnivores, that showed more appetite for salt in nature. Historically most herbivores (wild or domestic) have been found to have a natural liking for salt licks. In old times, the salt-ingesting behavior of the herbivores was used as a tool by the hunters to capture the wild animals by putting salt on the rocks. Leek (1917) has described how in the United States hunters would make a salt lick to lure the deer and other wild animals and lie in wait for them to come. Similar stories were heard from the high Himalayas in Himachal Pradesh—how the villagers in Kinnaur district used salt blocks to collect their yaks at one place every evening. We had also come across stories from shepherds that in Pin Valley alpine pastures located on the India-China border, the Chinese exchanged salt for pulses from Indian shepherds for their flocks. This barter practice for salt has also been reported between Chinese and Nepalese graziers (Shrestha and Pradhan 1995). Although craving for salt in herbivores is an innate character, it appears that the deciding factor is content of sodium in the herbage. Most plant species and forages are quite low in sodium but very rich in potassium content. Analytical reports confirmed that sodium content in the herbage/fodder species decreased with the increase in elevation and distance from the sea level (Denton 1982).

In our study on migratory sheep, we had observed an intense salt hunger in sheep after their prolonged stay in the alpine pastures. During upward journey with flocks, we had seen the animal attendants carrying the salt bags, but they were ignorant what the salt was required for. In the background of the salt-seeking stories of animals mentioned above, we decided to investigate the cause of salt hunger observed in the migratory sheep at the alpine pastures (Singh et al. 2000).

8.3.1 Methods of Diagnosis of Sodium Deficiency in Ruminants

Traditionally, critical level of a mineral element in blood plasma or serum forms the basis of its deficiency or sufficiency. However, blood sodium level measurement is not a good reference to diagnose sodium deficiency because body tries to maintain sodium concentration within the normal range through increased adrenocortical secretion of aldosterone, which effectively reduces the amount of sodium in saliva and urine and muzzle secretion (Singh and Rani 1999). Among other tests, parotid salivary sodium-to-potassium ratio (Na:K ratio) has been found to be the most sensitive index of sodium status of ruminants (Minson 1990). Kemp and Hartmans (1990) reported that cattle with saliva sodium concentration <87 mEq/L (or Na:K ratio <5 to 8) could be classified as sodium deficient. Skydsgaard (1967) observed that salivary Na:K ratio <10 was indicative of sodium deficiency.

Using urine sodium concentration as a basis, Morris (1980) suggested that cattle were probably sodium deficient when urine sodium concentration decreased to <7 mmol/L. McDowell (1997) claimed that concentrations of sodium in saliva and feces below 100–200 mg/ml and 1000 ppm, respectively, were indicative of sodium deficiency. Our studies in buffalo calves conclusively showed that Na:K ratio of <0.20 in muzzle secretion and Na:K ratio of <0.10 in urine were indicative of sodium deficiency (Singh and Rani 1999). In case of sheep, a Na:K ratio <2–4 or plasma aldosterone concentration >200 pg/ml was indicative of sodium deficiency (McSweeney et al. 1988; Denton 1982). A summarized view of the above proposed tests is presented in Table 8.4.

8.3.2 Salt Hunger in Migratory Sheep in North-Western Himalayas

The principal author of this book and his coworkers in the year 2000 undertook a detailed and systematic study with the objective to investigate whether the salt-seeking behavior of sheep flocks was as a result of their becoming sodium deficient during grazing at the alpine pastures (Singh et al. 2000), and if yes, what possible factors led to this status? For this purpose, twenty migratory crossbred sheep aged about two years belonging to the Himachal Pradesh State Government Farm, Tal, 1189 mamsl, were selected. The animals were adapted to an annual migratory cycle leaving the base camp for high hills in the month of June every year and returning to the base camp in the month of October. The sheep were stationed at the base

Table 8.4 Diagnostic criteria to detect sodium deficiency in small and large ruminants

Species	Body fluid	Parameter	Normal value	Critical level indicating Na deficiency	Reference
Cattle	Parotid saliva	Na concentration (mmol/L)	130–160	<87	Kemp and Hartmans (1990)
		Na:K ratio	12–15	<5 to 8	
Sheep	Parotid saliva	Na:K ratio	23–35	<4	McSweeney et al. (1988)
Cattle	Urine	Na concentration (mmol/L)	45–120	<7	Morris (1980)
Cattle, buffalo, sheep	Urine	Na:K ratio	>0.3–0.7	<0.1	Singh and Rani (1999)
Cattle, buffalo	Muzzle secretion	Na:K ratio	>0.3–0.5	<0.2	Singh and Rani (1999)
Sheep	Plasma	Aldosterone concentration (pg/ml)	20–25	>200	McSweeney et al. (1988)

camp for nearly eight months and migrated to Kara and Kanta pastures of Kinnaur district, Himachal Pradesh (3350 mamsl) in the month of June, where these animals stayed for about four months. Initial sampling was performed at the base camp and subsequent sampling at the Kara pastures after the flock had spent there a period of more than one month. At both the stations, samples of blood and urine were collected from each sheep. Besides, samples of mixed pasture grasses and water ingested by the animals were also collected. All samples were subjected to sodium and potassium analysis by using atomic absorption spectrometry (Singh et al. 2000).

The study revealed that the sheep flock did not exhibit the behavioral symptoms of salt hunger or geophagia during stay at the base camp Tal. The animals did not show any appreciable interest in the salt placed on the rocks except an occasional sniffing at the worker's hand carrying the salt. However, majority of sheep after prolonged stay at the high-altitude station, Kara, showed intense appetite for salt. Many a times, the animals were found chasing the animal attendants carrying the salt. As soon as the salt was placed on the rocks in the grazing area, sheep were found colliding with each other in the race to get to the salt spot. It was noticed that the animals not having an opportunity to consume salt exhibited continuous bleating and a tendency to refuse to retire for the day. The shepherds generally believed that if salt was not given to the animals, oral papillae became too large and painful (Om Prakash 1996, Personal communication).

We determined the sodium status of sheep flock at the base camp and high altitude station by monitoring Na and K concentrations in plasma, urine, pasture grasses, and water samples. Results showed that the salt-seeking behavior and intense appetite for salt observed in sheep on arrival and stay at the high altitude were in fact due to their being in poor sodium status which was confirmed by urine analysis. The mean urinary Na concentration at the base camp, Tal, was measured as 38.25 mmol/l with Na:K ratio of 0.28. However, after stay at the high-altitude station, Kara, the urinary sodium concentration and Na:K ratio dropped significantly to 14.23 mmol/l and 0.039, respectively (Fig. 8.3 and Table 8.5), and some sheep tested were voiding almost sodium-free urine.

The urinary potassium concentration showed an opposite trend. Contrary to a significant decline in urinary Na content at the high-altitude station, Kara, the mean urinary K concentration was significantly elevated at Kara (355.9 mmol/l) compared to the urinary K concentration at the base camp Tal (128.4 mmol/l, Table 8.5). Blood plasma Na concentration of sheep at high-altitude station, Kara, did not show any decline, but instead showed an increase from 125.42 mmol/l at the base station to 155.64 mmol/l at Kara. Similarly, mean plasma K concentration increased from 5.47 mmol/l at the base camp to 7.96 mmol/l at Kara. Consequent to these changes in plasma Na and K levels, the plasma Na:K ratio declined from 23.10 at the base camp to 19.87 at Kara.

The sodium content in the mixed grass samples collected from the two stations did not vary significantly, but the herbage at high-altitude station, Kara, had significantly higher K concentration (1.72 g%) than the corresponding value of 1.13 g% obtained at the base camp (Table 8.5). On the other hand, water samples collected

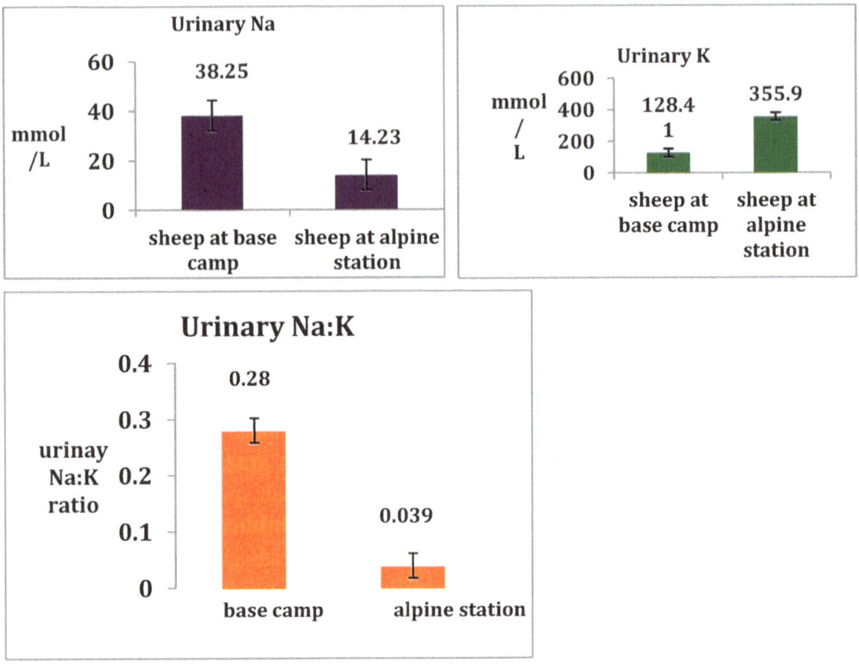

Fig. 8.3 Changes in sheep urinary Na and K concentrations (mmol/L) and Na:K ratio of sheep after prolonged stay at the Kanta pastures, Himachal Pradesh

from high-altitude station, Kara, had very low ($P < 0.05$) concentrations of sodium (0.85 ppm) and potassium (0.70 ppm) compared to the water samples analyzed for sodium (12.63 ppm) and potassium (1.87 ppm) at the base camp Tal.

Although the farm attendants accompanying the flock to the hills every summer were aware of the increased salt liking by the animals on arrival and during stay at the high-altitude station, they were ignorant about the actual sodium requirement and status of the animals. During the course of the study, it was informed that traditionally the shepherds offered some salt to their sheep biweekly but it was not sufficient to alleviate the salt appetite of these animals.

It was conclusively established that salt hunger in sheep was related to grazing on low-sodium and high-potassium herbage, and insufficient dietary sodium intake ultimately led to sodium deficiency in these animals. Based on the findings, it was recommended to the State Animal Husbandry Department of Himachal Pradesh that sheep at the alpine pastures must be provided extra salt supplementation at the rate of 3 to 4 g salt/animal/day and the shepherds be advised accordingly (Singh et al. 2000).

Table 8.5 Urinary and plasma sodium and potassium concentrations (mmol/l) and Na:K ratio in migratory crossbred sheep and Na and K concentrations in the pasture grasses (g%) and water samples (ppm) at the base camp, Tal, and high-altitude station, Kara (Singh et al. 2000)

	Base camp (Tal)			High-altitude station (Kara)		
Item	Na	K	Na:K ratio	Na	K	Na:K ratio
Urine	38.25 ± 10.1	128.41 ± 20.3	0.28 ± 0.064	14.23 ± 6.12*	355.9 ± 25.2*	0.039 ± 0.022*
Plasma	125.42 ± 2.95	5.47 ± 0.17	23.10 ± 0.69	155.64 ± 2.0*	7.96 ± 0.17*	19.87 ± 0.58*
Pasture grasses	0.0042 ± 0.0006	1.13 ± 0.20	0.0038 ± 0.0007	0.0053 ± 0.0009	1.72 ± 0.057*	0.0031 ± 0.0006
Water samples	12.63 ± 2.94	1.87 ± 0.66	7.80 ± 1.87	0.85 ± 0.086*	0.70 ± 0.11	1.24 ± 0.085*

*Significantly different ($P < 0.05$) from the corresponding value determined at base camp, Tal. Each value is a mean ± SEM of urine ($n = 12$), plasma ($n = 20$), pasture grasses ($n = 3$), and water ($n = 3$) samples

References

Aitken FC (1976) Sodium and potassium in nutrition of mammals, commonwealth agricultural Bureaux. Farnham Royal, pp 131–141. https://openlibrary.org/books/OL4590907M/Sodium_and_potassium_in_nutrition_of_mammals

Ashraf A, Sheikh GG, Ganai AM, Reshi PA (2021) Mineral status of dairy animals in Pulwama district of Jammu and Kashmir. Indian J Anim Nutr 38(4):353–359

Chatterjee A, Ghosh MK, Roy PK, Das SK, Santra A (2011) Macro and micro-mineral status of feeds and fodders in West Kameng district of Arunachal Pradesh. Indian J Anim Sci 81(10):1076–1079

De Waal HO, Baard MA, Engels EAN (1989) Effects of sodium chloride on sheep. 1. Diet composition, body mass changes and wool production of young Merino wethers grazing native pastures. S Afr J Anim Sci 19(1):27–33

Denton DA (1982) The hunger for salt. Springer, New York, pp 154–242. https://archive.org/details/hungerforsaltant0000dden

Dobson A (1963) Changes in composition of the saliva of cows on grazing heavily fertilized grass. Res Vet Sci 49(2):238–246

Dougherty CT, Wells KL, Mitchell GE (1995) Sodium in pasture species and grazing livestock. Agron Notes 28(5):1–4. https://uknowledge.uky.edu/pss_notes/37

Fan Q, Wanapat M, Hou F (2019) Mineral nutritional status of yaks (*Bos Grunniens*) grazing on the Qinghai-Tibetan Plateau. Animals (Basel) 9(7):468. https://doi.org/10.3390/ani9070468

Fan Q, Wang Z, Chang S, Peng Z, Wanapat M, Bowatte S, Hou F (2020) Relationship of mineral elements in sheep grazing in the highland agro-ecosystem. Asian Australas J Anim Sci 33(1):44–52

Freer M, Dove H, Nolan JV (eds) (2007) Nutrient requirements of domesticated ruminants. CSIRO Publishing, Clayton. https://www.researchgate.net/publication/303130298_Nutrient_requirements_of_domesticated_ruminants

Gowda NKS, Pal DT (2020) Mineral status in different agro-eco zones of India, implications and amelioration in livestock. Epashupalan 4(85) https://epashupalan.com/4016/animal-nutrition/mineral

Habib G, Saleem M, Hameed A (2013) Mineral composition of local tree leaves for feeding sheep and goats in Kohat district of Khyber Pakhtunkhwa. Sarhad J Agric 29(1):97–103

Hutton JT (1958) The chemistry of rain water with particular reference to conditions in South East Australia. In: The hunger for salt. Springer, Berlin/Heidelberg, p 11

Kemp A, Hartmans J (1990) Mineral stoffversorgung tiergesundheit. In: Minson DJ (ed) Forage in ruminant nutrition. Academic, San Diego, p 307

Khan SA, Bedi SPS, Sawhney PC, Ranjhan SK (1979) Zinc status of soil, plants and animals in Tarai Area, Uttar Pradesh, India. Indian J Anim Sci 49(8):612–617

Khan ZI, Ashraf M, Valeem EE (2006) Forage mineral status evaluation: the influence of pastures. Pak J Bot 38(4):1043–1054

Khan ZI, Ahmad K, Danish M, Mirzaei F (2012) Minerals profile of forages for grazing ruminants in Pakistan. Open J Anim Sci 2(3):133–141

Kumar V, Singh SP (2007) Determination of plasma macro-element levels in migratory and stationary sheep flocks of Himachal Pradesh. Paper presented at the, XVI Ann Conf of SAPI, Guwahati, India, 10–12 Feb, 2007

Kumar R, Verma MS, Sharma KB (2008) Plasma mineral status in animals of high altitude area of HP. Paper presented at 17th Ann Conf, Society of Animal Physiologists of India (SAPI), GB Pant University of Agriculture & Technology, Pantnagar, India, 7–9 Feb, 2008

Leek SN (1917) Salt licks and alkali springs for elk-Sci Am suppl 2179 IN: Denton DA 1982. In: The hunger for salt. Springer, New York. https://www.scientificamerican.com/article/salt-licks-and-alkali-springs-for-elk/

Martens H, Leonhard-Marek S, Röntgen M, Stumpff F (2018) Magnesium homeostasis in cattle: absorption and excretion. Nutr Res Rev 31(1):114–130

Masters DG, White CL (1996) Detection and treatment of mineral nutrition problems in grazing sheep. Australian Centre for International Agricultural Research (ACIAR) Monograph Series, pp 1–7

Masters D, Yu S, Gao F, Bin S, De-Xun LJ, Shao L, Guimet K, Xu W, Jiakong R, Chenglun R, Qinglei K, Fanjiang K, Harali S, Nuernisha LN, Fuben B, Lindsay J (1996) Identification of mineral elements limiting sheep production in Northern China. In: Proceedings of workshop on mineral problems in sheep in Northern China. Chinese Academy of Agricultural Science, pp 24–44

McDowell LR (1997) Minerals for grazing ruminants in tropical regions. Center for Tropical Agriculture, University of Florida

McSweeney CS, Cross RB, Wholohan BT, Murphy MR (1988) Diagnosis of sodium status in small ruminants. Aust J Agric Res 39(5):935–942

Minson DJ (1990) Sodium. In: Minson DJ (ed) Forage in ruminant nutrition. Academic, San Diego, pp 291–307. https://www.sciencedirect.com/book/9780124983106/forage-in-ruminant-nutrition

Morris JG (1980) Assessment of sodium requirements of grazing beef cattle: a review. J Anim Sci 50(1):145–152

Olson WG, Link KRJ, Otterby DE, Stevens JB (1989) Assessment of sodium deficiency and polyuria/polydipsia in dairy cows. Bov Pract 24:126–133

Om Prakash (1996) Personal communication, Farmers perception about migratory small ruminants farming system

Patkowska-Sokoła B, Dobrzański Z, Osman K, Bodkowski R, Zygadlik K (2009) The content of chosen chemical elements in wool of sheep of different origins and breeds. Arch Tierz 52(4):410–418

Rajora VS, Pachauri SP (1993) Micronutrient status of dairy cattle as influenced by soil-plant relationship in Tarai region. Indian J Anim Sci 63(8):882–890

Sharma KB, Kumar R (2007) Plasma mineral status in animals of cold desert area of HP. Paper presented at XVI Ann Conf of SAPI, 10–12 Feb, 2007, Guwahati, India

Sheikh FA, Ganie AM, Mir NA, Sheikh GG, Ahmad HA, Beigh YA (2019) Chemical composition and nutritional profile of feed and fodders fed to cattle of Anantnag district of Kashmir valley during winter season. Vet Res Int 7(2):104–109

Shen X, Chi Y, Huo B, Xiong K (2019) Studies on phosphorus deficiency in the Qianbei-Pockmarked goat. Asian Australas J Anim Sci 32(6):896–903

Shrestha HR, Pradhan DR (1995) Research priorities for animal agriculture agro-ecological zone in Nepal. In: Devendra C, P. Gardner Global P (eds) Agenda for Livestock Research- Proceedings of Consultation for South Asia Region, 6–8 June, 1995. International Crop Research Institute for Semi Arid Tropics (ICRISAT) Asia Centre, Patencheru

Singh SP (1989) Muzzle gland responses to potassium loading in buffalo calves. Indian J Anim Sci 59:28–31

Singh SP, Rani D (1999) Assessment of sodium status in large ruminants by measuring the sodium-to-potassium ratio in muzzle secretions. Am J Vet Res 60(9):1074–1081

Singh SP, Kumar R, Sharma KB, Kumar N (2000) Intense salt hunger in migratory crossbred sheep at high altitude pastures of North-West Himalayas. Oral presentation in 8th world salt symposium, Hague, Netherland. 2:817–821

Skydsgaard JM (1967) En metode til diagnosticering af latent natriummangel i kvaegbesaetninger. Nord Vet Med 19:346–349

Stewart AJ (2022) Hypomagnesemic tetany in cattle and sheep. MSD Veterinary Manual. https://www.msdvetmanual.com/metabolic-disorders/disorders-of-magnesium-metabolism/hypomagnesemic-tetany-in-cattle-and-sheep

Suttle NF (2010) Mineral nutrition of livestock, 4th edn. CABI, Oxfordshire

Suttle NF, Abrahams P, Thornton I (1984) The role of a soil × dietary sulphur interaction in the impairment of copper absorption by ingested soil in sheep. J Agr Sci Cambridge 103(1):81–86

Szigeti E, Kátai J, Komlósi I, Oláh J, Szabó C (2020) Newly grown wool mineral content response to dietary supplementation in sheep. Animals (Basel) 10(8):1390. https://doi.org/10.3390/ani10081390

Thomas DT, Rintoul AJ, Masters DG (2007) Increasing dietary sodium chloride increases wool growth but decreases *in vivo* organic matter digestibility in sheep across a range of diets. Aust J Agric Res 58(11):1023–1030

Thorndyke MP, Guimaraes O, Kistner MJ, Wagner JJ, Engle TE (2021) Influence of molybdenum in drinking water or feed on copper metabolism in cattle – a review. Animals (Basel) 11(7):2083. https://doi.org/10.3390/ani11072083

Tiwari SK, Kumar A, Tiwari DP, Mondal BC, Saxena PC (2012) Response to strategic dietary mineral mixture supplementation in cattle and buffaloes under field condition (hill region) of Nainital district of Uttarakhand. Indian J Anim Sci 82(11):1381–1385

Underwood EJ (1981) The mineral nutrition of livestock, 2nd edn. Commonwealth Agricultural Bureaux, Farnham Royal, p 1

Verma MS, Kumar R, Sharma KB (2008) Effect of specific mineral supplementation on the blood profile of dairy cattle of Himachal Pradesh. Indian J Anim Reprod 29(1):74–77

Warner AC, Stacy BD (1972) Intraruminal and systemic responses to variations in intake of sodium and potassium by sheep. Q J Exp Physiol Cogn Med Sci 57(2):89–102

Xin GS, Long RJ, Guo XS, Irvine J, Ding LM, Ding LL, Shang ZH (2011) Blood mineral status of grazing Tibetan sheep in the Northeast of the Qinghai–Tibetan Plateau. Livest Sci 136(2–3):102–107

Yatoo MI, Saxena A, Jhambh R, Nabi SU, Melepad DP, Kumar P, Dimri U, Sharma MC (2013) Status of trace mineral deficiency in sheep and goat in Kashmir Valley. Res J Vet Pract 1(4):43–45

Yatoo MI, Gopalakrisnan A, Gopinath D, Dimri U (2017) Evaluation of serum mineral status of pregnant and non-pregnant Changthangi ewes under different seasons. Indian J Anim Sci 87(12):1536–1538

Zhai B, Zhao K, Shen X (2021) Effects of sulphur fertilizer on copper metabolism in grazing Tibetan sheep in fertilized pasture. Pol J Environ Stud 30(6):5351–5356

Nutritional Status of Common Himalayan Pastures and Tree Foliages

9

Abstract

Livestock feed and fodder scarcity has been one of the major limiting factors for grazing animals in the hilly regions. Shrinkage of grazing land, degradation of pastures, increasing grazing pressure, and problems in cultivating fodder crops at variable heights due to adverse climatic conditions have resulted in huge gap between demand and supply of fodder in most Himalayan countries, especially in north-western Indian Himalayan states. The shortage of green and dry fodder in these states has been estimated to the extent of 36.5 and 50.8%, respectively. During summer, the animals thrive by grazing on alpine pastures and alpine tree fodder and leaves and for the remaining part of the year on, orchards, agroforestry, road side grasses, lower hill tree leaves, and crop residues. The current chapter lists the common species of Himalayan grasses, fodder trees, and crops and crop residues which comprise the main feed resources for grazing small ruminants. We also critically provide an in-depth analysis of nutritional and mineral profile of most common alpine grasses and tree foliage and present their salient characteristics including how altitude seems to affect the nutritional contents. It was observed that the mineral content as well as proximate profiles of grasses changed with the altitude. Samples collected from higher hills were characterized by substantially higher concentrations of iron, manganese, molybdenum, and potassium as compared to the grasses from mid and lower hills. Due to their interference in copper uptake from the soil, higher concentrations of iron and molybdenum in high altitude grasses could be a dominating factor in causing copper deficiency in farm animals. Some grasses from Lahaul and Spiti region of Himachal Pradesh, besides minerals, were found to be very rich in essential amino acids. Another salient feature was higher level of crude protein in alpine grasses and tree leaves as compared to the grasses grown at lower hills. Crude protein showed a rising trend with increase in elevation. Furthermore, as the cli-

S. P. Singh, J. K. Malik, *Migratory Small Ruminant Farming System in the Himalayas-Peculiarities, Problems, Prospects, Climate Effects, Nutrition and Health*, https://doi.org/10.1007/978-981-96-6691-1_9

mate changed from subtropical to alpine, the levels of nutritionally desired components in the grasses (crude protein and certain minerals) increased, whereas the levels of incriminating components (dry matter and total ash) decreased. Most crops and crop residues collected from the foot hill regions were severely deficient in sodium causing sodium deficiency in animals. Copper levels in straws and roughages were found to be below the critical level of the ruminant requirement. In neighboring Pakistan, many grasses from hilly regions were found to have sodium, calcium, magnesium, zinc, manganese, and cobalt below the critical levels of the requirement of small ruminants. In China, several sheep producing hilly regions had herbage that was found severely deficient in sodium, copper, phosphorus, and selenium having levels below the critical limits of ruminant requirement.

Keywords

Nutritional status · Himalayan grasses · Fodder trees and crop residues · Altitude and change in mineral content · Altitude and crude protein content

Abbreviations

amsl	above mean sea level
HP	Himachal Pradesh
J and K	Jammu and Kashmir
m	meter
UT	Union territory
UK	Uttarakhand

The Himalayas with diverse climatic conditions are well known for a variety of fauna and flora. The mountain ranges rise from subtropical heat region (Shivaliks) to the alpine zone between the foot hills and Tibetan plateau. As one moves from base of the hills to the alpine heights, the tree line changes from shesham, poplar, kail (800 to 1200 m) to sal, pine, oak, Deodar, and Junipers (1500 to 2000 m). The coniferous forests begin to thin around 2000 m and birches and junipers mark the end of the tree line. The grasslands found beyond the tree line on the southern face of the Greater Himalaya (approximately above an altitude of 2500 m and up to 5000 m) are generally referred as the Alpine meadows, which form the grazing grounds for small ruminants during summer (Joshi 2020). The meadows contain highly nutritious and palatable grasses. Most of the Himalayan pastures are spread over flat or undulated sloppy terrain and are called by various local names across the mountain range. They are known as Margs or Bahaks in Kashmir, Thach in Himachal Pradesh, and Bugyal in Uttarakhand. They are also commonly called as Pabo in some areas of Himachal Pradesh and Payar in Garhwal region of Uttarakhand (Chandran 2015; Mir et al. 2016).

In north-western Himalayan states of Himachal Pradesh, Uttarakhand, and the UT of Jammu and Kashmir, the area under permanent pastures and other grazing land has been estimated at 18140 Km^2 (ENVIS Bulletin 2010). Beyond certain altitude cultivation of fodder is not practical and, for this reason, in most of the Indian Himalayan regions, the farmers depend on forest-based fodder to feed their animals. Most fodder requirements are met through grasses, tree fodders, cultivated grasslands, orchards, agroforestry (trees planted on the edges of terraces), and grazing on open pastures. (Figs. 9.1, 9.2, and 9.3).

At lower elevations, crop residues of traditional crops like paddy, maize, pulses, and wheat also contribute to fodder requirements. Paddy hay meets the bulk fodder requirement for stall feeding during winter. However, most concentrates have to be procured from other states located in planes (Mir et al. 2016). In UT of Jammu and Kashmir, use of unconventional feeds like aquatic vegetation and leaves from willow tree is quite popular. In Lahaul region of Himachal Pradesh, about 16% of total fodders requirements is met from willow tree forestry (Rawat and Vishvakarma 2011).

Due to encroachment on pasture lands, increasing grazing pressure, contamination of grasslands with weeds, and poisonous plants and soil erosion, the feed and fodder resources have been depleting with time. Consequently, almost all the north-western states have been facing problem of inadequate fodder supply to sustain increasing livestock population (Dev et al. 2006). Shortage of concentrate and green fodder is chronic in north-western hilly states where as dry fodder availability vary. In a recent review, Dhamodharan et al. (2024) reported green fodder deficit in the union territory of Jammu and Kashmir and state of Uttarakhand to the extent of

Fig. 9.1 Himalayan alpine meadows of Tungnath area (3689 m amsl) of District Rudraparyag of Uttarakhand are well known (Photograph by the courtesy of social activist, Mr. Sankar Sridhar)

Fig. 9.2 Local sheep at Munsiyari pastures, District Pithoragarh, Uttarakhand, (above mean sea level 2200 m). (Photograph by the courtesy of social activist, Mr. Vipin kumar)

Fig. 9.3 Leh area of Ladakh in cold desert zone has very scanty grass cover for grazing (above mean sea level 3200 m). (Photograph by courtesy of Director, ICAR-CAZRI, Jodhpur with permission)

24.9%, but the State of Himachal Pradesh was surplus. There was overall surplus of dry fodder availability to the extent of 55% in most states.

What is so special about alpine pastures? What attracts the shepherds and grazers to reach to the highland pastures every summer? Do the alpine grasses differ in mineral content or proximate profile compared to the grasses in tropical or subtropical region? We have attempted to answer some of these queries. As the related work is random, scattered, and not along the systematic lines, sincere efforts have been made to unravel the nutritional analysis of fodder species from the available literature from different Himalayan zones.

During the migratory cycle, the animals spend summers in the alpine pastures and start the downward journey as the winter sets in. En route to the lower heights, they thrive on the meager grasses, weeds, and tree leaves. During winter, when they spend time around the base camps in the plains, they are maintained on partial grazing or fed on crops and crop residues or tree leaves.

The objective of the present chapter is to list the common alpine grasses, fodder trees and foliages, and crops and crop residues, which form the main feed resources of the migratory livestock and assess their nutritional status at mid hills and in and around base camps. In the lower sub-mountainous sub-alpine region, the grass cover is of Themeda-Arundinella type, whereas on the higher elevations above 2000–3000 meters, the grass cover belongs to temperate alpine category (Chandran 2015). Although the Himalayas extend through many Indian states from north west to the east, the focus of the present chapter would be north-western Himalayan states of Himachal Pradesh, Uttarakhand, and UT of Jammu and Kashmir including Ladakh, where a well-defined annual migratory cycle is operational. These three states/Uts represent more than 60% of the total geographical area (GA) of the Indian Himalayas.

9.1 Main Species of Himalayan Grasses

The most common fodder species of the Himalayan states constitute the "Ghasnees" (grasslands). Research on Himalayan grasses has received.

Limited attention in the past two decades, and several investigators have focused on the different available resources of the common grasses and tree species, which serve as a source of fodder in different hilly states. Some Important grass species of mentioned Himalayan States are presented in Table 9.1.

Besides the grasses mentioned in UT of Jammu and Kashmir in Table 9.1, a distinctive feature in the area is the use of aquatic vegetation, namely, *Nymphea tetragona*, *Cares* sp., *Typha angustata,* and *Phragmites elephantoides*, as a feed source for animals (Tomar and Sharma 2002; Mir et al. 2016).

Table 9.1 Important grass species of some Himalayan States

State/UT	Region	Grass species	Reference
Jammu and Kashmir	Budgam District	*Achillea milifollfolium, Chrysopogon echinulatus, Cynodon dactylone, Dichanthium annulatum, Fragaria nubicola, Imperata cylindrical, Pennisetum orieentale, Panicum antidotale, Poa annua, Taraxacum officinate, Themada anaethera, Thymus linearis, Trifolium pretense, Tulipa stellate*	Radotra et al. (2015) and Baba et al. (2017)
	Kashmir Himalaya (Pahalgam) and adjoining area)	*Agrostis* spp., *Bromus* spp., *Bothriochloa pertusa, Chrysopogon gryllus, Cymbopogon* spp., *Cynodon dactylon, Dactylus glomerata, Digitaria* spp., *Festuca* spp., *Lolium perenne, Lespedeza sericea, Medicago* spp., *Poa* spp., *Potentilla* spp., *Trifolium pratense, T. repens*	Mir et al. (2016)
Himachal Pradesh	Mid hill region	*Agropyron* species (wheat grass), *Agrostis stolonifera, Alopecurus arundinaceous, Arundinella setosa* (Arundinella), *Bothriochloa intermedia, Chrysopogon echinulatus, Cymbopogon jwarancusa, Cynodon dactylon* (Bermuda grass/Dub), *Dactylis glomerata* (Cocksfoot), *Dichanthium annulatum, Eragrostis pappora, Festuca rubra, Imperata cylindrical, Lotus corniculatus* (Birds foot trefoil), *Pennisetum orientale* (Bari jhan), *Poa pretense* (Kentucky blue grass), *Saccharum spontaneum* (Kai)	Katoch and Dogra (1998), Misri (2003), and NATP Ann Rpt (2004–2005)
	High hill region	*Agrostis Stolonifera, Agrostis micrantha, Andropogon ischaemun* (Veeran), *Artimisia vulgaris* (Dhudhal), *Bromus inermis, Dactylis glomerata* (Juara), *Festuca alpinum (Neeru), Glyceria tonglensis, Medicago falcate* (Lucerne/yellow flower), *Pennisetum flacidum* (Krash), *Phleum alpinum* (Timothy grass) *Pinus longifolia, Poa bulbosa* (Blue grass), *Poa alpine, Poa stirillis* (Blue grass), *Poa pretense* (Kentuky blue grass), *Trifolium repens* (White clover)	Katoch and Dogra (1998), Misri (2003), NATP Ann Rpt (2004–2005)
	Lahaul Spiti	*Festuca kashmiriana L., Medicago sativa L., Melilotus indica L.*	Kumar et al. (2022)
Uttarakhand	Almora region	*Brachiaria mutica, Brachiaria ruzizensis, Cetoria cajungula, Chloris gayana, Chrysopogon zizanioides, Cymbopogon citrates, Desmodium intortum, Digitaria eriantha, Digitaria sanguinalis, Megathyrsus maximus, Paspalum mandiocanum Penicum capillare, Penicum chlorotum, Pennisetum pedicellatum, Pennisetum purpureum, Setria sphacelata*	Das (2019)

Local names of some grasses are mentioned in parenthesis.

9.1.1 Mineral Profile of Important Grasses

Only a few studies seem to have been undertaken on mineral composition of alpine grasses, although tree foliage proximate analysis and mineral contents have been thoroughly covered. But in both cases the available data is scattered and unsystematic. Most of the recent publications on the subject are confined to the nutritional and mineral profile of alpine tree leaves and tree fodders which constitute a larger source of fodder for small ruminants. In early investigative work, Sawhney et al. (1977) and Dogra et al. (1984 and 1986) have reported the mineral profile of some important alpine grasses of Jammu and Kashmir and Himachal Pradesh. Their findings are presented in Table 9.2. Results revealed striking variations in mineral contents of grasses collected from different places in these two states.

Although the sample size was not very large and there was no uniformity in experimentation for obvious difficulties, some conclusions could be drawn about the mineral profile of grasses at different locations. The salient features of the investigative work were as follows:

(i) Majority of grass species from Chhrabra region, Shimla of Himachal Pradesh, had higher iron (mean, 768.2 mg/kg DM) and Manganese (mean 60.7 mg/kg DM) contents than the grasses from other regions of the State. (Figure 9.4).
(ii) The mean molybdenum content of grasses collected from Kupa/Kanda area of Kinnaur district was markedly higher (2.68 mg/kg DM) than the grasses from Chhrabra region (1.22 mg/kg DM) of Himachal Pradesh, and some grasses, namely, *Agrosstis stolonifera* and *Brosmis inermis*, contained molybdenum as high as 6.14–6.32 mg/kg DM.
(iii) The grasses collected from Jammu and Kashmir region were characterized by lower iron (266.3 mg/kg DM) and manganese contents (36.2 mg/kg DM) compared to the respective mineral concentrations in grasses from Himachal Pradesh.

In another study, Sahoo et al. (2016) and Das (2019) who analyzed grasses and tree leaves collected from Uttarakhand region revealed far lower iron content ranging from 119.8 to 153.0 (mean value 136.4) mg/kg DM compared to the iron concentrations in studied grasses of Himachal Pradesh and Jammu and Kashmir (Sahoo et al. 2016; Das 2019).

In some recent work on alpine grasses from Lahaul and Spiti region of Himachal Pradesh, Kumar et al. (2022) studied the nutritional quality of five high altitude forage grass species, namely, Medicago sativa L., Medicago falcate L., Trifolium pratense L., Festuca kashmiriana L., and Melilotus indica L. Elemental composition revealed that calcium (13.91–132.05 mg/kg DM) was the predominant mineral followed by magnesium (4.60–12.92 mg/kg DM), iron (2.04–76.13 mg/kg DM), and zinc (1.07–2.17 mg/kg DM). Three forage grass species were also rich in essential amino acids like histidine, tryptophan, valine, and isoleucine.

Table 9.2 Mineral profile of some alpine grasses of Himachal Pradesh and Jammu and Kashmir

Location	Grass species	Total Ash	Ca	P	Iron	Mn	Zn	Cu	Mo	Reference
		g % DM				mg/Kg DM				
Chhrabra region, Simla HP (> 2685 m)	*Agrostis canina*	6.4	0.59	0.22	390.3	48.6	34.8	16.1	1.1	Dogra et al. (1986)
	Chrysopogon pseudo-phragmite	7.9	0.51	0.23	906.3	33.6	39.8	0.0	2.8	
	Chrysopogon echinulatus	6.7	0.64	0.24	840.0	33.5	37.6	24.4	1.3	
	Digitalia sp.	11.1	0.42	0.21	1004.8	83.5	35.4	20.1	0.74	
	Festuca kashmiriana	6.4	0.72	0.22	1205.9	113.5	48.1	20.2	0.86	
	Hetropogon lanatus	8.5	0.41	0.22	667.7	120.2	93.5	12.0	0.85	
	Microstegium ciliatum	8.2	1.02	0.23	774.5	67.3	37.7	19.6	1.4	
	Pennisetum orientale	9.5	0.51	0.22	535.9	13.4	34.8	0.0	1.1	
	Themeda anathera	7.2	0.52	0.22	588.9	33.3	29.3	0.0	0.85	
Mean ± SE		7.9 ± 0.52	0.59 ± 0.6	0.2 ± 0.01	768.2 ± 84.2	60.7 ± 12.6	43.4 ± 6.5	12.4 ± 3.3	1.22 ± 0.2	

Kupa/Kanda, region, Kinnaur, HP (> 3356 m)	*Agrostis stolonifera*	7.52	0.45	0.21	479.7	54.8	46.6	23.3	6.14	Dogra et al. (1984)
	Bromus inermis	10.1	0.44	0.21	579.3	34.1	43.6	26.1	6.32	
	Carex sp.	7.71	0.62	0.21	480.5	26.8	29.5	6.1	0.42	
	Dactylis glomerata	7.43	0.40	0.21	337.8	32.7	35.8	14.9	2.21	
	Danthonia jacguemontii	4.42	0.75	0.20	232.9	26.6	29.6	23.9	0.64	
	Glyceria tonglensis	9.56	0.43	0.18	539.9	121.5	48.7	10.2	1.83	
	Koeleria Cristata	11.8	0.60	0.20	172.3	47.7	154.1	18.4	1.20	
Mean ± SE		8.3 ± 0.90	0.5 ± 0.04	0.20 ± 0.0	403.2 ± 59.5	49.2 ± 12.7	55.4 ± 16.7	17.5 ± 2.8	2.68 ± 0.95	
Jammu region J and K (> 750 m)	*Cynodon dectylon*	–	–	–	256.3	42.7	39.4	17.8	–	Sawhney et al. (1977)
	Dichanthium annulatum	–	–	–	311.8	51.5	20.4	9.8	–	
	Medicago sativa	–	–	–	219.5	33.8	27.1	13.8	–	
	Pennisetum purpureum	–	–	–	181.3	30.2	28.0	16.5	–	
	Trifolium alexandrinum	–	–	–	433.2	29.3	85.6	45.4	–	
	Trifoleum repens	–	–	–	195.5	30.1	27.4	20.5		
Mean ± SE		–	–	–	266.3 ± 15.7	36.2 ± 1.5	37.9 ± 4.0	20.6 ± 2.1	–	

Values in parentheses are height(m) from mean sea level

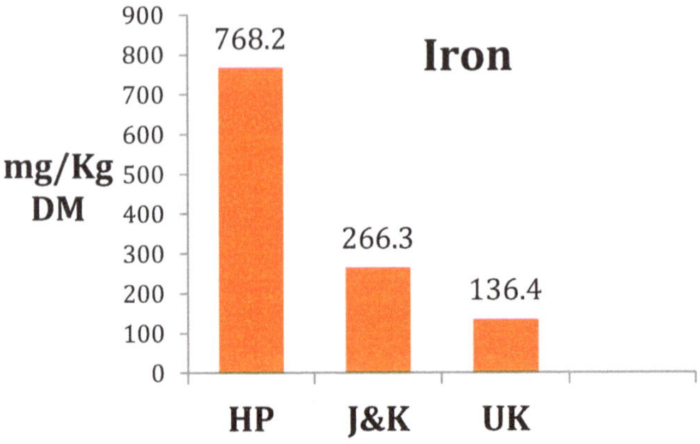

Fig. 9.4 Alpine grasses of HP contained higher mean iron content than the grasses from J&K and UK (Dogra et al. 1986, (Sahoo et al. 2016; Das 2019)

9.1.2 Variation in Mineral Profile of Grasses with Altitudes

The grasses, fodder trees, and shrubs on Himalayan ranges are distributed at different elevations. There are reports that the nutritive contents of plant species not only change from season to season but also on the elevations at which they are growing (Sharma and Puri 2022). Navale et al. (2022) who evaluated the nutritional characteristics of multipurpose fodder tree species in north-western mid hills of Himachal Pradesh reported that nutritive and mineral contents decreased as the leaves matured. As early as 1992, Morecroft et al. (1992) had reported that nitrogen concentration in plants increased with altitude gradient. In very recent findings, Singh et al. (2022) working on chemical composition of *Celtis australis* tree foliage in Garhwal region of Uttarakhand recorded a significant ($P < 0.05$) and positive correlation between altitude and crude protein. Foliage crude protein values at the elevation of 1980 m above mean sea level were recorded higher (14.2%) compared to the value at lower height of 550 m (12.66%). Potassium and calcium content of foliage were also found related with altitude. Minocha et al. (2021) investigating the foliar metabolism of Himalayan cedar and Himalayan Spruce trees in Himachal Pradesh reported that cedar foliage from higher elevation site had higher concentration of potassium, amino acids, soluble proteins, and some other minerals.

We have found that the mineral profile of some grasses, to a large extent, was affected by change in altitude. For this study, the samples were collected from places located at different elevations in the states of Himachal Pradesh and UT of Jammu and Kashmir. Salient results are presented in Table 9.3.

The grass species collected from places located at higher elevations were characterized by substantially higher concentrations of iron and manganese compared to the grasses from lower hills. The iron content in some grasses like F. kashmiria in Chhrabra region was recorded as high as 1205.9 mg/kg DM compared to the grasses from lower elevation of Jammu (203.9 mg/kg DM). Similarly, manganese

Table 9.3 Mineral status of grasses collected from different altitudes in HP and UT of J and K

State/place/ elevation	Grass species	Mean mineral content mg/Kg DM					
		Iron	Mn	Cu	Mo	Cu:Mo	
J and K Ravi/Indus river basin (450 m)	Avena sativa, Morus laevigata, Albizzia stipulate, Triticum aestivum, Butea monosperma, Grewia elastic, Pennisetum purpureum	198.1	38.8	12.7	–	–	Sawhney et al. (1977)
HP Bara Banghal, Kangra (3170 m)	Andropogon ischaemum, Agropuon spp., Agrostis stolonifera, Dactylis glomerata, Poa triandra, Pennesetum flacidum	542.1	94.7	6.0	–	–	Dogra et al. (1994)
HP Chhrabra region (2685 m)	A. canina, C. oseudophrgmites, C. echinulatus, Digitaria spp., F. kashimiria, H. lanatus, M. ciliatum, M. vagans, P. orientale, T. anathere	778.5	62.0	13.4	1.1	12.2	Dogra et al. (1986)
HP Gumshal, (3515 m)	Agrostis stolonfera, Bromis inermis, Carex spp., Dactylis glomerata, Koeleria cristata	409.9	38.1	17.6	3.7	5.0	Dogra et al. (1984)
HP Kandah area (3810 m)	Agropyron canaliculatus, A. schugnanicum, Dactylis glomerata, Danthonia jacguemontii, Glyceria tonglensis	476.9	49.0	16.7	2.4	12.9	Dogra et al. (1984)

concentration in the grasses grown on higher hills was of higher order than the values in grasses from lower elevations.

9.1.3 Altitude and Crude Protein (CP)

We have previously reported how the mineral contents and some other nutritive parameters of alpine grasses and foliages are affected by the altitude. It was interesting to observe similar pattern in case of crude protein. The CP level in the grasses also seemed to show a rising trend with the increase in elevation from sea level. Only limited investigative work seems to have been undertaken on the effect of altitude gradient on various nutritional components. Koutsoukis and Akrida-Demertzi (2020) reported that crude protein content significantly ($p < 0.05$) varied among the altitudinal zones. Singh et al. (2022) who worked on Celtis australis foliage also reported that altitude had significant influence on nutritive contents.

A few investigators in Himachal Pradesh have investigated the crude protein level in samples of grasses and foliages collected from lower hills and sub-temperate regions (Katoch et al. 1992). The crude protein (CP) levels were found to vary from 7.57% to 9.0 g% on DM basis. The individual values in some prominent grass species were recorded as follows: Agrostis canina (8.32%), Chrysopogon echinulatus

(8.13%), Themada ananthera (7.57%), M. vagus (9.0%), and H. lantus (8.9%) (Dogra et al. 1986; Katoch et al. 1992). Identical trend in crude protein has also been reported in the pasture and forage samples from lower and higher hills of north-western Himalayan region by Radotra et al. (2015). In their studies, crude protein level rose from 4.3 to 10.1 g%, whereas NDF and ADF declined from 69.2 to 58.3% and 43.1 to 35.1%, respectively, in the samples collected from lower and higher hill regions.

Most alpine grasses at higher elevations were found to have significantly higher crude protein content compared to the level in grasses from lower hill regions. On the other hand, the dry matter content was reported to be lower in the alpine grasses than the grasses grown in sub-temperate region. Aquatic vegetation, which is quite popular as a fodder source for livestock in Jammu and Kashmir, was found to be quite rich in crude protein content. Table 9.4 presents the crude protein level in tree leaves and grasses collected from different elevations from three north-western Himalayan UT/states.

Table 9.4 Crude protein level in sub-alpine and alpine grasses and tree leaves of some UT/Himalayan states

State/place/elevation	Name of grass/tree	Crude protein (g%DM)	Reference
UT of Jammu and Kashmir Anantnag/Baramula 1600 m amsl	*Alanthus*	15.9	Tomar and Sharma (2002)
	Robinia	18.5	
	Willow	15.8	
	Nymphaea[a]	19.4	
	Phragmites[a]	13.5	
UT of Ladakh >2700 m amsl	*Morus alba*	15.79	Sheikh et al. (2011)
	Salix tetragonal	16.40	
	Elaeagnus angustifolia	15.64	
	Ephedra gerardiana	16.29	
	Caragana vesicular	19.21	
	Lepidium latifolium	27.35	
Himachal Pradesh Kullu 1279 m amsl	*Bauhinia variegata*	15.6	Samant et al. (2007)
	Celtis australis	14.0	
	Pueraria tuberosa	23.8	
Lahaul spiti Six different sites 3283–3661 m amsl	Medicago falcate	12.96	Kumar et al. (2022)
	Medicago sativa	12.12	
	Trifolium pretense	14.71	
	Festuca kashmiriana	12.59	
	Medicago sativa	12.12	
Uttarakhand Almora, 1642 m amsl	*Pennisetum purpureum*	15.8	Das (2019)
	Quercus serrate	18.5	
	Cetoria cajungula 16.8		

[a] Aquatic grasses

Kumar et al. (2022) also reported that protein content in the high altitude forage species was very rich in essential amino acids like histidine, tryptophan, valine, phenylalanine, tyrosine, leucine, and to some extent tyrosine.

9.1.4 Studies from Neighboring Countries

There are very few reports on the mineral status of grasses and foliages in the mountainous regions of neighboring Pakistan and China. Many grasses from hilly regions of Pakistan were found to have sodium, calcium, magnesium, zinc, manganese, and cobalt below the critical levels of the requirement of small ruminants (Inam-ur-Rahim et al. 2008; Mirzaei 2012; Khan et al. 2012). Similarly, the grasses from hilly regions of China were found to be deficient in sodium and phosphorus having levels below the critical limits of ruminant requirement (Table 9.5). In China, several

Table 9.5 Mineral status of forages (grasses/tree leaves) in some sheep rearing areas of neighboring Pakistan and China

Mineral element	Pakistan			China
Macro minerals	Foot hills	High hill	Kohat region	High hills
	Macro minerals (g%/DM)			
Na	0.094[a] (0.006-0.021)	–	0.026[a] (0.02-0.04)	0.07–0.09[a]
K	1.43 (0.12-2.8)	0.53 (0.40–0.75)	1.03 (0.70–1.55)	0.6–2.2
Ca	0.41 (0.21–0.58))	0.32 (0.09–0.72)	2.32 (0.22–5.14)	0.44–2.2
P	–	0.041 (0.03–0.051)	0.102 (0.05–0.14)	0.16–0.20[a]
Mg	0.21 (0.17-0.29)	0.007 (0.002–0.01)	–	0.09–0.48
Micro minerals	*Micro minerals (mg/kg DM)*			
Iron	623.6 (265.0 —894.0)	–	186.17 (152.2–205.9)	4000. 0
Zn	41.3 (35.0–48.0)	4.30[a] (1.9-10.5)	25.74 (24.5–131.3)	14.6–16.1
Cu	6.7 (6.40–7.00)	18.4 (3.0–33.0)	3.32 (2.4–4.7)	1–5[a]
Mn	179.3 (106.0-217.0)	4.67[a] 1.5–9.4)	91.69 (42.7–134.4)-	–
Reference	Mirzaei (2012)	Inam-ur-Rahim et al. (2008)	Habib et al. (2013)	Shunxiang (1995)

Values in parentheses show ranges

[a] Concentration below the critical level of ruminant requirement. For Na, 0.7–0.9, for P, 0.9–3.8 (g/kg DM), for Cu, 4–14, for Mn, 20–25(mg/kg DM) ((Freer et al. 2007)

sheep producing regions were severely deficient in selenium, whereas iron was in excess with levels as high as 4000 ppm. In fact the high iron content was responsible for widespread copper deficiency (Shunxiang 1995).

9.2 Common Fodder Trees of North-Western Himalayan Region

Several species of fodder-yielding plants are present in the north-western Himalayan region, which are known to play a significant role to meet the deficiency of animal feeds. The nomads and local inhabitants extract the fodder by cutting the floor vegetation and lopping the trees. The trees, shrubs, and herbs contribute to the fodder supply to the extent of 51%, 12, and 29%, respectively. Table 9.6 presents the common tree species found in hills of north-western Himalayas whose leaves/foliage constitute the major fodder resource for the livestock.

9.2.1 Proximate Analysis and Mineral Content

The crude protein, calcium, phosphorus, dry matter, and total ash content of tree foliage collected from different regions of Himachal Pradesh and Uttarakhand varying in altitude from 560 to 3515 m are compared in Table 9.7.

The table reveals that most foliage from high hill trees were found to contain considerably high content of crude protein to the extent of 24.4 g% (in *Grewia optiva)*, whereas the dry matter was as low as 29.7%. Significant increase in the protein contents with elevation has also been shown in forage crop *Leymus secalinus* in Minhe County of the Qinghai-Tibetan Plateau (Cui et al. 2018). Singh et al. (2022) have also reported that values of nutritive parameters of foliage are influenced by altitude. The crude protein value in foliage of C. australis was recorded as 12.66% (at the elevation of 560 m) and 14.02% at (1980 m asl), respectively. It has been suggested that increase in soluble proteins with altitude may raise the functional protein content and ensure cell metabolism (Cui et al. 2018). Obviously, forages containing high percentage of crude protein may serve as source of rich nutrition for grazing sheep.

Recently, Sahoo et al. (2016) reported the chemical and mineral profile of foliage of 11 high hill tree species in the state of Uttarakhand including *Grewia oppositifolia*, *Ficus* species, *Celtis austrails*, and *Bahunia variegate.* The foliage from high hill trees were characterized by very high crude protein level varying from 13.3 to 21.1 g% DM. It was interesting to note that foliage of trees collected from Uttarakhand contained significantly lower concentrations of iron, zinc, copper, and molybdenum but higher manganese content compared to the tree foliage from Himachal Pradesh (Table 9.8).

Table 9.6 Common fodder trees of north-western Himalayan states

UT/State	Region	Tree species	Reference
Jammu and Kashmr and Ladakh	Kashmir Himalaya(near pahalgam)	Willow, Alanthus, Robinia, Leaves of lotus plant, Azola, Lemna, water chestnut, hydrilla, pistia, aquatic weeds, etc.	Mir et al. (2016)
	Kathua area	*Aleizia lebbeck,Bauhinia variegata, Dalbergia sissoo, Dendrocalamus strictus, Dendrocalamus strictus, Ficus palmate, Grewia optiva, Leucaena leucocephala, Quercus leucotrichophora, Robinia pseudo-acacia, Salix alba, Terminalia bellirica*	Gupta (2014)
	Ladakh area	*Chenopodium alvum, Popalus tremula, Elaeagnus angustifolia, Hippophae rhamnoides, Prunus armeniaca, Ephedra gerardiana, Poa pratensis, Lepidium lati*	Sheikh et al. (2011)
Himachal Pradesh	Sirmour	Acacia catechu, A. nilotica, *Albezia lebbeck, A. procera, Bauhinia variegata, B. malabarica, B. vahlii, Butea monosperma, Cassia fistula*	Saini and Sood (2018)
	Kullu	Aesculus indica, Albizzia stipulate, Bauhinia vahlii, B. variegata, Bothriochloa intermedia, Celtis australis, Grewia oppositifolia, Morus serrate, M. alba, Pueraria tuberosa, *Quercus leucotrichophora, Q. glauca, Robinia pseudacasia, Salix tetrasperma, T. Viburnum cotonifolium*	Samant et al. (2007) and Katoch et al. (1992)
	Solan	Acasia catechu, Leucaena leucocephala, Melia composita, M. serrate, O. glandulifera, Albizia chinensis, Celtis australis, Ficus roxburghii	Navale et al. (2022)
Uttarakhand	Mukteshwar	*Bauhinia variegate, Ficus nemoralis; Ficus palmate; Ficus roxburghii; Grewia oppositifolia; Quercus leucotricophora; Quercus floribunda; Quercus glauca; Celtris australis*	Sahoo et al. (2016)
	Garhwal	*Celtis australis L*	Singh et al. (2022)
	Almora (ICAR hill Institute)	Alnus nepalensis, Bauhinia variegate, Celtis australis, Grewia optiva, Moras alba, Prunus cerasoides, *Quercus leucotricophora; Quercus floribunda; Quercus glauca,* Robinea seudoacacia	Das (2019)
	Almora (mid hill villages of Almoda District)	Melia azedarach L., Debregeasia longifolia, Alnus nepalensis, Bauhinia variegata, Bauhinia retusa L Celtis australis, Grewia optiva, Morus alba, Prunus cerasoides, *Quercus leucotricophora; Quercus glauca, Quercus serrata Murray,* Robinea pseudoacacia	Mukherjee et al. (2018)

Table 9.7 Values of proximate analysis and macro mineral profile of some fodder tree leaves in the state of Himachal Pradesh and Uttarakhand

State/location (Elevation in m from msl)	Name of fodder tree/foliage	Crude Proteing%	Dry Matter%	Total Ash%	Ca g % in DM	P g% in DM	Reference
Himachal Pradesh (Sangla, Keylong) (3515 m)	*Salix dephnoides* (Beuns)	16.8	29.7	9.65	2.07	0.23[a]	Dogra et al. (1985)
	Desmodium tiliaefolium	18.2	30.5	6.46	0.63	0.22[a]	
	Indigofera gerardiana	20.1	35.4	7.84	1.52	0.22[a]	
Himachal Pradesh (Palampur, Kangra) 1350 m asl	*Grewia optiva*	24.4	–	12.8	3.27	0.20[a]	NATP (2004–2005)
	Albizia stipulate	22.3	–	5.93	1.03	0.14[a]	
	Leucaena leucocephala	21.1	–	11.0	2.41	0.19[a]	
	Acacia catechue	21.0	–	9.66	2.46	0.14[a]	
	Bauhinia verigata	20.6	–	7.40	1.76	0.22[a]	
	Dentdrocalamus Hamiltonii	19.8		13.7	0.76	0.17[a]	
	Artocarpus lakocha	19.4	–	13.3	2.19	0.18[a]	
	Morus alba	17.8	–	16.9	3.62	0.23[a]	
	Quercus incana	12.9	–	6.43	1.65	0.11[a]	
Himachal Pradesh (Chitkul, Simla) 3400 m asl	*Acer pictum Thunb.*	6.57	–	6.13	7.71	0.12[a]	Radha et al. (2021)
	Euphorbia pilosa L.	4.34	–	4.6	6.46	0.30[a]	
	Urtica dioica L.	4.74	–	4.73	3.29	0.21[a]	
	Quercus ilex L.	5.74	–	8.3	4.17	0.22[a]	
	Berberis asiatica Roxb. ex. DC	5.28	–	5.4	4.39	0.26[a]	
Uttarakhand, Almora (1250 m)	Robinia pseudoacacia L	26.7	4.7452.2	5.61	–	–	Mondal et al. (2016)
	Quercus glauca	13.7	60.0	6.88	–	–	
	Tushar	19.2	39.9	16.5	–	–	
	Bahuhinia retusa L	17.6	58.8	14.1	–	–	
	Melia azedarach L	20.3	39.8	19.06	–	–	
Uttarakhand, Garhwal region (560 m)	Celtis australis	12.66	41.28	16.36	2.87	0.11[a]	Singh et al. (2022)

[a]Values below critical level of ruminant requirement. For P, 0.9- 3.0 g/kg DM ((Freer et al. 2007)

Table 9.8 Mineral composition of foliage collected from Himachal Pradesh and Uttarakhand

State/location (elevation in m from msl)	Minerals (mg/kg DM) Iron	Manganese	Zinc	Copper	Molybdenum	Reference
Himachal Pradesh keylong, Kupra, Sangla Division, (3510 m)	425.1 (324.3–478.6)	52.8 (27.3–87.7)	81.8 (41.0–129.6)	15.9 (10.1–20.4)	2.4 (0.43–3.7)	Dogra et al. (1985)
Himachal Pradesh Kinnaur, Chitkul range	216.3 (161.7–322.7)	62.7 (46.5–87.0)	42.9 (18.0–38.7)	52.1 (34.3–69.9)	–	Radha et al. (2021)
Uttarakhand Mukteshwar, (1000–2500 m)	153.0 (121.0–177.0)	144.0 (42.0–402.0)	52.0 (32.0–83.0)	6.7 (5.2–7.9)	0.61 (0.51 0.71)	Sahoo et al. (2016)

Table 9.9 Mineral status of common crops, crop residues, and grass leaves offered to small ruminants in foot hill areas

Item	Macro mineral (g%)					Micro mineral (mg/kg)				Reference
	Na	K	Ca	P	Mg	Fe	Zn	Cu	Mn	
Wheat	0.070[a]	2.05	0.34	–	0.23	712.0	48.0	6.4	217.0	Mirzaei (2012)
Barley	0.060[a]	1.57	0.53	–	0.19	894.0	41.0	6.8	215.0	
Oat hay	0.017[a]	1.64	0.18[a]	–	0.16[a]	265.0	35.0	7.0	106.0	
Grasses Leaves	0.094[a]	14.2	0.41	–	0.21	563.0	36.8	6.5	168.5	
Wheat straw	–	–	0.40	0.14	0.10[a]	271.3	43.0	3.5[a]	33.6	Tiwary et al. (2007)
Green fodder	–	–	0.32[a]	0.26[a]	0.11[a]	317.5	64.1	6.7	60.4	
Roughage	–	–	0.18[a]	0.27	0.15[a]	437.5	108.5	6.8	117.1	

[a] Values are below the critical level of ruminant requirement. For Na 0.7–1.0, Ca 1.4–7.0, P 0.9–3.0, Mg 0.9–1.2(g/kgDM), and Cu 4–14(mg/kgDM) (Freer et al. 2007)

9.3 Mineral Status of Crops and Crop Residues in Foot Hill Regions of Himalayas

Winter months are very hard for the shepherds as the fodder supply becomes very scarce. The migratory sheep and goats thrive on crops, crop residues, tree leaves, weeds, and whatever other fodders are available during winter months after their return from alpine areas. There are a very few systematic studies on the mineral status of these stray feeds and fodders, and whatever information is available is scattered and incomplete. Table 9.9 summarizes the mineral status of crops, crop residues, tree leaves, and weeds collected from foot hill mountain areas of India (Tiwary et al. 2007) and Pakistan (Mirzaei 2012) .

As is evident from the available studies, most crop residues and hay were severely deficient in sodium, calcium, and magnesium. Copper level in wheat, barley, and oat hay were on borderline, whereas its concentration in wheat straw was below the critical level of the ruminant requirement. Other minerals such as potassium, iron, zinc, and manganese were optimum or in excess of the ruminant requirement.

References

Baba AA, Geelani SN, Saleem I, Husain M (2017) Pytosociological status of the selected sites (Protected site) for assessing the effect of grazing in Kashmir Valley, India. J Pharmacogn Phytochem 6(4):388–393

Chandran M (2015) Grassland vegetation of India: an update. In: Rawat GS, Adhikari BS (eds) Ecology and management of grassland habitats in India ENVIS bulletin: wildlife and protected areas, vol 17. Wildlife Institute of India, Dehradun, pp 12–27. https://www.researchgate.net/publication/293098333

Cui G, Li B, He W, Yin X, Liu S, Lian L, Zhang Y, Liang W, Zhang P (2018) Physiological analysis of the effect of altitudinal gradients on *Leymus secalinus* on the Qinghai-Tibetan Plateau. PLoS One 13(9):e0202881. https://doi.org/10.1371/journal.pone.0202881

Das SK (2019) Qualitative evaluation of fodder trees and grasses in hill region. J Krishi Vigyan 7(2):276–279

Dev I, Misri B, Pathania MS (2006) Forage demand and supply in western Himalaya: a balance sheet for Himachal Pradesh. Indian J Anim Sci 76(9):720–726

Dhamodharan J, Bhuvaneshwari S, Sowmiva S, Chinnadurai R (2024) Revitalizing fodder production: challenges and opportunities. Int J Res Agron 7(1):208–217

Dogra KK, Khatta VK, Katoch BS (1984) Macro and trace mineral status of some indigenous forages of sub-alpine regions of Himachal Pradesh. Paper presented at Proceedings of the symposium on recent advances in mineral nutrition, Oct 29–31, 1984, Department of Animal nutrition, Haryana Agriculture University, Hissar, pp 220–226

Dogra KK, Khatta VK, Katoch BS (1985) Nutrient composition, trace minerals and in vitro dry-matter digestibility of some top foliages of sub-alpine regions of Himachal Pradesh. Indian J Anim Sci 55(3):220–222

Dogra KK, Khatta VK, Katoch BS (1986) Nutrient composition, trace minerals and in vitro dry matter digestibility of some indigenous grasses of Chhrabra region of Himachal Pradesh. Forage Res 12(1):27–32

Dogra KK, Katoch BS, Singh B (1994) Botanical composition, nutritional status and carrying capacity of native alpine pastures of Bara Bangahal for sheep. Indian J Anim Sci 64(5):489–492

ENVIS Bulletin (2010) Compendium of statistics for the states of Indian Himalayan Region, official web portal of GB Pant National Institute of Himalayan Environment and sustainable development (http://www.gbpihad.gov.in), Kosi-Katamal, Almora

Freer M, Dove H, Nolan JV (eds) (2007) Nutrient requirements of domesticated ruminants. CSIRO Publishing, Victoria

Gupta SK (2014) Fodder trees of district Kathua. J&K GJBAHS 3(3):171–177

Habib G, Saleem M, Hameed A (2013) Mineral composition of local tree leaves for feeding sheep and goats in Kohat district of Khyber Pakhtunkhwa. Sarhad J Agric 29(1):97–103

Inam-ur-Rahim SJI, Yaqoob M, Nawaz H, Javed I, Hameed M (2008) Mineral profile, palatibility and digestibility of marginal land grasses of Trans-Himalayan grasslands of Pakistan. Pak J Bot 40(1):237–248

Joshi H (2020) The Bugyals or Himalayan alpine meadows: A treasure trove endangered. india.mongabay.com

Katoch BS, Dogra KK (1998) Present scenario and strategies for improving the nutrional status and carrying capacity of the natural grasslands/pastures of Himachal Pradesh. Monograph, Department of animal nutrition, College of veterinary and animal science, HPKV, Palampur

Katoch DC, Dogra KK, Bhandari JC (1992) Present status, future perspectives and strategies for pasture improvement in Himalayas. Monogr Strat Forage prod Imp. pp 99–118

Khan ZI, Ahmad K, Danish M, Mirzaei F (2012) Minerals profile of forages for grazing ruminants in Pakistan. Open J Anim Sci 2(3):133–141

Koutsoukis CA, Akrida-Demertzi K (2020) The effect of altitude on the botanical composition of forage and the variation of its chemical composition in a subalpine grassland. IOSR-JAVS 13(7):36–48

Kumar R, Joshi R, Kumar S, Srivatsan V, Satyakam CA, Patial V, Kumar S (2022) Nutritional quality evaluation and proteome profile of forage species of Western Himalaya. Grassl Sci 68(3):214–225

Minocha R, Contosta AR, Lawrence GB, Kohli RK, Minocha SC, Long S (2021) Changes in soil chemistry and foliar metabolism of Himalayan Cedar (*Cedrus deodara*) and Himalayan Spruce (*Picea smithiana*) along an elevational gradient at Kufri, HP, India: The potential roles of regional pollution and localized grazing. Forests 12(4):400. https://doi.org/10.3390/f12040400

Mir NH, Ahmad S, Verma DK (2016) Livestock rearing–sectorial status and fodder-feed strategies in Kashmir Himalaya. Ann Biol 32(2):253–259

Mirzaei F (2012) Minerals profile of forages for grazing ruminants in Pakistan. Open J Anim Sci 2(3):133–141. http://www.scirp.org

Misri B (2003) Migratory goats and sheep rearing in Himachal Pradesh, India. In: Suttie JM, Reynolds SG (eds) Transhumant grazing systems in Temperate Asia, Plant production and protection series No. 31. Food and Agriculture Organization of the United Nations, Rom

Mondal T, Yadav RP, Bisht JK, Meena VS, Kumar S (2016) Nutritional analysis of locally preferred tree leaves in mid hill of Himalayas. Presented at Indian Ecological Society International Congress, Shere Kashmir University of Agriculture Sciences, Jammu. http://www.researhgate.net

Morecroft MD, Woodward FI, Marrs RH (1992) Altitudinal trends in leaf nutrient contents, leaf size and S13 C *Alchemilla alpina*. Funct Ecol 6:730–740

Mukherjee A, Mondal T, Bisht JK, Pattanayak A (2018) Farmers' preference of fodder trees in mid hills of Uttarakhand: a comprehensive ranking using analytical hierarchy process. Range Manag Agrofor 39(1):115–120

NATP (National Agricultural Technology Project) (2004–2005). Hill and mountain Agroecosystem. Final Report Central soil and water Research and Training Institute, Dehradun, Indian Council of Agriculture Research(ICAR), Uttarakhand. www.icar.gov.in

Navale MR, Bhardwaj DR, Bishist R, Thakur CL, Sharma S, Sharma P, Kumar D, Probo M (2022) Seasonal variations in the nutritive value of fifteen multipurpose fodder tree species: A case study of north-western Himalayan mid-hills. PLoS One 17(10):e0276689. https://doi.org/10.1371/journal.pone.0276689

Radha KM, Puri S, Pundir A, Bangar SP, Changan S, Choudhary P, Parameswari E, Alhariri A, Samota MK, Damale RD, Singh S, Berwal MK, Dhumal S, Bhoite AG, Senapathy M, Sharma A, Bhushan B, Mekhemar M (2021) Evaluation of nutritional, phytochemical, and mineral composition of selected medicinal plants for therapeutic uses from cold desert of Western Himalaya. Plants (Basel) 10(7):1429. https://doi.org/10.3390/plants10071429

Radotra S, Dev I, Ahmad S, Kannan A (2015) Pasture and forages in north western Himalayan region: current status and future strategies. Paper presented at Grassland conference "Status and future prospects of Animal Production in Himalayan Region". http://www.researchgate.net/publication/286363506

Rawat YS, Vishvakarma SCR (2011) Pattern of fodder utilization in relation to sustainability under Indigenous agroforestry systems, north-western Himalaya, India. Environ We Int J Sci Tech 6:1–13

Sahoo B, Garg AK, Mohanta RK, Bhar R, Thirumurgan P, Sharma AK, Pandey AB (2016) Nutritional value and tannin profile of forest foliages in temperate sub-Himalayas. Range Manag Agrofor 37(2):228–232

Saini JS, Sood SK (2018) Ethnobotanical enumeration of forage plant species in and around Colonel Sher jung National Park, Simbalbara (CSJNPS), Sirmour (H.P.). India Int J Adv Sci Res Manag 3(8):133–140

Samant SS, Singh M, Lal M, Pant S (2007) Diversity, distribution and prioritization of fodder species for conservation in Kullu District, Northwestern Himalaya, India. J Mt Sci 4(3):259–274

Sawhney PC, Bedi SPS, Khan SA (1977) A note on trace element contents in the fodders of Jammu and Kashmir State. Indian J Anim Sci 47(4):229–233

Sharma S, Puri S (2022) Nutrients composition changes in leaves of *Quercus leucotrichophora* from different seasons and altitudes in the North-West Himalaya. Preprints 2022070293. https://doi.org/10.20944/preprints202207.0293.v1

Sheikh GG, Ganie M, Ganie AA (2011) Nutritional evaluation of some tree leaves, feeds and fodders of Ladakh. Indian J Anim Nutr 28(4):427–431

Shunxiang Y (1995) Review of mineral research in grazing sheep in China. In: Mineral problems in sheep in northern China and other regions of Asia. Proceedings of a workshop held in Beijing. pp 13–17. https://ageconsearch.umn.edu/record/134748?ln=en&v=pdf

Singh B, Kumar M, Cabral-Pinto MMS, Bhatt BP (2022) Seasonal and altitudinal variation in chemical composition of *Celtis australis* L. tree foliage. Land 11:2271. https://doi.org/10.3390/land11122271

Tiwary MK, Tewari DP, Mondal BC, Kumar A (2007) Macro and micro mineral profile in soil, feeds and animals in Haridwar district of Uttarakhand. Anim Nutr Feed Technol 7(2):187–195

Tomar SK, Sharma RL (2002) Fodders and feeding practices of cattle and sheep in Kashmir (India). TARE (Tropical Agricultural Research and Extension) 5(1–2):48–52

Part V

Parasitic Infestations and Disorders of Small Ruminants on Indian Himalayas

10 Helminths and Other Diseases Affecting Migratory Small Ruminants

Abstract

Heavy parasitism, parasitic or verminous bronchitis, peste des petits ruminants (PPR), and chlamydiosis have been recognized as serious productive constraint in migratory small ruminants of the Indian Himalayan states. Studies have revealed that over 90% of gastrointestinal helminth infections belonged to the order *Strongylida* and the remaining to the other orders. When ranked for infectivity, *Haemonchus* species was found to be the most prevalent nematode in all the studied states and union territories (UT). The second prevalent Nematodes in the infesting order in Jammu and Kashmir and the states of Himachal Pradesh, Uttarakhand, Sikkim, and Meghalaya were recorded as *Chabertia*, *Ostertagia*, *Trichuris*, *Bunostomum*, and *Ostergia*, respectively. Overall, the prevalence rate of infestation in different states varied from 61.9% to 96%. In north-western states of Himachal Pradesh and Uttarakhand and union territories of Jammu and Kahmir and Ladakh, parasitic or verminous bronchitis caused by *Dictyocaulus filaria* ranged from 18.7–43.5% in sheep and goats, respectively, leading to heavy productive losses. Other serious problems in the migratory animals were persistent abortions caused by *Chlamydia* infection and peste des petits ruminants. Goats were found to be more susceptible than sheep and infection rate was higher in males than females. As compared to the north-western states, the north-eastern states of Sikkim and Meghalaya had some peculiarities. For instance, in these states, the infection of *Moniezia* and *Eimeria* was very high and no incidence of *Fasciola* infestation was reported. Besides, two new species of *Oesophagostomum*, namely, *O. venulosum* and *O. aspersum* were reported from the north-eastern states.

S. P. Singh, J. K. Malik, *Migratory Small Ruminant Farming System in the Himalayas-Peculiarities, Problems, Prospects, Climate Effects, Nutrition and Health*,
https://doi.org/10.1007/978-981-96-6691-1_10

Keywords

Helminthiasis in migratory small ruminants · Features of parasitic infestations · Dominant parasites · Parasitic bronchitis · Persistent abortions · Peste des petits ruminants (PPR)

Abbreviations

EPG	Egg per gram
FMD	Foot-and-mouth disease
OP	Overall prevalence
PPR	Peste des petits ruminants
UK	Uttarakhand
UT	Union territory

Small ruminants such as sheep and goats are the main source of livelihood for small and marginal farmers of both Indian north-western states of Himachal Pradesh and Uttarakhand and union territories of Jammu and Kashmir and Ladakh and north-eastern states of Sikkim and Meghalaya. As stated earlier, out of several rearing constraints, shortage of feed and fodder has been the main impediment forcing these animals for seasonal migration to alpine grasslands during summer and back to the plains in the winter season. The migration pattern is well established for generations especially in the north-western mountain states. This peculiar ruminant production system in the hills has a few special features which also reflect on the health and disease status of migratory livestock in the long run. Some of these unique characteristics include the following:

(a) Exposure of sheep and goats to long harsh environment, variety of natural vegetation, unhygienic water resources, companionship with large farm ruminants and wild animals during upward and downward migration to and from alpine pastures.
(b) Exposure of migratory animals to infections during grazing in alpine pastures and during winter stay at the base camps.
(c) Seasonal migration and constant movement of animals facilitate the transmission of infectious agents across the migratory routes and thus the infectious agents are introduced to the newer territories.
(d) Inability or difficulty to adopt or implement appropriate disease preventive and control measures/treatments due to constant movement and long migratory routes.
(e) Exposure of animals to toxic and poisonous plants and vegetation during en route journeys and while grazing on the high-altitude pastures.

Apart from traditional infectious and noninfectious diseases and occasional disease outbreaks, heavy parasitism is a serious and constant problem in migratory animals leading to heavy economic losses. Diseases prevailing in the migratory sheep and goats including heavy gastrointestinal parasitism have been well recognized but have not been systematically studied with respect to specific causative agent (region wise), prevalence rate (overall and parasite wise), factors affecting the prevalence, magnitude of economic losses, and problems related to control measures. Besides heavy parasitism, which accounts as a major cause of high morbidity (Wani 2015), some other diseases such as peste des petits ruminants (PPR), brucellosis, chlamydiosis, foot-and-mouth disease (FMD), rinderpest, pasturellosis, scabies, and alopecia and poisoning due to toxic plants have seriously affected health and growth performance of migratory sheep and goats. Also there is hardly any systematic work on wild vegetation–related poisonings which are of common occurrence in the migratory livestock. One such incident has been reported due to *Rhododendron* poisoning leading to heavy mortality in sheep in Palampur area of Himachal Pradesh (Uniyal et al. 2005).

The purpose of this chapter is not to offer general description of various diseases affecting the migratory livestock or suggest treatment or control measures in detail but to document the important disease outbreaks and other ailments in the migratory flocks especially the parasitic infestation including their prevalence, most common internal parasites, extent of infection, factors affecting the prevalence (season, age, sex, species), infestation rate, and host species.

Among Himalayan states of India, major annual migratory activity of small ruminants is confined to the north-western states of Himachal Pradesh and Uttarakhand and union territories of Jammu and Kashmir and Ladakh. Seasonal migration of livestock of lesser order is also prevalent in the north-eastern states of Sikkim and Meghalya. But in these states, goats, rather than sheep, are the main migratory species. And there does not seem to be a well-defined migratory cycle. In view of the above, it was considered expedient to present the state-wise information. It is worth mentioning that considerable migration of small ruminant species also takes place in neighboring countries, namely, Pakistan, Afghanistan, Nepal (Ghimire and Bhattarai 2019), and Bhutan (Namgay et al. 2014), and the problems encountered relating to migratory animals in these countries are more or less identical to the migratory livestock of Indian states. Wherever necessary, specific mention of the problem has been made at appropriate place.

10.1 Helminthiasis and Other Diseases Affecting Migratory Small Ruminants in Indian Himalayan States and Union Territories

10.1.1 The Union Territories (UT) of Jammu and Kashmir and Ladakh

(a) The union territory of Jammu and Kashmir has a geographical area of 42,241 km^2 and has two distinct regions, namely, Kashmir valley (15,948 km^2) and Jammu region (26,293 km^2) comprising of 20 districts each. The elevation in the territory varies from as low as 247 m to as high as 7135 m. As per the 20th Livestock census data of Govt of India, the region has 1,768,400 of sheep and 383,000 of goats.

(b) The union territory of Ladakh constitutes the eastern portion of Kashmir region and has a geographical area of 59,146 km^2 with elevation ranging from 2550 to 7742 m. The total population of sheep and goats has been estimated at 569,000 (AUTL 2024).

Owing to favorable agro-climatic conditions, sloppy terrains, and rich alpine pastures, sheep and goats rearing is the prime occupation of small and marginal farmers and pastoral nomads in both the union territories. In fact, sheep and goats are unique to the economy of regions particularly to the communities such as Bakerwals, Gujjars, Gaddis, and Chopans who have been following a traditional transhumance profession for generations (SHDK, 2024). Alpine pastures are the natural habitat of sheep and goats migrating from lower ranges to graze in high-altitude areas from May to October.

The main indigenous sheep breeds in Jammu and Kashmir are Gurez, Gaddi, Kashmir valley, Karnah, Bhakarwal, Changthangi, and Poonchi. In order to promote fine wool and mutton quality, the Government of Jammu and Kashmir introduced crossbreeding program in sheep to boost the pastoral economy of the region (Hamadani et al. 2022). As a result, Kashmiri Merino was developed by crossing Kashmir valley ewes with Tasmanian Merino rams. The dual-purpose breeds were produced by crossing local ewes with Corriadale and Dorset rams. As regards goats (Changthangi and Pashmina), they have been improved by grading up and selection. The large-scale seasonal migration of sheep and goats during the summer season from lower plains to the alpine pastures and back is a regular feature in Jammu and Kashmir. As pointed out earlier, the nomadic tribes are involved in this transhumance activity.

Parasitic infections, especially due to helminths, are proving as main deterrent to productive life of sheep and goats in union territory regions of Jammu and Kashmir. Of the three agro-climatic divisions of Jammu and Kashmir, the majority of studies on parasitic burden in sheep have been reported from Kashmir valley. Area-wise details of prevalence of infections are presented below.

(a) *Kashmir Valley*

Gastrointestinal parasitism is a major limiting factor affecting productivity of small ruminants in Jammu and Kashmir owing to diverse climatic conditions and variable terrain. Prevalence of helminths varies from region to region depending upon the local ecological conditions and management practices. For instance, Wani et al. (2013) reported prevalence of *Trychostrongylus* as low as 3.6% from Ganderbal area, whereas it was recorded as high as 33.9% by Tariq et al. (2008b) in sheep from government and private farms. Prevalence of *Moniezia* infection in Kashmir valley has been reported to vary from 3.5% (Shah et al. 2018) to 18.9–29.1% (Lone et al. 2012). Attempts have been made to develop resistant breeds of sheep without any remarkable success (Bhat et al. 2014). Status of parasitic infestation in sheep and goats has been reported in several studies on the basis of collection of samples from different areas of the valley. It was shown that prevalence of helminths infection was higher in goats (83.6%) than in sheep (64.1%) (Lone et al. 2012). Table 10.1 presents the main nematode species infecting small ruminants in different areas of Kashmir valley.

Among trematodes, *Fasciola*, Dicrocoelium, *Amphistomum*, and *Paramphistomum* were reported to be the most prevalent (Table 10.2), whereas the Cestode species occurring frequently were *Moniezia* and *Aviteliana*. Most prevalent protozoan species was *Eimeria* (Table 10.3).

Tariq et al. (2008a) conducted an epidemiological study exclusively on Bhakarwal and Kashmiri goats ($n = 1267$) in different areas of Kashmir valley comprising of Anantnag, Pahalgam, and Srinagar to ascertain the prevalence of *Paramphistomum* infection. Gut and fecal samples were collected from slaughtered goats, and it was found that 6.7% goats were infected with the parasite. Highest infection was found during summer season, in lower age group and in males compared to their counterparts. Goats of Bhakarwal breed which are migratory in nature had higher infection rate and with the increase in age from <1 year onward the infection level was decreased.

(b) *Jammu Region*

Goats occupy greater importance in Jammu region, as majority of the state's total goat population is reared in this region. Recent work on prevalence of helminths in both stationary and nomadic animals from various locations of Jammu region showed that they harbor a variety of different species of nematodes, trematodes, and cestodes. The overall prevalence of infection varied from as low as 58.2% (Singh et al. 2014) to as high as 91.0% (Saleem et al. 2018). In all the studies, *Haemonchus contortus* was observed to be the most dominant parasite. Studies on sheep and goats of wandering nomads revealed an overall prevalence of 68.5% with *Haemonchus* being most prevalent and *Trichuris* being least (Javaid et al. 2018). In an exclusive study on tapeworm in sheep and goats from Jammu district, Ahmad et al. (2018) reported a prevalence rate of 34.6% with *Cysticercus tenuicollis* and hydatid cysts as the main metacestodes. The findings of various studies on the prevalence of gastrointestinal (GI) nematodes in small ruminant species in the Jammu region are summarized in Table 10.4.

Studies conducted in Jammu region revealed that the main trematode species were amphistomes (8.3%), *Fasciola* (5.3–8.2%), paramphistomes (6.7%), and *Dicrocoelium* (3.9–18.9%), whereas *Stilesia* (3.5%), *Moniezia* (2.1%), and anoplocephalids comprised the main cestode species. Ahmad et al. (2018) also reported presence of metacestode (larval) stages in sheep and goats from Jammu district (Table 10.5). High coccidial infection (58.3–68.3%) in these animals was also reported from various farms of Jammu (Saleem et al. 2018).

Table 10.1 Main species of parasitic nematodes prevalent in sheep and goats in different regions of Kashmir valley

Study area	Number tested	Number infected	Main nematode species	Infected (%)	OP (%)	Reference
Different districts	5361[a]	2503	Strongyles *Strongyloides*	34.9 0.69	46.7	Shah et al. (2018)
Bandipura	1016[a]	758	*Haemonchus* spp. *Nematodirus* *Trichostrongylus* *Chabertia* *Ostertagia* *Strongyloides* *Trichuris*	71.05 68.3 12.86 8.47 7.62 5.9 4.3	74.7	Bihaqi et al. (2017)
Budgam district	1200[a]	924	*Strongyle*[b] *Haemonchus* *Trichostrongylus* *Ostertagia* *Chabertia* *Strongyloides* *Trichuris*	57.7 55.0 17.5 9.17 6.67 1.67 1.5	77	Tramboo et al. (2015)
Various farms	750[a]	466	*Strongyle* spp. *Strongyloids* *Nematodirus* spp.	– – 9.0	62	Bhat et al. (2014)
Ganderbal district	55[c]	40	*H. contortus* *Trichuris* *Chabertia* *Dictyocaulus* *Mecistocirrus* spp. *Oesophagostomum* *Ostertagia* *Bunostomum* *Trichostrongylus* *Nematodirus* *Gaigeria*	72.2 30.9 20.0 16.4 12.7 12.7 9.0 7.3 3.6 1.8 1.8	72.7	Wani et al. (2013)
Various farms	1477[a]	1076	Trichostrongyle eggs		72.8	Allaie et al. (2018)
Various farms	1035[a], 498[c]	945	*H. contortus* *O. circumcincta* *B. trigonocephalum* *C. ovina* *Trichostrongylus* spp. *N. spathiger* *Oesophagostomum* spp. *M. marshalli* *Trichuris*	59.6 38.0 37.7 37.7 33.9 29.4 28.4 22.1 23.5	61.6	Tariq et al. (2008b)

(continued)

Table 10.1 (continued)

Study area	Number tested	Number infected	Main nematode species	Infected (%)	OP (%)	Reference
Various farms	5836[a]	2989	*Strongyle* spp.		51.2	Lone et al. (2013)
Anantnag, Pulwama, Srinagar	310[d]	198	*H. contortus* (abomasum)		63.8	Tak et al. (2013)
Banihal, Doda, Kishtwar	295[e]	137	*Trichuris ovis* *Chabertia ovina*	21.4 16.3	46.4	Amin and Wani (2012)
Ganderbal	284[c] (sheep), 318[c] (goats)	182 (sheep), 266 (goats)	*Haemonchus* *Trichuris* *Nematodirus* *Trichostrongylus* *Chabertia* *Oesophagostomumm* Strongyloids	33.5–48.4 32.2–41.8 27.6–32.4 25.4–32.0 24.0–28.8 20.0–26.0 20.4–21.5	74	Lone et al. (2012)

OP overall prevalence
[a]Fecal sample
[b]Samples found positive for strongyle eggs after subjected to coproculture
[c]GI tract necropsy sample
[d]Gut sample and abomasum sample
[e]Viscera sample

Table 10.2 Main species of trematodes prevalent in sheep and goats in different regions of Kashmir valley

Study area	Number tested	Main trematode species	Prevalence (%)	Reference
Bandipura	1016[a]	*Dicrocoelium* *Paramphistomum* *Fasciola*	3.4 3.05 2.75	Bihaqi et al. (2017)
Budgam district	1200[a]	*Dicrocoelium* *Paramphistomum* *Fasciola*	11.5 4.8 3.5	Tramboo et al. (2015)
Ganderbal	284[b] (sheep), 318[b] (goats)	*Dicrocoelium* *Fasciola* *Paramphistomum*	20.0–32.0 29.3–30.6 22.4–23.5	Lone et al. (2012)
Various farms	5836[a]	*Fasciola* Amphistomes *Paramphistomum* *Dicrocoelium* *Dendriticum*	–	Lone et al. (2013)
Various farms	1035[a], 498[c]	*Paramphistomum*	7.3	Tariq et al. (2008c)

[a]Fecal sample
[b]Gut sample and abomasum sample
[c]GI tract necropsy sample

Table 10.3 Main species of cestodes and protozoa prevalent in sheep in different regions of Kashmir valley

Study area	Number tested	Main cestode/protozoa species	Prevalence (%)	Reference
Different districts	5361[a]	*Moniezia* *Eimeria* spp.	3.49 7.59	Shah et al. (2018)
Bandipura	1016[a]	*Moniezia*	9.4	Bihaqi et al. (2017)
Various farms	310[b]	*Eimeria*	9.7	Tak et al. (2013)
Banihal, Doda, Kishtwar	295[c]	*Moniezia*	13.9	Amin and Wani (2012)
Budgam district	1200[a]	*Moniezia*	7.92	Tramboo et al. (2015)
Ganderbal	284[c]	*Avitellina* *Thysaniezia*	17.7–24.2 10.1–17.8	Lone et al. (2012)
Various farms	5836[a]	*Moniezia* *Avitellina*	– –	Lone et al. (2013)
Various farms	750[a]	*Moniezia*	3.3	Bhat et al. (2014)

[a]Fecal sample
[b]Abomasum sample
[c]Viscera sample

(c) *Ladakh Region*

Species, sex, and age of the animal, season, and agro-ecology seem to be important risk factors associated with helminth parasitism in the Ladakh region. Kuchai et al. (2011a) studied the prevalence of helminth parasites and various associated risk factors in small ruminants of Ladakh region during 2007–2008. Samples of GI tract, heart, and lungs from 313 sheep and 268 goats were collected from slaughterhouses for larval and adult worm identification. Of 581 small ruminants examined, 405 (69.7%) were found infected with one or more helminth parasite species. The parasites detected included *Trichuris ovis*, *Haemonchus contortus*, *Dictyocaulus filaria*, *Chabertia ovina*, *Fasciola hepatica*, *Dicrocoelium dendriticum*, *Moniezia expansa*, and *Stilesia globipunctata*. The prevalence of parasites was found to be 68.3% in sheep and 71.26% in goats. Higher prevalence was reported during the wet season (76.5%) compared to dry season (58.1%). Female and young sheep and goats were found to be more infected than the male and adult animals. Furthermore, prevalence of helminth parasites was relatively higher in lowland (Kargil) areas compared to highland (Leh) areas.

In another study, Kuchai et al. (2011b) examined a total of 313 sheep in different areas of Ladakh for nematode parasites and 136 (43.4%) animals were found infected. Among four nematode species found, *Trichuris ovis* was the most dominant followed by *Haemonchus contortus*, *Dictyocaulus filaria*, and *Chabertia ovina*. They also evaluated the effect of season, sex, age, and agro-ecology on the prevalence of nematode parasites. The prevalence was higher in wet season (61.7%) than in the dry season (38.2%). Higher infection was reported in females (53.6%) and adults (48.5%) as compared to the males (46.3%) and young ones (51.4%).

Table 10.4 Seasonal prevalence of GI nematodes in sheep and goats in Jammu region

Study area	Number tested	Main nematode species	Infected (%)	OP (%)	Reference
Four farms	240[a] (sheep), 240[a] (goats)	*Strongyle* *Strongyloides* *Trichuris*	62.1 16.9 14.6	91.04	Saleem et al. (2018)
Wandering nomads	103[a] (sheep), 180[a] (goats)	*Haemonchus* *Ostertagia* *Chabertia* *Bunostomum* *Trichostrongylus* *Nematodirus* *Oesophagostomum* *Trichuris*	73.8 70.8 45.9 33.1 34.8 38.6 24.1 32.9	68.5	Javaid et al. (2018)
RS Pura region	150[a] (sheep)	*H. contortus* *Oesophagostomum* *Trichostrongylus* Others	76 12 9 3	–	Vohra et al. (2016)
Four selected villages	960[a] (sheep), 960[a] (goats)	*Strongylus* *Trichuris*	50.1 12.1	67.2 68.5	Khajuria et al. (2012)
Jammu area	257[b] (sheep)	*Haemonchus contortus* *Trichuris*	23.7 12.4	58.4	Mir et al. (2013a)
Jammu area	280[c] (goats)	*Haemonchus contortus* *Trichuris ovis*	28.8 14.4	72.9	Mir et al. (2012)
R.S. Pura, Akhnoor, Negrota plain region	680[a] (goats)	*Haemonchus* *Ostertagia* *Chabertia*	84.9 70.6 59.8	72.5	Mir et al. (2013c)
Kathua, Sambha, Udhampur districts	500[a] (sheep), 500[a] (goats)	*Haemonchus*	59.8	58.2 (sheep), 61.4 (goats)	Singh et al. (2014)

OP overall prevalence
[a]Fecal sample
[b]GI tract, lung and trachea samples
[c]Gut (stomach, intestine), lung, and liver samples

Among trematodes, *Dicrocoelium*, amphistome, paramphistome, and fasciola, and among cestodes *Avitellina*, *Anoplocephala*, *Thysaniezia*, and *Moniezia* were found to be the dominant infesting parasites.

Comparative Analysis of Nematode Infectivity in Small Ruminants in Different Divisions of Jammu and Kashmir

It was revealed that in all the zones of Jammu and Kashmir, most dominant nematode parasites were *Haemonchus* followed by either of the following parasites, namely, *Ostertagia*, *Chabertia*, *Bunostomum*, *Trichostrongylus*, *Oesophagostomum*, and *Nematodirus* (Fig. 10.1).

Table 10.5 Seasonal prevalence of trematodes and cestodes in sheep and goats of Jammu region

Study area	Number tested	Main nematode species	Infected (%)	OP (%)	Reference
Trematodes					
Four selected villages	960[a] (sheep), 960[a] (goats)	Amphistomes *Fasciola* spp. *Dicrocoelium*	8.3 8.2 5.4	21.9	Khajuria et al. (2012)
Jammu area	280[b] (goats)	*Paramphistomum* *Fasciola* spp. *Dicrocoelium*	6.7 5.3 3.9	61.2	Mir et al. (2012)
Jammu area	284[c] (goats)	*Fasciola* spp.	5.3	72.8	Mir et al. (2013b)
Kandli, R.S. Pura sectors	228[c] (goats)	*Dicrocoelium dendriticum*	18.9[d]	18.8	Godara et al. (2014)
Cestodes					
Jammu area	280[b] (goats)	*Stilesia* *Moniezia expansa* Anoplocephalids	3.5 2.1 1.7	72.8	Mir et al. (2012)
Jammu district	300[e] (sheep), 300[e] (goats)	*Cysticercus tenuicollis* Hydatid cysts	31.8 2.8	34.6	Ahmad et al. (2018)

OP overall prevalence
[a]Fecal sample
[b]Gut (stomach, intestine), lung, and liver samples
[c]Liver sample
[d]Highest prevalence during December followed by October and January
[e]Visceral organs in pelvic, abdominal thoracic

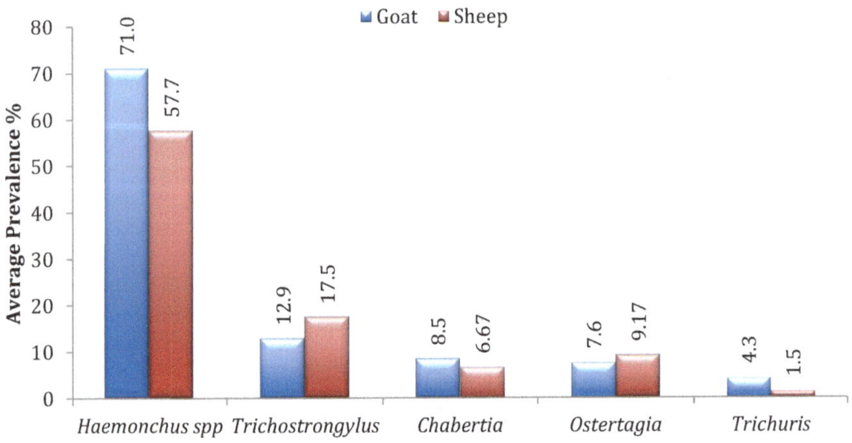

Fig. 10.1 Main nematode species infesting goats (Bihaqi et al. 2017) and sheep (Tramboo et al. 2015) in UT of Jammu and Kashmir and Ladakh

10.1.2 State of Himachal Pradesh

As discussed earlier, sheep and goat rearing is generally the sole source of livelihood for various pastoral nomadic tribes such as Gaddis in high hills of Himachal Pradesh. Gaddi sheep and goats are most prominent and are mainly distributed in the districts of Kangra, Kullu, Chamba, Shimla, and Kinnaur.

Parasitic Infestations

Information on prevalence of GI parasites in the state of Himachal Pradesh is based on long- and short-term field studies as well as from sudden disease outbreaks in migratory ruminants. We have, therefore, preferred to present the main parasites/pathogens identified during experimental and field studies and cases of disease outbreaks.

Gastrointestinal (GI) Parasitic Infestations

Even though Gaddi sheep and goats form the main small ruminant species in Himachal Pradesh, studies on their GI tract helminths status has been stray and unsystematic (Jithendran 1998, 2000; Sharma et al. 2007, 2023; Moudgil et al. 2017; Singh et al. 2023).

Jithendran (1998) randomly collected 493 fecal samples from migratory sheep (n = 335) and goats (n = 158) over a period of 12 months of which a total of 458 animals (93%) were found positive for eggs of one or the other worms. Among the nematodes, prevalence of strongyle was highest in sheep (91.6%) as well as in goats (100%) followed by *Trichuris* which was prevalent to the extent of 14.3% and 1.3% in sheep and goats, respectively (Table 10.6). The major proportion of larvae were *Trichostrongylus*, *Haemonchus*, *Oesophagostomum*, *Bunostomum*, and *Chabertia*. The overall mean monthly fecal eggs per g (EPG) ranged from 236 to 3400 in sheep and 325 to 5908 in goats with the highest peak during the months from June to August. Among trematodes, prevalence of *Fasciola* was highest (9.6% in sheep and 8.9% in goats) followed by *Dicrocoelium* (7.2% in sheep and 2.5% in goats). Among cestodes, the prevalence of *Moniezia* was 2.7% and 1.3% in sheep and goats, respectively. Protozoan parasite *Eimeria* prevalence was of the order of 29.9% in sheep and 15.2% in goats (Table 10.6).

In another study by Jithendran (2000) lasting for more than a decade, fecal samples were collected from migratory sheep and goats during necropsy from abattoirs of 11 out of 12 districts of the state. The samples were subjected to qualitative and quantitative examination for GI parasites, and a representative number of fecal samples were subjected to coproculture at 27 °C and infective larvae were harvested and used for larval identification. Sixteen species of nematodes, 24 species of trematodes, and 4 species of cestodes were identified. Main species identified and their prevalence are presented in Table 10.6.

In yet another study exclusively on migratory Gaddi sheep of Palampur valley, Sharma et al. (2007) collected 240 rectal fecal samples at monthly intervals and subjected them to qualitative and quantitative examination. Two hundred thirty six fecal samples (98.3%) were found positive for the parasites. The strongyle eggs were subjected to monthly coproculture for species identification. It was found that *Haemonchus contortus* (52.5 ± 3.48%) was the most dominant species followed by *Ostertagia* spp. (24.46 ± 2.46%), *Trichostrongylus* spp. (18.25 ± 2.82%), *Oesophagostomum* spp. (4.41 ± 0.84%) and *Strongyloides* spp. (1.5 ± 0.35%). Overall mean EPG value of 1017.5 ± 205.6 with a range of 100–8600 was recorded with highest EPG of feces in the rainy season.

Among trematodes, *Fasciola* (14.16%) and amphistomes (12.08%) and, among cestodes, *Anoplocephala* (20.41%) were the main infecting species. Moudgil et al. (2017) have reported *Fasciola* and *Dicrocoelium* infection to the extent of 15.9% in liver of Gaddi goats of Kangra valley of Himachal Pradesh.

Besides Gaddi goats, incidence of GI parasite infection of Chegu goats of Lahaul and Spiti region has been reported by Sharma et al. (2012). Of the 30 fecal samples collected, 23 (76.7%) were found positive for single or mixed infection. Interestingly, incidence of coccidia was highest (71.4%) followed by strongyles (17.8%) and amphistomes (3.6%). Three species of *Eimeria* identified included *E. parva*, *E. ovina*, and *E. ninakohlyakimovae*. The main nematode species infesting the small ruminants in the state of Himachal Pradesh are shown in Fig. 10.2.

Table 10.6 Prevalence of gastrointestinal parasites in migratory sheep and goats of Himachal Pradesh

	Prevalence (%)			
Parasite species	Sheep	Goats	OP (%)	Reference
Strongyle[a]	91.6	100	94.3	Jithendran (1998)
Strongyle[b]	98.3	–	–	Sharma et al. (2007)
Strongyloides spp.[a]	4.8	5.1	4.9	Jithendran (2000)
Strongyloides spp.[b]	16.66	–	–	Sharma et al. (2007)
Dictyocaulus[a]	1.2	1.3	1.2	Jithendran (1998)
Trichuris[a]	14.3	1.3	15.2	
Trichuris[b]	25.83	–	–	Sharma et al. (2007)
Fasciola[a]	9.6	8.9	9.3	Jithendran (1998)
Fasciola	14.16	–	–	Sharma et al. (2007)
Amphistome[a]	3.8	2.5	3.4	Jithendran (1998)
Amphistome	12.08	–	–	Sharma et al. (2007)
Dicrocoelium dendriticum[c]	8.1	4.1	–	Jithendran and Bhat (1996)
Dicrocoelium dendriticum[d]	24.1	12.3	–	
Dicrocoelium[a]	7.2	2.5	5.7	Jithendran (1998)
Schistosoma[a]	1.2	0.6	1.0	
Anoplocephala spp.	20.41	–	–	Sharma et al. (2007)
Moniezia[a]	2.7	1.3	2.2	Jithendran (1998)
Eimeria[a]	29.9	15.2	25.2	

OP overall prevalence
[a]Based on random fecal samples collected over a period of 12 months
[b]Fecal samples positive for strongyle eggs were coprocultured for larval identification
[c]Based on fecal samples
[d]Based on necropsy examination

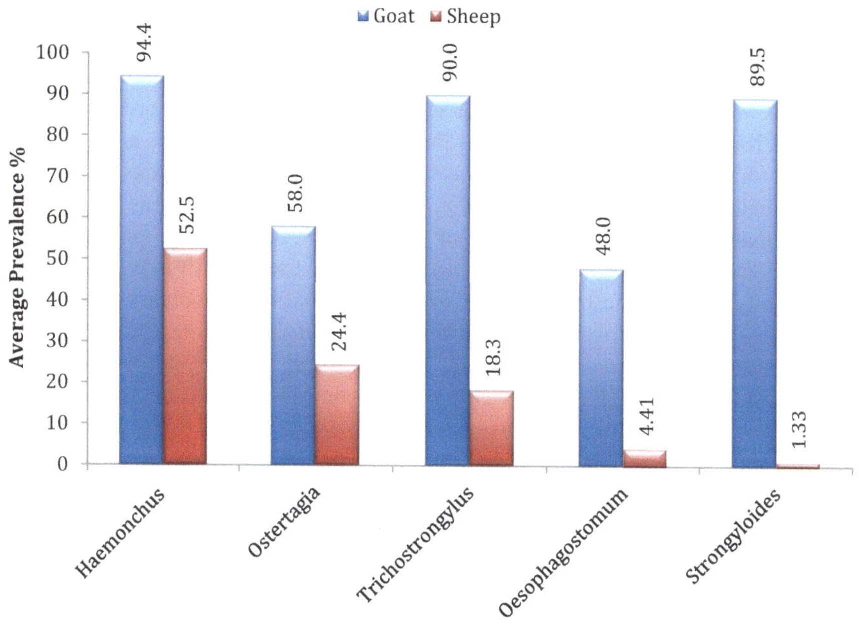

Fig. 10.2 Main nematode species infesting sheep (Sharma et al. 2007) and goats (Sharma et al. 2023) in Himachal Pradesh

Results of some recent studies on gastrointestinal nematodes of migratory sheep of Himachal Pradesh have been reported by Sharma et al. (2023). The authors studied the prevalence of gastrointestinal parasites in small ruminants of Sirmaur area of Himachal Pradesh. *Haemonchus contortus* was found to be predominant parasite both in goats and sheep. The highest prevalence of gastrointestinal parasites was recorded in the rainy season. The copro-parasitological evaluation of the animals revealed the prevalence of ten genera of gastrointestinal nematodes. Out of these, four genera, namely, *Haemonchus contortus*, *Trichostrongylus* spp., *Strongyloides* spp., and *Trichuris* spp. had very high prevalence of the order of 94.4%, 90%, 89.5%, and 87%, respectively. Among the animal species the group-wise prevalence of gastrointestinal nematode parasites were 95.3%, 90.9%, 93.7%, and 94.4% in cattle, buffalo, goats, and sheep, respectively.

In yet another recent study on migratory sheep in Himachal Pradesh Singh et al. (2023) reported higher prevalence (93.3%, 28 out of 30) of GIT parasites on direct smear examination and 100% by floatation method. This study was conducted on a total of 40 migratory Gaddi sheep, aged above 2 years, belonging to shepherd Suresh Kumar V.P.O Gugga Saloh, Palampur, Distt., Kangra, temporarily stationed at Nearby Thakurdwara town. Fecal eggs of *Strongyle*, *Trichuris*, and *Moniezia* were identified. The authors also observed reduction in hemoglobin and packed cell volume in the infected animals compared to the healthy ones.

Gastrointestinal Parasitic Transmission in Grazing Animals
Khanyari et al. (2022) have reported an interesting observation of transmission of parasitic infection from wild Asiatic Ibex (*Capra sibirica*) to the animals of migratory flock when both were sharing the same grazing range in the pastures of Pin Valley of Kinnaur District of Himachal Pradesh. The Kinnaura flocks are generally known to share pastures with Ibex. It was found that the migratory flocks that shared pastures had mean *Nematodirus* and *Moniezia* prevalence to the extent of 15% and 55% compared to the prevalence value of 6% and 34%, respectively, for these parasites in flocks that had no opportunity to share the pastures with Ibex. In another recent study on parasitic transmission in the Bani Wildlife Sanctuary of western Himalayas, Wani and co-workers (2024) documented the parasites infecting wild ungulates and emphasized the potential for cross-transmission via the fecal-oral route. Results suggested elevated risk of gastrointestinal parasitic transmission between the animals sharing the pastures.

10.1.3 State of Uttarakhand

Uttarakhand (UK), another north-westeren Himalayan state of India, is located between 28° 43′–31° 27′ N latitude and 77° 34′–81° 02′ E longitude. Of the total area of 53,483 km^2, 44.76% has a forest cover including 4.04% under permanent pasture. Around 92% of the UK is mountainous and the altitude varies from 200 m to more than 8000 m (ENVIS 2016). The state from north to south comprises of the following subdivisions: Greater Himalayas, Lesser Himalayas, Sub-Himalayan belt, and Tarai and Bhabhar.

More than two thirds of the population of the state depends on agriculture of which majority of the farmers (88%) are small and marginal and depend on livestock and meager land. The total population of goats and sheep in the state spread over in 13 districts is estimated at 1,656,500 of which 1,371,900 are goats and 284,600 sheep (20th Livestock Census, Govt. of India). According to a study, goat numbers are increasing, whereas the number of sheep is on decline (Rawat 2010). To-and-fro migration of animals between alpine pastures and base camps in the plain is widely practiced by shepherds. In general, they migrate with some members of family. The author has blamed the graziers/nomads for unscientific grazing practices leading to degradation of forests and deterioration in quality of natural pastures and cautioned that if adequate arrangements for the supply of feed and fodder were not made, there would be grave consequences to the pastures and the environment.

In spite of large population of small ruminants in the state and heavy burden of gastrointestinal helminths, only a few scientific studies seem to have been undertaken to determine the region-wise situation of parasitism. Whatever published reports currently available are stray and unsystematic. Obviously, there is a need to undertake detailed survey of parasitic load in small ruminants in all the 13 districts of the state.

In perhaps the first study in UK, Meena et al. (2016) ranked the status of general diseases in goats of Uttarakhand placing parasitic infestation on top of the list followed by diarrhea/dysentery (31.5%, 18/57), ticks/lice infestation (14.0%, 8/57), and abortions (3.5%, 2/57).

Extent of Parasitic Infestation
In a study on Pashmina and local hilly goats belonging to the high-altitude experimental farm of Indian Veterinary Research Institute (IVRI) at Mukteshwar, Ram et al. (2007) reported nematode infection to the extent of 98.3%. Fecal samples from 366 goats were collected per rectum and parasitic eggs were counted, and the samples were also subjected to coproculture for larval identification. The eggs per gram were found to be higher during monsoon season (1717) followed by summer (838) and winter (775). Higher nematode egg output during monsoon was related to high humidity. Larval identification indicated presence of *Haemonchus*, *Trichostrongylus*, *Oesophagostomum*, and *Teladorsagia* (*Ostertagia*) species. Among cestodes, *Moniezia* spp. infection was found to be quite high to the extent of 9.8–12.7%.

In Tarai region of the state, Pant et al. (2009) conducted a survey in surrounding villages and organized farms. Overall prevalence of gastrointestinal parasites was found to be 96%. Among nematodes, trematodes, and cestodes, the major species identified were *Haemonchus* (85.9%), *paramphistomum* (29.2%), and *Moniezia* (9.5%). Infection rate was higher in samples collected from villages compared to the organized farms.

In another elaborate study, Chaudhary et al. (2014) and Maitra et al. (2014) carried out detailed investigation on the epidemiology and prevalence of paramphistomiosis in Garhwal and Kumoan regions of Uttarakhand, respectively. Comparison of the prevalence of paramphistomes in the two regions and also in the union territory of Jammu and Kashmir is presented in Table 10.7.

In both the studies by Chaudhary et al. (2014) and Maitra et al. (2014), the peak infection was observed in the monsoon and post-monsoon seasons. Lower prevalence was recorded both in sheep and goat flocks from Kumaon region compared to Garhwal region. Infectivity of paramphistomes was found higher in Tarai region compared to hill region. It was clarified that lower infectivity in the hills was perhaps due to lower prevalence of snails in the region.

10.1.4 North-Eastern States

Sikkim and Meghalaya are the prominent Himalayan states of eastern India with geographical areas of 7299 km^2 and 83,743 km^2, respectively. The terrain and topography vary considerably from state to state. Both states are hilly and covered with varied vegetation and forest. Mountain slopes are covered by sub-tropical and alpine temperate forest and grasses having *Rhododendron*, oak, pine, maple, and Sal trees with meager cultivable area. The elevations range from 150 to 1960 m (Meghalaya) and 300 to 8570 m (Sikkim) (Department of Animal Husbandry and

Table 10.7 Prevalence of paramphistomosis in sheep and goats in Uttarakhand and union territory of Jammu and Kashmir

Study area (period)	Species	Season	Number tested	Number infected	Worm count	Reference
Jammu (July 2004–June 2005)	Sheep		45	2 (4.44%)		Shabih and Juyal (2006)
	Goat		30	4 (13.3%)		
Nainital, Almora, Bageswar, Pithoragarh districts of Uttarakhand (2005–2007)	Sheep		599	15 (2.5%)		Maitra et al. (2014)
	Goat		1638	33 (2%)		
Jammu (March 2010–February 2011)	Sheep	Summer	48	22.9	12.8 ± 3.8	Godara et al. (2014)
		Rainy	51	35	28.8 ± 4.4	
		Post-rainy	55	47.3	41.1 ± 7.1	
		Winter	45	33.3	20.0 ± 5.3	
	Goat	Summer	53	22.6	11.1 ± 3.6	
		Rainy	52	32.7	25.3 ± 3.9	
		Post-rainy	51	43.1	37.6 ± 6.9	
		Winter	51	25.5	19.5 ± 4.9	
Jammu (July 2004–June 2005)	Sheep		45	2 (4.44%)		Shabih and Juyal (2006)
	Goat		30	4 (13.3%)		
Jammu region and Kashmir (November 2020–October 2021)	Sheep	Summer	390	101 (25.9%)	27.6 ± 0.41 EPG	Sofi et al. (2022)
		Rainy	305	152 (49.8%)	46.5 ± 0.42 EPG	
		Post-rainy	225	96 (42.7%)	22.0 ± 0.17 EPG	
		Winter	290	91 (31.4%)	12.7 ± 0.42 EPG	

Veterinary Science, Sikkim 2016; Department of Animal Husbandry and Veterinary Science, Meghalaya 2016).

Majority of the people depend on agriculture and allied activities and over 75% of rural households own livestock. The total small ruminant population in both the states has been estimated at 505,700 which include 488,000 of goats and 17,700 of sheep showing that goat is the main rearing species (GOI 2019). The animals are generally reared on free grazing on mountain slopes without any scientific inputs such as feeding, breeding, housing, vaccination, disease control, and management.

Status of Gastrointestinal Parasitic Infestation

There are only a few studies on the GI parasitic infection in the north-eastern states. The published work is stray rather than systematic or region-wise. In a few published research findings, investigators have not clarified whether the work reported was on the migratory or sedentary or stray undefined animals. State-wise status of gastrointestinal parasitism in north-eastern states is presented in the following discussion.

10.1.4.1 Sikkim

Rahman et al. (2013) investigated the incidence of GI parasites in ruminants of organized farms in Sikkim. The fecal samples were randomly collected from 158 goats and sheep and examined for presence of GI parasites. Fecal egg counts were also done. The adult worms identified during necroscopic examination of GI tracts of sheep and goats were strongyle and *Strongyloides*. Parasitic eggs recovered from feces belonged to strongyle, *Strongyloides*, *Nematodirus*, amphistomes, and *Trichuris* spp. Coproculture studies showed that *Haemonchus* was the most prevalent nematode followed by *Oesophagostomum* and *Bunostomum* spp. In another study but exclusively on goats belonging to seven villages of Sikkim, Pal and Bandyopadhyay (2004) examined the fecal samples collected from 970 goats and also by necropsy of slaughtered animals for the presence of GI tract worms. A representative quantity of fecal sample was subjected to coprological culture and processed for larval identification. Out of 970 goats, 750 (77.3%) were found positive for eggs of various parasites. Seven species of worms were found. Highest infection was observed during the summer season. The monthly mean fecal EPG counts ranged from 150 to 3500 with a high value during July to November (1000–3500). Table 10.8 and Fig. 10.3 present the summarized view of the helminths infection in Sikkim.

10.1.4.2 Meghalaya

Similar to Sikkim, goat is the main small ruminant species in Meghalaya and sheep is hardly a priority. There does not seem to be a definite animal migratory cycle in the state as practiced in north-western states. Besides heavy parasitic load, cases of peste des petits ruminants (PPR) have been reported in goats in Meghalaya although its prevalence was low (5.88%) (Balamurugan et al. 2014b).

Despite few reports on the prevalence of GI parasites in the state, Bandyopadhyay et al. (2010) investigated the burden level of GI parasites in goats slaughtered for human consumption in Shillong town from 2001 to 2006 and observed 92.4% prevalence. In a later study, Laha et al. (2013) reported GI parasitism prevalence rate of 46.5% based on manual fecal sample collected from 636 goats in Barapani area of north-eastern hill region of Meghalaya. Das et al. (2017) collected a total of 834 fecal samples manually from goats in two districts of Meghalaya and reported an overall prevalence of GI parasitic infection to the extent of 28.65%. In all these studies, egg count data showed that highest infection was prevalent in rainy and summer seasons and lowest in cold season. The coproculture of fecal samples revealed the presence of *Haemonchus contortus* and other species. In East khasi hill area of

Meghalaya, several species of *Eimeria* were reported. Among the trematodes, amphistome parasites were the most predominant while among cestodes, *Moniezia expansa* and *Moniezia benedeni* were found to be common in sheep and goats (Table 10.9).

Helminthological studies in north-eastern states revealed quite a few interesting findings. Two new genera and species, namely, *Oesophagostomum venulosum* and *O. aspersum* were reported which do not seem to have been reported in small ruminants of north-western states (Bandyopadhyay et al. 2010). Yet another interesting finding was the prevalence of *Haemonchus contortus* in reticulum. Das et al. (2017) for the first time reported seven different species of *Eimeria*, namely, *E. christenseni*, *E. hirci*, *E. caprina*, *E. jolchijevi*, *E. ninakohlyakimovae*, *E. arloingi*, and *E. kocharii* in goats of Meghalaya.

10.2 Salient Features of the Parasitic Infections in Small Ruminants of Himalayan States

Studies showed that over 90% of the gastrointestinal helminth infection of small ruminants in north-western and north-eastern Himalayan states belonged to the sub-order Strongylida. Majority of the infective parasites belonged to four families, namely, Trichostrongylidae (*Haemonchus*, *Ostertagia*, *Trichostrongylus*, and *Nematodirus*), Ancylostomatidae (*Bunostomum*, *Gaigeria*), Strongylidae (*Oesophagostomum*, *Strongyloides*, *Chabertia*), and Dictyocaulidae (*Dictyocaulus* or lungworm).

10.2.1 Dominant Parasites

In all the Himalayan states, *Haemonchus* spp. was the most predominant nematode infecting sheep and goats followed by other variable genera as presented in Table 10.10.

10.2.2 Two Most Predominant Helminth Species

Based on the available published data we have ranked the two most predominant nematodes, trematodes, and cestodes each infecting sheep and goats in the Himalayan states/UTs with a view to draw appropriate conclusions as far as possible (Figs. 10.4, 10.5, and 10.6).

Table 10.8 Prevalence of gastrointestinal parasitic infections in sheep and goats of Sikkim

Study area	Number tested/ number infected (OP)	Endoparasite	Genus/species	Infection rate (%)		Reference
				Sheep	Goat	
Organized farms	158[a]/106 (67%)	Nematode	Strongyles, *Haemonchus*, *Oesophagostomum*, *Bunostomum*	76.5	58.9	Rahman et al. (2013)
			Strongyloides	52.9	34.8	
			Trichuris	29.4	7.8	
			Nematodirus	–	15.6	
			Ascaris	–	9.9	
		Trematode	Amphistomes	17.6	5.7	
		Cestode	*Moniezia*	47.1	26.2	
			Eimeria	64.7	36.2	

(continued)

Table 10.8 (continued)

Study area	Number tested/ number infected (OP)	Endoparasite	Genus/species	Infection rate (%)		Reference
				Sheep	Goat	
Several villages	970[b]/750 (77.3%)	Nematode	*Haemonchus contortus*	–	70.0	Pal and Bandyopadhyay (2004)
			Bunostomum trigonocephalum	–	51.6	
			Oesophagostomum columbianum	–	43.0	
			Trichostrongylus columbriformis	–	28.3	
			Nematodirus filicolis	–	21.6	
			Trichuris globulosa	–	10.0	
			T. ovis	–	6.7	
Thirteen different places (600–1600 amsl)	1921[a]/1258 (65.5%)	Nematode	*Strongyles*	–	63.5	Pal et al. (2017)
			Strongyloides spp.	–	34.2	
			Nematodirus spp.	–	22.3	
			Trichuris spp.	–	11.7	
				–		
		Trematodes	*Amphistomes*	–	8.9	
			Dicrocoelium spp.	–	10.2	
		Cestods	*Moniezia* spp.	–	32.1	
			Eimeria		54.5	

Phalung valley (5250–6000 msl)	213[a]/71 (33.3%)	Nematode	*Haemonchus* spp.	24.5	–	Pal et al. (2020)
			Strongyles	31.5	–	
			Strongyloides spp.	19.2	–	
			Nematodirus spp.	11.7	–	
			Trichuris spp.	3.7	–	
		Trematodes	*Amphistomes*	5.6	–	
			Dicrocoelium spp.	4.2	–	
		Cestodes	*Moniezia* spp.	17.4	–	

OP overall prevalence
[a]Fecal sample. Eggs subjected to coproculture for larval identification
[b]Fecal sample and GI tract necropsy sample. Eggs subjected to coproculture

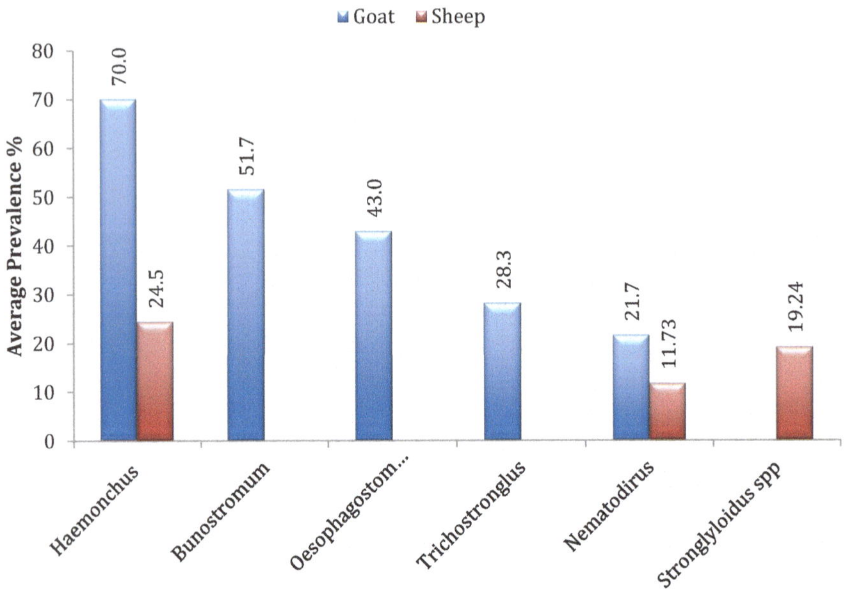

Fig. 10.3 Main nematode species infesting goats (Pal and Bandyopadhyay 2004) and sheep (Pal et al. 2020) in Sikkim

10.2.3 Peculiarities of Parasitic Infestations in North-Eastern and North-Western States

The published reports generally point to certain distinct peculiarities between north-eastern and north-western states in relation to the prevalence of parasitic infestations in small ruminants. In the north-western states, the major focus is on sheep, whereas in the north-eastern states, goats are more important mainly as a source of mutton. Also the annual cycle of migration of animals practiced in north-western states is not strictly followed in the north-eastern states. It was interesting to note that although *Moniezia* infection has been reported from all the states, its prevalence was highest in Sikkim and Meghalaya. Similarly, *Eimeria* infection was most predominant in Sikkim and Meghalaya followed by Himachal Pradesh.

Some other distinctive features are presented in Table 10.11.

The occurrence of some parasites in small ruminants has been found to be restricted to certain states. The prevalence of *Marshallagia marshalli*, *Chabertia*, *Avitellina*, and *Thysaniezia* has been reported only from Jammu and Kashmir, whereas *Anoplocephala* and *Schistosoma* have been reported only from Himachal Pradesh. On the other hand, *Oesophagostomum venulosum* and *O. aspersum* have been reported only from Meghalaya.

Table 10.9 Prevalence of gastrointestinal parasitic infections in goats of Meghalaya

Study area	Number tested/ number infected (OP)	Endoparasite	Genus/species	Infected (%)	Reference
Shillong	250[a]/231 (92.4%)	Nematode	*Haemonchus*	68.4	Bandyopadhyay et al. (2010)
			Bunostomum	22.4	
			Oesophagostomum	70.8	
			O. venulosum	92.4	
			O. aspersum	19.6	
			Trichostrongylus	3.1	
			Gaigeria	29.6	
			Trichuris spp.	66.8	
		Trematode	Amphistomes	16.8	
		Cestode	*Moniezia expansa*	40.0	
			M. benedeni	38.0	
Barapani	636[b]/296 (46.5%)	Nematode	*Haemonchus*	73.2	Laha et al. (2013)
			Oesophagostomum	17.7	
			Trichostrongylus	3.1	
			Strongyle	67.5	
			Strongyloides	6.1	
		Cestode	*Moniezia* spp.	1.5	
		Protozoa	*Eimeria*	16.8	
East Khasi hills	834[b]/239 (28.6%)	Nematode	*Haemonchus contortus*	72.2	Das et al. (2017)
			Oesophagostomum spp.	14.4	
			Strongyloides spp.	8.9	
			Trichostrongylus spp.	4.5	
		Cestode	*Moniezia expansa*	10.04	
		Protozoa	*Eimeria*	23.02	

OP overall prevalence
[a]GI tract necropsy sample
[b]Fecal sample subjected to coproculture

10.2.4 Factors Affecting Prevalence of Gastrointestinal Helminths

A number of factors have been found to significantly affect the extent of parasitic infestation and prevalence rate both in sheep and goats in various Himalayan states. The major factors include season, sex and age of the animal, type of livestock production system (whether migratory or sedentary), and the nature of the host species. Although breed-wise infective or prevalence rates have not been systematically reported, data suggest that as compared to sheep, goats were relatively more susceptible to gastrointestinal parasitism.

Table 10.10 Nematode infections in sheep and goats in order of predominance in different Himalayan states

	North-western states/UT			North-eastern states	
Prevalence[a]	UT of Jammu and Kashmir	Himachal Pradesh	Uttarakhand	Sikkim	Meghalaya
I	*Haemonchus*	*Haemonchus*	*Haemonchus*	*Haemonchus*	*Haemonchus*
II	*Bunostomum*	*Ostertagia*	*Dictyocaulus*	*Bunostomum*	*Oesophagostomum*
III	*Ostertagia*	*Trichostrongylus*	Trichuris	*Oesophagostomum*	*Bunostomum*
IV	*Trichostrongylus*	*Gaigeria*	–	*Trichostrongylus*	*Trichostrongylus*
V	*Oesophagostomum*	*Dictyocaulus*	–	*Nematodirus*	*Gaigeria*
VI	*Dictyocaulus*	*Trichuris*	–	*Trichuris*	*Trichuris*
VII	*Nematodirus*	*Strongyloides*	–	*Strongyloides*	*Strongyloides*
VIII	*Chabertia*	–	–	–	–
IX	*Trichuris*	–	–	–	–
X	*Strongyloides*	–	–	–	–

[a]Prevalence in descending order ranging from highest (I) to lowest (X)

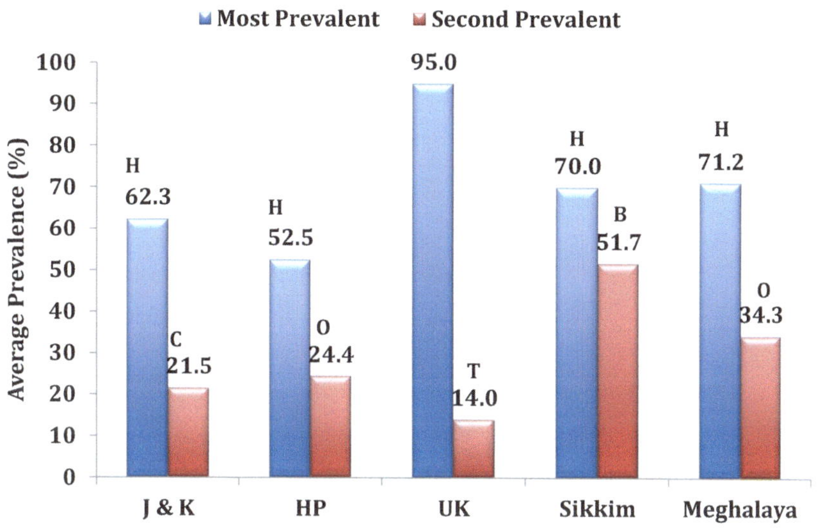

Fig. 10.4 Ranking (prevalence) of the first two major nematodes infesting sheep (Tramboo et al. 2015; Wani et al. 2013; Tariq et al. 2008c; Sharma et al. 2007) and goats (Pant et al. 2009; Ram et al. 2007; Pal and Bandyopadhyay 2004; Bandyopadhyay et al. 2010; Laha et al. 2013; Das et al. 2017) in the Himalayan states. H = *Haemonchus* spp., C = *Chabertia*, O = *Ostertagia*, T = *Trichuris*, B = *Bunostomum*

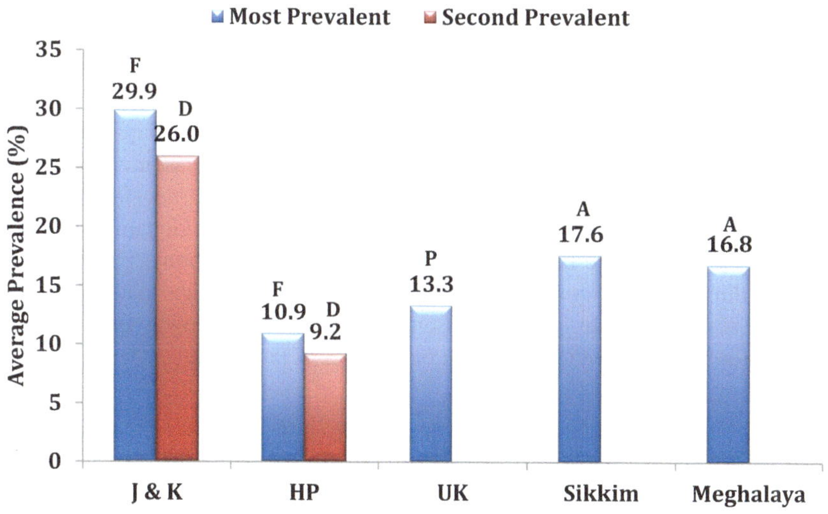

Fig. 10.5 Ranking (prevalence) of the first two major trematodes infesting sheep (Tramboo et al. 2015; Lone et al. 2012; Sharma et al. 2007), goats (Bihaqi et al. 2017; Bandyopadhyay et al. 2010), and goats and sheep (Jithendran 1998; Maitra et al. 2014; Rahman et al. 2013). F = *Fasciola*, D = *Dicrocoelium*, P = *Paramphistome*, A = *Amphistome*

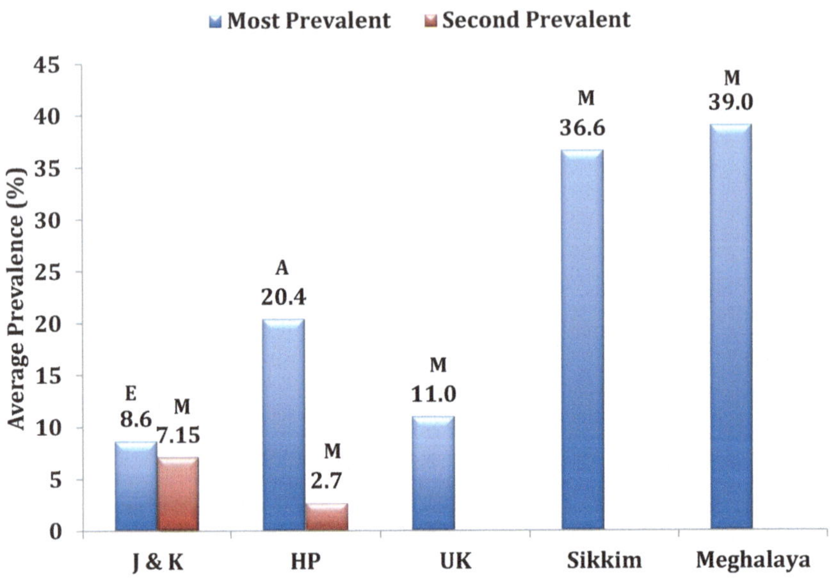

Fig. 10.6 Ranking (prevalence) of the first two major cestodes infesting sheep (Shah et al. 2018; Amin and Wani 2012; Tramboo et al. 2015; Bhat et al. 2014; Tak et al. 2013; Sharma et al. 2007) and goats (Ram et al. 2007; Bandyopadhyay et al. 2010; Jithendran 1998; Rahman et al. 2013) in the Himalayan states. E = *Eimeria*, M = *Moniezia*, A = *Avitellina*

(a) *Season*

Majority of the studies have revealed seasonal prevalence of gastrointestinal parasitic infections both in sheep and goats. Among helminth groups, nematodes were found to have higher prevalence during summer/rainy season and lowest in winter. On the contrary, trematodes especially *Fasciola* and *Dicrocoelium dendriticum* had the highest prevalence during winter (Mir et al. 2013b; Godara et al. 2014). These observations were based on liver necropsy. The maximum egg production occurred during January to March. It appears that meterological parameters in a particular season provide ideal conditions for the herbage growth and necessary moisture for the optimum development of the eggs of parasites.

(b) *Sex*

Studies on parasitic nematodes have shown that the infection rate was significantly higher in males than the females. Interestingly, this pattern was not evident in the occurrence of trematode infection, where no significant difference was observed in the prevalence rate between male and female animals.

Table 10.11 Salient features of parasitic infestations in the north-eastern and north-western states

S. no.	Prevalence in north-western states	Prevalence in north-eastern states
1	Low to moderate infection of *Strongyloides*, *Oesophagostomum*, *Moniezia*, *Eimeria*, and amphistomes	Very heavy infection of *Strtongyloides*, *Oesophagostomum*, *Moniezia*, *Eimeria*, and amphistomes
2	Among trematodes, *Fasciola* was the main species	Among trematodes, amphistome was the main helminth. We have not come across any published report of *Fasciola* in these states
3	Among cestodes, besides *Moniezia* prevalence of *Anaplocephala*, *Avitellina*, and *Thysaniezia* has been reported	Only *Moniezia* has been reported but the degree of infection was very heavy
4	*Dictyocaulus filaria* occurring commonly in sheep and goats in higher altitude regions	No report of *Dictyocaulus* seems to be available in these states
5	*Oesophagostomum columbianum* reported as highly prevalent species	Two new species *O. venulosum* and *O. aspersum* reported in goats of Meghalaya
5	–	Seven new species of *Eimeria* reported in goats of Meghalaya

(c) *Age*

Without exception, the prevalence of infection was recorded higher in younger animals/kids compared to older/adults animals (>2 years) (Amin and Wani 2012).

(d) *Livestock Production System Whether Migratory or Stationary*

Tariq et al. (2008c) studied the prevalence *of Paramphistomum* infection in various Kashmiri breeds and reported higher infection rate in migratory Bakerwal goats. It was reported that local Kashmiri breeds were resistant to *Paramphistomum* and had lower prevalence of infection, but the values were statistically not significant.

(e) *Nature of Host Species*

Various studies have indicated significant differences in the susceptibility of migratory sheep and goats to infective organisms including helminths. Compared to sheep, goats were relatively more susceptible to infection of *Dictyocaulus filaria* (lungworm infection) (Dar et al. 2012) and *Chlamydia* (Sharma et al. 2008). Similarly, fatality caused by PPR was significantly higher in kids than in lambs (Jithendran et al. 2000). On the other hand, sheep were found be more susceptible to *Brucella* and goats to *Chlamydia* in causing abortions in the migratory flocks (Sharma et al. 2008).

10.3 Other Disease Conditions Affecting Productivity of Small Ruminants in North-Western Himalayan States of India

Besides heavy gastrointestinal parasitism, some other disease conditions such as pneumonia, parasitic or verminous bronchitis, and persistent abortions due to *Chlamydia* and peste des petits ruminants have been found responsible for causing heavy losses in migratory sheep and goats. The prevalence of these conditions is presented below.

10.3.1 Parasitic or Verminous Bronchitis

Parasitic or verminous bronchitis has been found widely prevalent in flocks of small ruminants in the north-western Himalayan region of India. The first report of the disease in Indian goats has been traced in early 1930s (Bhalerao 1932). Subsequently, the occurrence of parasitic bronchitis was reported from different parts of the country (Sadana et al. 1979; Banerjee and Gupta 1980). Although the most common lungworm species causing the disease has been identified as *Dictyocaulus filaria* (Dar et al. 2012), involvement of other worm species such as *Protostrongylus rufescens*, *Varestrongylus pneumonicus*, and occasionally, *Muellerius* spp. have also been reported in India and other developing countries. Some early reports pointed to a lower prevalence rates in Himachal Pradesh compared to the union territory of Jammu and Kashmir (Dhar and Dash 1982; Krishna et al. 1987). Jithendran (1998) reported the prevalence of *Dictyocaulus* infection in Himachal Pradesh to the extent of 1.2–1.3%. Asrani et al. (1999) reported five outbreaks of verminous bronchopulmonitis in peak winter season among migratory Gaddi sheep (1204) and goats (627) at the foothills of Sirmaur district of Himachal Pradesh. The morbidity and mortality were reported as 52.1% and 9.4% in sheep and 35.7% and 7.5% in goats, respectively. Two species of lungworms, namely, *Dictyocaulus filaria* and *Protostrongylus rufescens* were isolated from trachea and bronchi. The microbiological examination of lung pieces from affected animals revealed the occurrence of *Escherichia coli*, *Corynebacterium* spp., *Pasteurella haemolytica*, *Pasteurella multocida*, and *Streptococcus pyogenes* indicating secondary complications due to these microorganisms.

Parasitic bronchitis has been reported in migratory flocks of goats in the north-western Himalayan regions. Sharma (1994) carried out a detailed study on its prevalence in 5554 goats spread over in 31 villages in different agro-climatic regions of Himalayan belts including Himachal Pradesh, Uttarakhand, and union territory of Jammu and Kashmir. The coprological examination of fresh fecal samples from these goats revealed the presence of lungworm larvae.

In another study exclusively on indigenous sheep from different regions of Kashmir valley, Dar et al. (2012) collected lung samples from 1455 slaughtered sheep from Sambha region of the Jammu area to find out the prevalence of lungworm infection. Most adult lungworms were identified as *Dictyocaulus filaria*. The prevalence was significantly higher in emaciated sheep than those with relatively good body conditions. Results of both the studies (Sharma 1994; Dar et al. 2012) have been presented in Table 10.12. The prevalence varied from 18.7% to 43.5% in goats and 26.8% in Kashmiri indigenous sheep.

The cold and humid climate was found to be most favorable for the development of the parasitic larvae. As many as 27.6% of goats tested positive belonged to the high-altitude area Kargil in UT of Ladakh (2700–3900 m amsl). Goats were more susceptible to lungworm infection than sheep and susceptibility to infection was higher, and the characteristic signs were more severe in kids than in the adult goats.

Sharma (1994) also developed a vaccine against *D. filaria* and tested the immunological response of goats. It was found that, in contrast to the control group, the vaccinated animals had a negligible number of *D. filaria* larvae in their feces from day 30 post-challenge. Also clinically, the vaccinated animals did not show any characteristic symptoms of the disease.

10.3.2 Pneumonia

Pneumonia is common during winter when the animals are grazing at high-altitude pastures. Asrani et al. (1999) and Kumar et al. (2004) have reported two severe outbreaks of pneumonic disease in migratory sheep and goats of Himachal Pradesh. The histopathological findings of lung tissues collected from the affected animals showed characteristic pneumonic lesions and varying degrees of emphysema, congestion, hemorrhages along with edema, and polymorphonuclear infiltration. The microbiological examination revealed involvement of several bacterial species like *Pasteurella* and *Corynebacterium*. In a recent study, Bhardwaj et al. (2017) investigated the involvement of *Chlamydia* in pneumonitis in sheep and goats grazing on four different tracks in Himachal Pradesh and found that in 72.7% of the samples ($n = 139$) the causative agent was *Chlamydia psittaci*.

Table 10.12 Prevalence of parasitic bronchitis in goats and sheep in Uttarakhand state, union territories of Jammu and Kashmir and Ladakh

Species	Survey area	Number tested	Number infected	Prevalence (%)	Reference
Goat	Kashmir (Handwara and Lolab)	400	174	43.5	Sharma (1994)
	Ladakh (Kargil)	754	208	27.6	
	Uttarakhand (Garhwal)	4400	822	18.7	
Sheep	Kashmir	2910	773	26.8	Dar et al. (2012)

10.3.3 Persistent Abortions and Other Reproductive Disorders

In migratory sheep and goats, the lambing and kidding normally occurs during winter months after their return from alpine pastures and when the flocks are confined to limited housing spaces at the base camps in the plains. Confinement increases the risk of disease transmission. The reproductive losses are numerous which generally include abortions, repeat breeding, and various kinds of infections and mortality.

Investigations to identify the possible pathogens causing persistent abortions in migratory animals of Himachal Pradesh and UT of Jammu and Kashmir were initially started at the Himachal Pradesh Agriculture University (HPAU), Palampur, and the findings indicated that *Chlamydia psittaci*—a bacterial pathogen—was the potential causative agent for abortions, metritis, and mastitis in migratory sheep and goats of Himachal Pradesh. Batta et al. (1996) screened sheep and goats both from Himachal Pradesh and Jammu and Kashmir regions. Of the 149 flocks comprising of 27,240 sheep and 20,400 goats, 119 flocks were found to have history of diseases including outbreaks of abortions in 28 flocks and other sporadic illnesses such as pneumonia, polyarthritis, enteritis, mastitis, and conjunctivitis in 91 flocks. The vaginal swabs from the suspected animals were collected and processed for isolation of *Chlamydia psittaci*. Out of 49 vaginal samples collected from sheep, 7 (14.3%) were found positive for the presence of *Chlamydia psittaci*, whereas 38 (35.5%) vaginal samples from 107 goats were found to be positive for the bacteria.

In a subsequent study from the same laboratory, Sharma et al. (2008) showed that, besides *Chlamydia*, four other bacterial pathogens were involved in causing abortions and other reproductive problems in migratory flocks. The field study on bacterial etiology of abortions was carried out in 203 flocks comprising of 33,790 sheep and 27,737 goats. Abortion outbreaks occurred in 37 flocks and 56 flocks having sporadic abortions. Vaginal swabs from aborted animals were processed for isolation of pathogens, and bacteria known to cause abortions were identified. An important finding of the study was that goats were found to be more susceptible to *Chlamydia* spp., whereas sheep were more susceptible to *Brucella melitensis*. This increased susceptibility of goats to *Chlamydia* had also been reported by an earlier investigator (Rodolakis et al. 1984).

Chahota et al. (2015) studied the chlamydial seroprevalence in sheep and goats in Jammu and Kashmir and Himachal Pradesh. Of the 439 serum samples collected from sheep, 47 (10.7%) were found positive. In goats, a prevalence of 22.5% was recorded, and 62 samples out of 276 serum samples collected were found to be positive.

10.3.4 Peste des Petits Ruminants (PPR)

PPR, also known as "goat plague" or "ovine rinderpest," is turning out to be most threatening viral disease to the grazing small ruminants causing heavy economic losses to small and marginal farmers. It is highly contagious and acute condition is caused by *Morbillivirus*. The condition is characterized by high fever, conjunctivitis, necrotizing stomatitis, mucopurulent nasal and ocular discharge, and bronchopneumonia followed by recovery or mortality from the disease. In India, the disease was first reported from Tamil Nadu in the year 1987 (Shaila et al. 1989) and has since been reported by several investigators (Nanda et al. 1996; Dhar et al. 2002; Balamurugan et al. 2010). Due to endemic nature of the disease and high mortality rate (30–40%), the condition has been considered as serious impediment to rearing of small ruminants. Economic losses due to PPR have been estimated to the tune of Rs. 1800 million annually (Bandyopadhyay 2002).

The disease outbreaks have been found to occur most commonly during March to June, and goats have been found more susceptible compared to sheep. Based on a survey across the country, Balamurugan et al. (2014b) reported that in northern regions of India, PPR outbreaks were most frequent in goats, whereas in southern region, they were most common in sheep.

(a) *Prevalence of PPR in the Himalayan States/Region*

Among the north-western Himalayan region, many outbreaks of PPR have been reported from the union territory of Jammu and Kashmir and the state of Uttarakhand (Balamurugan et al. 2012). Mahajan et al. (2013) reported the disease in nomadic sheep and goats in Jammu region of the union territory of Jammu and Kashmir. Samples from the affected animals were collected from field outbreaks of six different districts of Jammu region. Out of 34 animals suspected of PPR infection (13 sheep and 21 goats), 19 were found positive using sandwich ELISA with overall prevalence rate of 55.88%. The prevalence was higher in goats (66.66%) as compared to sheep (35.71%) and higher in the younger animals (83.33%) than in the adults (31.35%). The highest number of PPR cases was reported in the month of February.

Several investigators have reported the prevalence of PPR in the state of Himachal Pradesh. News of heavy economic losses due to large-scale mortality of migratory sheep and goats has also appeared repeatedly in the print media (Tribune Newspaper, Chandigarh June 21, 2001 and August 10, 2014). Joshi et al. (1996), who confirmed PPR in migratory sheep and goat flocks in Himachal Pradesh, reported the morbidity and mortality rate of 58–83% and 17–53%, respectively. Jithendran et al. (2000) encountered the outbreak of the disease in several flocks in villages of Kangra district of Himachal Pradesh during their downward journey. Besides other typical symptoms of PPR, abortion cases were also reported. The tissues and buccal swabs collected from affected animals were found positive for PPR/RP antigens. Eighteen sera samples were also tested positive for the virus by ELISA. The average mortality was found significantly higher in goats (22.9%) compared to sheep (2.05%).

In a 2-year study on PPR covering main Himalayan states, Singh et al. (2004) reported the outbreaks and prevalence of antibodies to PPR virus in 4407 serum samples of small ruminants from 58 laboratories. Indigenously developed monoclonal antibody-based diagnostic kits were used for the detection of PPR virus antigen by enzyme-linked immuno sorbent assay (ELISA). Based on the limited data, it was observed that prevalence of PPR was highest in Jammu and Kashmir. The prevalence of PPR in migratory small ruminants in Himalayan states/region is presented in Table 10.13.

Table 10.13 Seroprevalence of PPR in sheep and goats in Himalayan states/region of India

		Sheep		Goats		
State/UT	Period	Number tested	Number positive (%)	Number tested	Number positive (%)	Reference
Jammu and Kashmir	1998–2003	27	11 (40.7%)	21	8 (38.1%)	Singh et al. (2004)
	2003–2009	129	63 (48.8%)	201	103 (51.2%)	Balamurugan et al. (2011)
	June 2017–March 2018	632	219 (34.6%)	633	195 (30.8%)	Balamurugan et al. (2020a)
Himachal Pradesh	1998–2003	97	15 (15.5%)	84	34 (40.5%)	Singh et al. (2004)
	2003–2009	87	8 (9.2%)	230	57 (24.8%)	Balamurugan et al. (2011)
	June 2017–March 2018	646	215 (33.3%)	640	262 (46.9%)	Balamurugan et al. (2020a)
Uttarakhand	1998–2003	44	12 (27.3%)	630	128 (20.3%)	Singh et al. (2004)
	June 2017–March 2018	132	79 (46.6%)	655	152 (41.2%)	Balamurugan et al. (2020a)
Arunachal Pradesh	2013–2014	–	–	32	5 (15.6%)	Balamurugan et al. (2014b)
Meghalaya	2013–2014	–	–	34	2 (5.9%)	
	April 2017–March 2018	–	–	655	31 (4.7%)	Balamurugan et al. (2020b)
Sikkim	September 2017–February 2018	99	1 (1.01%)	677	8 (1.2%)	Balamurugan et al. (2020c)

(b) *Prevention and Control Measures for PPR*

The most effective preventive step is not to allow contact between infected and susceptible animals. If necessary and possible, the animals or flocks should be put under quarantine for a month or so. Unfortunately, none of these conditions are truly implementable in case of migratory animals. The movement of animals in fact helps in transmission and maintenance of virus in nature. Mixing of migratory flock returning from the pastures with other animals en route or in the village or with newly purchased animals is a dangerous situation from the point of view of control of PPR. Obviously, animal movements have to be strictly controlled in the area of infection. As all such measures are difficult to implement, the only effective way to control PPR is by way of mass vaccination and quarantine measures. The flock owners and livestock keepers should be properly educated in all aspects of prevention and control of the disease. Availability of attenuated cell culture vaccine and various diagnostic techniques/kits for diagnosis of PPR should be ensured in advance for proper implementation of the program. The detailed aspects of prevention and control of PPR have been discussed by several investigators (Singh et al. 2004; Balamurugan et al. 2014a).

References

Administration of Union Territory of Ladakh (AUTL) (2024) Department of Animal/Sheep Husbandry, Administration of Union Territory of Ladakh. Developed and hosted by National Informatics Centre, Ministry of Electronics & Information Technology, Govt of India. Available: https://ladakh.gov.in/animal-sheep-husbandry/

Ahmad PA, Sharma R, Katoch R, Godara R, Yadav A (2018) Prevalence of larval tapeworm in sheep and goats of Jammu, India. Indian J Small Ruminants 24(2):304–307

Allaie IM, Shahardar RA, Tramboo SR, Bulbul KH, Wani ZA, Khan AA (2018) Prevalence of gastro-intestinal nematodes in small ruminants of Kashmir valley. J Entomol Zool Stud 6(2):2554–2559

Amin A, Wani SA (2012) Gastrointestinal helminth parasites of slaughtered sheep with special reference to age, sex and seasonal distribution in District Doda, Jammu and Kashmir state of India. Indian J Sci Res Technol 1(1):25–28

Asrani RK, Batta MK, Katoch RC, Jithendran KP, Sharma M, Singh SP, Gupta VK (1999) Outbreak of verminous bronchopulmonitis among sheep and goat in Himachal Pradesh. Indian J Anim Sci 69(4):207–210

Balamurugan V, Sen A, Venkatesan G, Yadav V, Bhanot V, Riyesh T, Bhanuprakash V, Singh RK (2010) Sequence and phylogenetic analyses of the structural genes of virulent isolates and vaccine strains of peste des petits ruminants virus from India. Transbound Emerg Dis 57(5):352–364

Balamurugan V, Saravanan P, Sen A, Rajak KK, Bhanuprakash V, Krishnamoorthy P, Singh RK (2011) Sero-epidemiological study of peste des petits ruminants in sheep and goats in India between 2003 and 2009. Rev Sci Tech 30(3):889–896

Balamurugan V, Saravanan P, Sen A, Rajak KK, Venkatesan G, Krishnamoorthy P, Bhanuprakash V, Singh RK (2012) Prevalence of peste des petits ruminants among sheep and goats in India. J Vet Sci 13(3):279–285

Balamurugan V, Hemadri D, Gajendragad MR, Singh RK, Rahman H (2014a) Diagnosis and control of peste des petits ruminants: a comprehensive review. Virusdisease 25(1):39–56

Balamurugan V, Das S, Raju DSN, Chakravarty I, Nagalingam M, Hemadri D, Govindaraj G, Singh NI, Ltu K, Devi M, Sharma K, Gajendragad MR, Rahman H (2014b) Prevalence of peste des petits ruminants in goats in North-East India. Virusdisease 25(4):488–492

Balamurugan V, Varghese B, Kumar KV, Muthuchelvan D, Dheeraj R, Govindaraj G, Suresh KP, Hemadri D, Roy P (2020a) Seroprevalence study of peste des petits ruminants in sheep and goats in the northern region of India. Vet World 13(8):1573–1580

Balamurugan V, Varghese B, Muthuchelvan D, SowjanyaKumari S, Vinod Kumar K, Dheeraj R, Govindaraj G, Suresh KP, Hemadri D, Roy P (2020b) Seroprevalence of peste des petits ruminants in small ruminants in the North Eastern Region of India. Vet Ital 56(1):55–60. https://doi.org/10.12834/VetIt.1985.10631.2

Balamurugan V, Varghese B, Muthuchelvan D, Kumar KV, Chettri K, Suresh KP, Govindaraj G, Hemadri D, Roy P (2020c) Serosurvey of peste des petits ruminants in small ruminants in the hilly terrain north-eastern state of Sikkim in India. Adv Anim Vet Sci 8(12):1421–1426

Bandyopadhyay SK (2002) The economic appraisal of PPR control in India. Paper presented at the14th Annual conference and national seminar of Indian Virological Society on management of viral diseases with emphasis on global trade and WTO regime (2002) held at Bangalore, 18–20 January

Bandyopadhyay S, Devi P, Bera A, Bandyopadhyay S, Bhattacharya D (2010) Prevalence of gastrointestinal parasite in goats in Shillong, Meghalaya, India. Webmed Cent Parasitol 1(9):WMC00777

Banerjee M, Gupta PP (1980) Incidence and pathology of verminous pneumonia in sheep and goat in Ludhiana area (Punjab). J Res Punjab Agric Univ 17:100–103

Batta MK, Joshi VB, Asrani RK, Sharma M, Katoch RC, Nagal KB (1996) *Chlamydia psittaci* infectons among migratory sheep and goats in Himachal Pradesh and Jammu & Kashmir, India. J Appl Anim Res 9:45–49

Bhalerao GD (1932) On some nematode parasites of goat and sheep at Mukteshwar. Ind J Vet Sci Anim Husb 2:242–254

Bhardwaj B, Chahota R, Gupta S, Malik P, Sharma M (2017) Identification of chlamydial strains causing abortions and pneumonia in sheep and goat flocks during trans Himalayan seasonal migration in the northern region of India. Vet Arhiv 87(2):157–170

Bhat AS, Reshi AA, Mir RM, Husain I, Sheikh BA, Khan HM (2014) Assessment of prevalence of parasitic infections in sheep reared in Kashmir valley, India. J Anim Prod Adv 4(5):416–421

Bihaqi SJ, Allale IM, Banday MAA, Wani ZA, Shahardar RA (2017) Prevalence of caprine GI helminths in temperate areas of Jammu & Kashmir. J Parasit Dis 41(3):843–849

Chahota R, Gupta S, Bhardwaj B, Malik P, Verma S, Sharma M (2015) Seroprevalence studies on animal chlamydiosis amongst ruminants in five states of India. Vet World 8(1):384–387

Chaudhary S, Vatsya S, Yadav CL (2014) Epidemiology of paramphistomosis in domestic ruminants of Garhwal region of Uttarakhand, India. Vet Res Int 2(1):12–14

Dar LM, Darzi MM, Mir MS, Kamil SA, Rashid A, Abullah S, Hussain SA, Bhat AA, Rashi PA (2012) Prevalence and pathology of lung worm infection in sheep in Kashmir valley, India. J Anim Sci Adv 2(8):678–685

Das M, Laha R, Goswami A, Sen A (2017) Gastrointestinal parasitism of goats in hilly region of Meghalaya, India. Vet World 10(1):81–85

Department of Animal Husbandry and Veterinary Science, Meghalaya. 2016. www.megahvt.gov.in

Department of Animal Husbandry and Veterinary Science, Sikkim. 2016. www.sikkim-ahvs.gov.in

Dhar DN, Dash PK (1982) Lung worms and strangulate nematode infections in sheep in the Kullu valley of Himachal Pradesh. Indian J Anim Sci 43(1):53–55

Dhar P, Sreenivasa BP, Barrett T, Corteyen M, Singh RP, Bandyopadhyay SK (2002) Recent epidemiology of peste des petits ruminants virus (PPRV). Vet Microbiol 88(2):153–159

ENVIS (2016) The Indian Himalayan states. Official web portal of GB Pant National Institute of Himalayan Environment and Sustainable Development, Almora, India. www.gbpihad.gov.in. Accessed 5 Sept 2016

Ghimire TR, Bhattarai N (2019) A survey of gastrointestinal parasites of goats in a goat market in Kathmandu, Nepal. J Parasit Dis 43(4):686–695

Godara R, Katoch R, Yadav A, Borah MK (2014) Dicrocoeliosis in goats in Jammu, India. J Parasit Dis 38(2):201–204

Government of India (GOI) (2019) 20th Livestock census: all India report. Department of Animal Husbandry and Dairying, Ministry of Fisheries, Animal Husbandry & Dairying, Govt of India. Available: https://dahd.nic.in/sites/default/filess/20thLivestockCensus2019AllIndiaReport.pdf

Hamadani A, Ganai NA, Rather MA, Shanaz S, Ayaz A, Mansoor S, Nazir S (2022) Livestock and poultry breeds of Jammu and Kashmir and Ladakh. Indian J Anim Sci 92(4):409–416

Javaid M, Rashid M, Nabi B, Mudasir M, Bhat MA, Wani NM, Bhat MA (2018) Prevalence of gastrointestinal nematodes in sheep and goats reared by nomads of Jammu region of Jammu and Kashmir. Int J Livest Res 8(8):219–225

Jithendran KP (1998) Epidemiology of gastrointestinal nematodes in migratory sheep and goats in north-western humid Himalayan region. Indian J Anim Sci 68(9):894–896

Jithendran KP (2000) Helminth parasites – a constraint in animal health management in Himachal Pradesh. ENVIS Bull Himal Ecol 8(2):7–20

Jithendran KP, Bhat TK (1996) Prevalence of dicrocoeliosis in sheep and goats in Himachal Pradesh, India. Vet Parasitol 61(3–4):265–271

Jithendran KP, Sharma AK, Kurade NP, Katoch VC (2000) Probable peste des petits ruminants among migratory sheep and goats in the sub-Himalayan region of India. Indian Vet J 77(5):435–437

Joshi VB, Nagal KB, Sharma M, Katoch RC, Batta MK, Sharma AK (1996) Peste des petits ruminants (PPR) among Gaddi sheep and goats in Himachal Pradesh. Indian J Anim Sci 66:1126–1127

Khajuria JK, Katoch R, Yadav A, Godara R, Gupta SK, Singh A (2012) Seasonal prevalence of gastrointestinal helminths in sheep and goat of middle agro-climatic zones of Jammu province. J Parasit Dis 37(1):21–25

Khanyari M, Robinson S, Milner-Gulland EJ, Morgan ER, Rana RS, Suryawanshi KR (2022) Pastoralism in the high Himalayas: understanding changing practices and their implications for parasite transmission between livestock and wildlife. Pastoralism 12:44. https://doi.org/10.1186/s13570-022-00257-1

Krishna L, Gupta VK, Katoch RC (1987) Studies on the outbreaks of verminous pneumonia in Gaddi sheep and goats in Himachal Pradesh. Indian J Comp Microbiol Immunol Infect Dis 8(4):170–172

Kuchai JA, Chishti MZ, Zaki MM, Ahmad J, Rasool M, Dar SA, Tak H (2011a) Prevalence of nematode parasites in sheep of Ladakh-India. J Agric Ext Rural Dev 3(13):229–231

Kuchai JA, Chishti MZ, Zaki MM, Ahmad J, Rasool M, Dar SA, Tak H (2011b) Epidemiology of helminth parasites in small ruminants of Ladakh, India. Online J Anim Feed Res 1(5):239–242

Kumar R, Katoch RC, Sharma M, Mahajan A, Dhar P (2004) Pathoetiology of pneumanic Gaddi goats in Himachal Pradesh. Indian J Anim Sci 74(6):569–570

Laha R, Das M, Goswami A (2013) Seasonal dynamics of gastrointestinal parasitic infections and prediction of *Haemonchus contortus* infections of goats in a subtropical hilly region of north-eastern India. Indian J Anim Res 47(4):309–314

Lone BA, Chishti MZ, Ahmad F, Tak H (2012) A survey of gastrointestinal helminth parasites of slaughtered sheep and goats in Ganderbal, Kashmir. Glob Vet 8(4):338–341

Lone IM, Shah MM, Iqbal I, Baba MA, Sakina A, Wafai B, Nasreen S (2013) Seasonal occurrence of gastrointestinal parasites in sheep of Kashmir division. Int J Livest Res 3(2):135–138

Mahajan S, Agrawal R, Kumar M, Mohan A, Pande N (2013) Incidence of peste des petits ruminants in nomadic sheep and goat of Jammu region. Vet World 6(7):384–387

Maitra A, Yadav CL, Sanjukta RK (2014) Seasonal prevalence of paramphistomosis in domestic ruminants in different agro-climatic zones of Uttarakhand, India. Asian Pac J Trop Dis 4(Suppl 2):S748–S753

Meena HR, Ram H, Seth P (2016) Livestock diseases in sub-Himalayan temperate region – a Garrett's ranking analysis. Int J Livest Res 2(1):160–172

Mir MR, Chishti MZ, Dar SA, Rashid M (2012) Bionomics of helminth parasites in goats of sub-tropical Jammu area of J&K, India. Agric Biol J N Am 4(3):316–319

Mir MR, Chishti MZ, Rashid M, Dar SA, Katoch R, Khajuria JK, Mehraj M, Dar MA, Rasool R (2013a) Incidence of gastrointestinal nematodosis in sheep of Jammu. Trends Parasitol Res 2(1):1–4

Mir MR, Chishti MZ, Rashid M, Dar SA, Katoch R, Mehraj M, Dar MA, Rasool R (2013b) The epidemiology of caprine fascioliasis in Jammu (J&K) – India. Int J Food Agric Vet Sci 3(1):233–237

Mir MR, Chishti MZ, Rashid M, Dar SA, Kuchay DA, Dar JA (2013c) Prevalence of gastrointestinal nematodes in goats of Jammu region. Int J Recent Sci Res 4(3):208–210

Moudgil AD, Sharma A, Verma MS, Kumar R, Dogra PK, Moudgil P (2017) Gastrointestinal parasitic infections in Indian Gaddi (goat) breed bucks: clinical, hemato-biochemical, parasitological and chemotherapeutic studies. J Parasit Dis 41(4):1059–1065

Namgay K, Millar JE, Black RS, Samdup T (2014) Changes in transhumant agro-pastoralism in Bhutan: a disappearing livelihood? Hum Ecol 42:779–792

Nanda YP, Chatterjee A, Purohit AK, Diallo A, Innui K, Sharma RN, Libeau G, Thevasagayam JA, Brüning A, Kitching RP, Anderson J, Barrett T, Taylor WP (1996) The isolation of peste des petits ruminants virus from northern India. Vet Microbiol 51(3–4):207–216

Pal P, Bandyopadhyay S (2004) Prevalence of gastrointestinal nematodosis in goats in Sikkim. J Vet Parasitol 18(2):127–130
Pal P, Chatlod LR, Avasthe RK (2017) Seasonal prevalence of gastrointestinal parasites of goats in North-East Himalayan region of Sikkim, India. Indian J Anim Sci 87(5):558–561
Pal P, Kumar B, Avasthe RK, Bhutia KT, Chatlod LR (2020) Prevalence of gastrointestinal parasites in Tibetan sheep in alpine zone of North Sikkim, India. Indian J Anim Sci 90(12):1567–1571
Pant K, Rajput MKS, Kumar J, Sahu S, Rajkumari V, Gangwar P (2009) Prevalence of helminthes in small ruminants in Tarai region of Uttarakhand. Vet World 2(7):265–266
Rahman H, Pal P, Chatlod LR (2013) Incidence of gastrointestinal parasites in ruminants of organized farm in Sikkim. Indian J Anim Sci 83(5):484–487
Ram H, Sharma AK, Singh SK, Meena HR, Rasool TJ (2007) Seasonal nematode egg output in goats at high altitude of Uttarakhand. J Vet Parasitol 21(1):73–74
Rawat VS (2010) Problems of animal grazing in Uttarakhand Himalaya. Researcher 2(5):81–85
Rodolakis A, Boullet C, Souriau A (1984) *Chlamydia psittaci* experimental abortion in goats. Am J Vet Res 45(10):2086–2089
Sadana JR, Purohit VD, Kalra DS (1979) Incidence and pathology of verminous pneumonia in sheep and goats. Haryana Vet 18:111–116
Saleem T, Katoch R, Godara R, Yadav A, Khursheed A (2018) Epidemiology of gatro-intestinal parasites in small ruminants of Jammu Province. Indian J Small Ruminants 24(2):357–359
Shabih HS, Juyal PD (2006) Epidemiological observations of paramphistomosis in ruminants in endemic regions of Punjab and adjoining state (India). In: Proceedings of the 11th international symposium on veterinary epidemiology and economics. www.sciquest.org.nz
Shah MM, Maqbool I, Manzoor A, Shbnum N, Rather T, Ahanger SA (2018) Seasonal prevalence of gastrointestinal parasites in sheep of Kashmir valley. J Entomol Zool Stud 6(6):534–536
Shaila MS, Purashothaman V, Bhavsar D, Venugopal K, Venkatesan RA (1989) Peste des petits ruminants of sheep in India. Vet Rec 125(24):602
Sharma RL (1994) Parasitic bronchitis in goats and the possible use of *Dictyocaulus filaria* vaccine for its control. Vet Parasitol 51(3–4):255–262
Sharma D, Katoch R, Agnihotri RK (2007) Gastrointestinal helminths in Gaddi sheep. J Vet Parasitol 21(2):141–143
Sharma M, Batta MK, Katoch RC, Anderson AA (2008) A field investigation of bacterial etiology of abortions among migratory sheep and goats in north-west hill states of India. Vet Arhiv 78(1):65–71
Sharma D, Agnihotri RK, Dogra PK, Sen D, Katoch A (2012) Incidence of gastrointestinal parasites among Chegu goats in Lahaul and Spiti districts of Himachal Pradesh. Himachal J Agric Res 38(1):61–63
Sharma A, Sharma S, Kour S, Avatsingh AU, Perveen K, Alsulami JA, Singh N (2023) Gastrointestinal nematodes and protozoa in small and large ruminants from rural agro-climatic regions of northern India. Diversity 15(11):1131. https://doi.org/10.3390/d15111131
SHDK (2024) Sheep Husbandry Department Kashmir, Livestock Statistics, Directorate of Sheep Husbandry Department. Available: www.jksheephusbandrykashmir.nic.in
Singh RP, Saravanan P, Sreenivasa BP, Singh RK, Bandyopadhyay SK (2004) Prevalence and distribution of peste des petits ruminants virus infection in small ruminants in India. Rev Sci Tech 23(3):807–819
Singh G, Rajeev S, Singh R, Katoch R, Wazir VS, Khajuria JK, Tikko A, Anand A (2014) Prevalence of haemonchosis in sheep and goats of Jammu. Indian Vet J 91(1):52–53
Singh J, Wadhwa DR, Sharma A, Katoch A, Vinay HB (2023) Insight into hematobiochemical profiling and gastrointestinal parasite prevalence among migratory Gaddi sheep in Himachal Pradesh. Pharma Innov SP-12(11):558–560
Sofi OMUD, Godara R, Katoch R, Yadav A, Nazim K, Chakraborty D (2022) Epidemiology and economic impact of amphistomosis in sheep of north-western Himalayas, India. Res Sq:1–10. https://doi.org/10.21203/rs.3.rs-1783347/v1
Tak IR, Chishti MZ, Ahmad F (2013) Epidemiological studies of abomasal nematodes of sheep of Kashmir valley with particular reference to *Haemonchus contortus*. Nat Sci 11(10):34–39

Tariq KA, Chishti MZ, Ahmed F, Shawl AS (2008a) Epidemiological studies on paramphistomum infection in goats in Kashmir valley. World J Agric Sci 4(1):61–66

Tariq KA, Chishti MZ, Ahmad F, Shawl AS (2008b) The epidemiology of paramphistomosis of sheep (*Ovis aries* L.) in the north west temperate Himalayan region of India. Vet Res Commun 32(5):383–391

Tariq KA, Chishti MZ, Ahmad F, Shawl AS (2008c) Epidemiology of gastrointestinal nematodes of sheep managed under traditional husbandry system in Kashmir valley. Vet Parasitol 158(1–2):138–143

Tramboo SR, Shahardar RA, Allaie IM, Wani ZA, Bushra MS (2015) Prevalence of gastrointestinal helminth infections in ovine population of Kashmir valley. Vet World 8(10):1199–1204

Tribune Newspaper (2001) Migratory cattle hit by virus. Tribune News Service, June 21

Tribune Newspaper (2014) Virus attacks sheep and goats in Kinnaur district of Himachal Pradesh—250 dead. Tribune on line edition, August 10

Uniyal SK, Lal B, Kumar A (2005) Mass mortality in livestock in the high altitude areas of western Himalayas. Curr Sci 88(4):534–535

Vohra S, Katoch R, Yadav A, Khajuria JK, Gupta S (2016) An outbreak of gastrointestinal parasitic infection in sheep of R. S. Pura region of Jammu. Int J Sci Environ Technol 5(4):2057–2061

Wani SA (2015) Small ruminant rearing among Bakerwal pastoralists of Jammu and Kashmir – an exploratory study. PhD thesis, ICAR-National Dairy Research Institute (Deemed University), Karnal, India

Wani ZA, Shahadar RA, Shahnawaz M (2013) Prevalence of nemathelminth parasites in locally reared sheep of Ganderbal district of Kashmir valley based on necropsy examination. Int J Adv Vet Med Sci 7(5):142–148

Wani, Quyoom I, Bhat BA, Zasay J, Tanveer S (2024) Does the presence of livestock in protected areas pose a risk of gastrointestinal parasite transmission to wild ungulates? A study from Bani Wildlife Sanctuary, Western Himalayas. J Wildl Biodivers 8(2):246–259

Vaccines Against Gastrointestinal Parasitic Infections for Sustained Health and Production of Small Ruminants

11

Abstract

Migratory sheep and goats in north-western Himalayan region are raised for meat, wool, and milk production. Factors adversely affecting animal health and performance are responsible for economic losses in animal production. Parasitic infections particularly those affecting gastrointestinal tract cause morbidity and mortality in farm animals and contribute substantially to the spread of zoonotic diseases. Thus, proper control and management of parasitic infections are crucial for sustainable and profitable sheep and goat production. The most common and efficacious approach to control parasitic infections is chemotherapy, but it has disadvantages such as development of drug resistance and chemical residues in the animal products. Management-related strategies include grazing management, biological control of parasites, and breeding for genetically resistant sheep and goats. Vaccination is likely to be part of integrated parasite management schemes. A gamma radiation-attenuated live parasite vaccine against *Dictyocaulus filaria* was developed in India. Helminth and protozoal vaccines currently available are those against *Haemonchus contortus*, *Echinococcus granulosus*, *Taenia ovis*, and *Toxoplasma gondii*. Vaccination of farm small ruminants against chronic and subclinical parasitic infections would be an effective, sustainable, and economical approach to control parasitic infections for improvement of animal health and production.

Keywords

Vaccines · Parasitic infestations · Chemotherapy · Biological control · *Haemonchus contortus*

S. P. Singh, J. K. Malik, *Migratory Small Ruminant Farming System in the Himalayas-Peculiarities, Problems, Prospects, Climate Effects, Nutrition and Health*, https://doi.org/10.1007/978-981-96-6691-1_11

11.1 Introduction

Sheep and goats with their multi-faceted utility for wool, meat, milk, skins, and manure form an integral component of hill farming system in the north-western Himalayan states. Small ruminants require low input and provide a consistent source of livelihood to the marginal farmers and shepherds through sale of animal products. Migratory system is highly relevant as a large sheep and goat population is reared under this system and many tribal communities are involved in migratory sheep and goat husbandry (Sankhyan and Thakur 2019). Parasitic infections impact huge economic losses in terms of negligible production, reduced weight gain, reduced wool and milk production, and reduced reproductive performance. In view of large small ruminant flocks, heavy annual economic losses due to roundworms parasites have been reported to the tune of $103 million in India alone (McLeod 2004). The control of parasitic diseases is still based on the application of anthelmintic drugs; however, their inadvertent and excessive use in the past decades has led to the widespread emergence of drug resistance by the parasite and the presence of chemical residues in the animal products. High levels of resistance in gastrointestinal nematodes of small ruminants have been reported for all classes of anthelmintic drugs worldwide (Claerebout and Geldhof 2020; Arsenopoulos et al. 2021; Charlier et al. 2022; Alaro et al. 2023; Babják et al. 2023). Prevalence surveys conducted in the north-western Himalayan states have indicated the widespread emergence and spread of anthelmintic resistance in gastrointestinal (GI) nematodes of both sheep and goats (Kumar et al. 2017; Bihaqi et al. 2020; Kalkal et al. 2020). The emergence of resistance to anthelmintic drugs poses a major constraint to future production and welfare of grazing small ruminants. Thus, effective integrated approaches for control of parasitic infections like grazing management, biological control, chemotherapy, genetically resistant strains, and effective parasitic vaccination strategies are vital for sustained productivity and performance of animals. This chapter deals mainly with the role of vaccines against important gastrointestinal parasitic infections of sheep and goats as an integrated approach for controlling helminth and potozoan parasites.

11.2 Approaches to Control Parasitic Infections in Small Ruminants

Strategies to control parasitic infections in sheep and goats should be formulated with the aims to maintain financial stability, sustainability, and their efficacy to reduce the parasite load. An approach for integrated parasite control in animals generally includes grazing management, breeding small ruminants for resistance to parasites, biological control, strategic chemotherapy, and prophylactic vaccination (Ramaswamy 2015; Sharma et al. 2015; Vercruysse et al. 2018; Verma et al. 2018; Arsenopoulos et al. 2021).

11.2.1 Grazing Management

In grazing management, the animal grazing pattern is manipulated to attain the optimum results by reducing the parasite burden. The major source of spread of parasitic infection in grazing animals is through grazing of pastures containing eggs and larvae of parasites, and its decontamination is very difficult. The strategies to prevent pasture infection generally comprise of controlled rotational grazing or pasture rotation system, alternate or mixed grazing, pasture resting, zero grazing, and use of bioactive forages (Verma et al. 2018; Haenlein and Park 2020).

11.2.2 Breeding Sheep and Goats for Resistance to Parasites

Among the various methods to control parasitic infections in farm animals, genetic resistance seems to be the best method to sustainable parasite control. At present, research efforts are being made to preferably select animal breeds which are genetically resistant for GI nematode infections for both sheep and goats. Sheep belonging to some breeds such as Morada Nova, Red Masai, Florida Native, St. Croix, Galore, and Barbados Backbelly are more resistant against GI nematodes. Animals of resistant breeds harbor fewer worms and fecal eggs excreted are also few and do not contribute to intense infection of pastures (Verma et al. 2018; Arsenopoulos et al. 2021). The selective crossbreeding is desirable for the development of breeds for enhanced resistance to parasitic infections which should ultimately lead to sustained improvements in health and production of sheep and goats.

11.2.3 Biological Control

Nematophagous fungi such as *Duddingtonia flagrans*, *Drechmeria coniospora*, *Harposporium anguillulae*, *Verticillium chlamydosporium*, *Arthrobotrys oligospora*, and *Paecilomyces lilacinus* have the ability to infect and damage parasite eggs, thus preventing the development of eggs to infective larvae on pasture. These fungi are naturally present in feces of various ruminant species (Larsen 1999). These nematophagous fungi remain unaltered during the GI passage and subsequently trap and destroy the developing helminth parasite larvae in the deposited feces. These fungi are very efficient in controlling *Haemonchus* spp., *Trichostrongylus* spp., *Ascaris* spp., *Fasciola* spp., and *Ostertagia* spp. of parasites (Sharma et al. 2015; Arsenopoulos et al. 2021). *D. flagrans* is one of the most studied nematophagous fungi, and there are already commercial products which are being employed for biological control of parasites in livestock production (Araújo et al. 2021). *D. flagrans* is known to produce a three-dimensional adhesive network responsible for capturing and eliminating nematodes (Wang et al. 2019). Similarly, *A. oligospora* is also the most widely investigated fungus which can make three-dimensional network traps that capture nematodes. The trapping process of this fungus usually involves attraction, recognition, trap formation, adhesion, penetration, and digestion (Wang et al. 2023).

11.2.4 Chemotherapy

Chemotherapy is one of the most economical, effective, and extensively used approach to control and treat parasitic infections in farm animals. At present, several broad-spectrum and highly efficacious anthelmintics are available to control parasitic infections. Administration of anthelmintic drugs in appropriate dose during the top season of parasitic infections can reduce the cost and maximize the benefit (Sharma et al. 2015; Ramaswamy 2015; Verma et al. 2018).

In spite of the fact that anthelmintic therapy has a great impact on animal health and production, there are several disadvantages. Resistance to anthelmintic agents and presence of their residues in meat and milk of treated animals are the major disadvantages of using chemotherapy for control of parasitic diseases. Undiscriminating and overuse of anthelmintics is supposed to be the cause of emergence of drug resistance. Anthelmintic resistance is currently a major problem in the production of small ruminant species in many countries worldwide (Kumar et al. 2017; Bihaqi et al. 2020; Claerebout and Geldhof 2020; Kalkal et al. 2020; Arsenopoulos et al. 2021). The problem of acquired anthelmintic resistance in parasitic nematodes has surfaced in nearly all drugs presently available and is likely to continue due to the high levels of genetic diversity among this group of parasites, mainly as a result of often improper dosing, and the suboptimal mode of application to many infected domestic animals (Rosa et al. 2023). For reducing the selection pressure for anthelmintic resistance in helminth parasite population, it has been recommended to decrease drench frequency, use of optimal doses, alteration of anthelmintic classes, combination drug strategy, and preferred application of narrow-spectrum anthelmintics whenever possible depending on the parasitic species. The target-selective treatment is another recommended strategy in which only those animals that will most benefit from treatment are given anthelmintic drug (Verma et al. 2018; Kalkal et al. 2020). The development of new anthelmintic classes with novel mechanisms of action could potentially solve this problem, but the discovery of new drugs is exceedingly expensive (Kaplan and Vidyashankar 2012). Drugs that are rapidly metabolized into harmless metabolites should be used to reduce drug residues in meat and milk (Waller 2006).

11.3 Vaccines Against Gastrointestinal Parasitic Infections

Parasitic vaccines are most effective and dependable method to control parasitic diseases and are valuable in reducing reliance on anthelmintic drugs. Besides improving animal health and production by controlling parasitic infections, there are other advantages of using veterinary parasite vaccines such as improving public health by controlling zoonoses and foodborne pathogens in animals, solving problems relating to anthelmintic resistance and keeping animals and the environment free of drug residues and preserving biodiversity (Innes et al. 2011; Sharma et al. 2015; Matthews et al. 2016; Claerebout and Geldhof 2020).

Three major families of parasitic helminths are recognized: nematodes (roundworms), trematodes (flatworms), and cestodes (tapeworms). Being the multicellular parasites having genetic complexity, and their physical size, helminths cannot be internalized by phagocytic cells of the immune system or killed by classical cytotoxic T cells. It has been documented that helminth parasites are recognized by dendritic cells which have capability to induce a Th2-type response. Other phagocytes are essential to generate an intricate network of cytokines, soluble factors, and danger signals originating from the damage produced by parasite migration, which promotes a Th2-type response. This response leads to alternative activation of macrophages, and the activation of eosinophils, basophils, and mast cells, which ultimately contribute to the expulsion of intestinal nematodes (Motran et al. 2018). The current status of vaccines for the gastrointestinal parasites of sheep and goats is discussed below.

11.3.1 *Haemonchus contortus*

Gastrointestinal nematodosis is a common parasitic infection of domestic small ruminant species in India including sub-Himalayan region of northern India. The salient features of the gastrointestinal helminth infections in small ruminants of the Indian Himalayan states have been extensively reviewed in Chap. 10 of GI nematodes. *Haemonchus contortus* is an economically important and highly pathogenic nematode parasite that leads to anemia, primarily due to the blood-sucking activity, thereby adversely affecting the health and productivity of animals and in few cases resulting in the death of the infected animals (Arsenopoulos et al. 2021). The financial losses associated with GI nematodes are considerable as most grazing animals are exposed to infection while at pasture (Matthews et al. 2016).

There are two types of main antigens produced by helminth parasites namely soluble excretory/secretory products and hidden gut membrane antigens. Hidden antigens are found in the intestinal cells of *H. contortus* and these are not normally exposed to immune system of the host during infection. Excretory/secretory products and hidden gut membrane antigens appear to be the potent candidates for development of vaccines against *Haemonchus* spp. (Bassetto and Amarante 2015). Two gut antigens namely H11 and *Haemonchus* galactose-containing glycoprotein complex (H-gal-GP) have been demonstrated to confer high degree of protection against larval challenge in sheep. Both H11 and H-gal-GP antigens may induce strong protective immunity in sheep and goats when used separately (Adduci et al. 2022). The H11 is a microsomal aminopeptidase expressed on the microvilli of intestine of adult parasites. The H-gal-GP contains aspartyl, metallo- and cysteine proteinases (Smith et al. 1999). The major component of this gut membrane glycoprotein complex is a family of four zinc metalloendopeptidases, designated as MEPs 1–4 of which MEP3 appears to be the most abundant. H-gal-GP also contains a family of galectins which particularly bind β-galactoside sugars and have varied roles in the cell adhesion, immune function, and apoptosis (Maqbool et al. 2018). In a recent study, Scarff et al. (2020) solved the structure of H-gal-GP and showed that it acts as an efficient digestion machine capable of trapping hemoglobin and channeling it to different enzymes for processing. A number of trials showed that immunization of lambs and sheep with either H11 or H-gal-GP antigens afforded reduction in fecal egg counts and worm burdens (Smith et al. 1994, 1999; Cachat et al. 2010; Roberts et al. 2013). Although native proteins are highly protective, the recombinant forms expressed in bacteria, yeast, or insect cells are far less effective, which is likely due to incorrect folding or inappropriate glycosylation (Newton and Meeusen 2003; Matthews et al. 2016; Maqbool et al. 2018). As adult parasites cannot be obtained by in vitro cultivation, the native antigens are required to be extracted and purified locally from adult *H. contortus* derived from the abomasa of infected sheep for production of the vaccine (Adduci et al. 2022).

A vaccine (Barbervax, Wormvax Australia Pty Ltd) containing gut membrane glycoprotein antigens from adult *H. contortus* is now marketed exclusively for sheep in Australia and South Africa (Table 11.1). In South Africa, it is named as Wirevax (Wirevax, Afrivet Business Management Pty Ltd). In the United Kingdom, the vaccine is available only under the veterinarian's prescriptions. By the year 2018, its respective commercial registration was obtained in New Zealand and Europe (Vargas et al. 2022). The vaccine contained 5 μg/ml of integral glycoprotein complexes (H11 and H-gal-GP), 1 mg saponin (Quil A) in 1 ml of Tris-buffered saline pH 7.4, and 0.1% Triton X-100 (Smith et al. 2000; Teixeira et al. 2019; Claerebout and Geldhof 2020). Antibodies present in the blood of sheep administered with the vaccine are directed against antigens expressed on the intestinal enterocytes of the parasite. As the parasite ingests blood of the vaccinated sheep, the specific antibodies present in the blood bind to and damage intestinal proteins of the

Table 11.1 Commercialized helminth and protozoal vaccines in small ruminants

Brand name	Parasite	Animal species	Vaccination schedule	Level of protection	Reference
Barbervax, Wirevax	*Haemonchus contortus*	Sheep	Basic scheme: three vaccinations with 4-week interval before turnout. Revaccination: single dose before turnout	93–95% (egg counts) 72–94% (worm burden)	Claerebout and Geldhof (2020)
Providean Hidatil EG95	*Echinococcus granulosus*	Sheep, goats	Two vaccinations with 1-month interval	95–100% (cysts)	Claerebout and Geldhof (2020)
Toxovax	*Toxoplasma gondii* (Toxoplasmosis abortion)	Sheep	One vaccination in pre-breeding ewe lambs	Enduring protection	Wilkins et al. (1988); Buxton et al. (1993); Mcallister (2014)

parasite ultimately leading to interference with the augmentation of *H. contortus*. This results in marked reduction in egg output and a decrease in numbers of parasite (Matthews et al. 2016; Teixeira et al. 2019).

Three priming doses of the vaccine are administered subcutaneously 3–4 weeks apart to induce immunity before the high *Haemonchus*-risk season and boosts are administered at 6-week intervals to maintain resistance. The vaccine produces 93–95% decrease in egg counts and reduction of 72–94% in worm burden. No withdrawal time after vaccination is recommended (Claerebout and Geldhof 2020). In a recently conducted trial, it was shown that immunization of hair sheep experimentally exposed to higher artificial challenge with Barbervax vaccine had high antibody levels and reduced egg counts and worm burdens by 87% and 79%, respectively (Teixeira et al. 2019). Goats are more susceptible than sheep to the nematode *H. contortus*. Attempt has been made to control haemonchosis with Barbervax vaccine in dairy goats. The average efficacy over the trial period was found to be 69.8 ± 2.11% and 57.4 ± 2.76% for the Anglo Nubians and Saanens goats, respectively. The protection induced by vaccination was considered to be useful for management of haemonchosis in dairy goats by reducing egg counts (de Matos et al. 2017).

11.3.2 *Dictyocaulus filaria*

Parasitic bronchitis is one of the most important diseases of small ruminants. The disease is widely prevalent in migratory sheep and goats in the northwest Himalayan and other hilly regions of India and is responsible for heavy mortality in young animals. Animals that survive the acute infection may become chronic carriers and serve as a source of contamination of pastures. Among the species of trichostrongyloid and metastrongyloid nematodes causing parasitic bronchitis, *Dictyocaulus filaria* is the most pathogenic and most widely prevalent constituting a major economic threat to the health and production of sheep and goats (Dhar et al. 1981; Kumar et al. 2018). In India, a gamma radiation-attenuated live parasite vaccine against *D. filaria* named DIFIL was developed at the Nuclear Research Laboratory, Indian Veterinary Research Institute. In this vaccine, third-stage larvae were exposed to 50 krad gamma radiation. The vaccine contains irradiated L3-larvae that do not mature to adult worms. In addition to *D. filaria*, the vaccine also appeared to induce a degree of acquired resistance against other nematode lungworms such as *Varestrongylus pneumonicus*, *Protostrongylus rufescens*, and *Muellerius capillaries*. It has been suggested that the vaccination should be done in very young lambs (6 weeks old) before they receive a natural infection. The vaccine is given orally to lambs between 6 and 10 weeks of age. Two doses of the vaccine are administered at an interval of 4 weeks and it produces effective immunity for about 1 year in the absence of further challenge. Due to repeated natural exposure to infection in endemic areas, good immune status may be maintained for even longer period (Dhar and Sharma 1981; Dhar et al. 1981; Kumar et al. 2018).

11.3.3 *Echinococcus granulosus*

Echinococcus granulosus is the causal agent of cystic echinococcosis or hydatidosis, one of the most important zoonotic parasitic disease in humans, livestock, and wildlife worldwide (Pourseif et al. 2018). Sheep are the most important reservoir of this small cestode parasite which can be spread directly or indirectly to human beings. Echinococcosis produces important economical losses in the livestock industry by affecting leather quality, milk production, and animal fertility. Oncospheres or secretory products from cultured oncospheres may provide protection against *E. granulosus* in sheep. EG95, an outer membrane protein from activated oncospheres, seems to be very potent immunogen. EG95 has been cloned into *Escherichia coli* and linked to an enzyme glutathione-S-transferase which helps in the purification of EG95 (Woollard et al. 1998). At present, a recombinant protein vaccine (Providean Hidatil EG95, Tecnovax) is available on the commercial market in parts of South America for the control of *E. granulosus* in sheep and goats (Table 11.1). The vaccine is marketed in Argentina, China, and Peru (Vargas et al. 2022). This vaccine is based on a single antigen EG95 with Montanide and saponin as adjuvants. Two vaccinations with 1-month interval are given by subcutaneous or intramuscular routes. There is no withdrawal period and the level of protection is 95–100% (cysts) (Matthews et al. 2016; Claerebout and Geldhof 2020).

11.3.4 *Taenia ovis*

Taenia ovis is an economically important cestode parasite infecting primarily sheep or goats as intermediate hosts. The parasite causes cysticercosis which is an important public health issue and a cause of considerable economic losses in small ruminants particularly in the sheep meat industry. The recombinant antigen vaccine against *T. ovis* in sheep was the first effective, defined antigen vaccine against a parasitic infection. The recombinant vaccine developed against *T. ovis* in sheep is based on oncosphere antigen which induces immune response. Three separate host-protective antigens namely To16, To18, and To45W have been cloned from the oncosphere of the parasite (Jabbar et al. 2010).

Considerable progress has been made with regard to vaccination against cysticercosis caused by *T. ovis* in sheep. Native oncosphere antigens were recognized as a source of host-protective antigens. Furthermore, specific host-protective molecules have been identified for the cestode, and recombinant DNA techniques have been employed successfully to clone and express the genes encoding three different antigens. It has been documented that each of these was able to induce almost complete immunity against experimental challenge infection in sheep (Johnson et al. 1989; Harrison et al. 1996; Lightowlers et al. 2000). Although the vaccine has been registered for commercial use in New Zealand, it is yet to be successfully marketed (Rickard et al. 1995). Immunization with the 45 W protein in saponin was shown to reduce cysticerci by 94% in vaccinated lambs as compared to control animals following a challenge infection (Johnson et al. 1989). Following two vaccinations with the 45 W protein in both naturally and artificially challenged lambs at 6 and 12 weeks of age, the vaccine was shown to reduce cysticerci counts by 98% compared to controls under field conditions (Lawrence et al. 1996; De Wolf et al. 2014). Two other highly immunogenic recombinant oncosphere antigens To16 and To18 have also been found to elicit protective immunity against *T. ovis* infection in lambs. In clinical trials, To16 and To18 provided a 92% and 99% reduction in the number of cysticerci in vaccinated lambs, respectively, as compared to control animals (Harrison et al. 1996; De Wolf et al. 2014).

11.3.5 *Toxoplasma gondii*

Toxoplasma gondii is an obligate, single-celled intracellular protozoan parasite which is of great economic importance and significantly impact productivity of sheep and goats. The parasite seems to be one of the main causes for reproductive disorders in the small ruminants and is the major cause of abortion in sheep and goats. The live attenuated vaccine Toxovax (Animal Health Ltd., New Zealand) is currently the only commercial vaccine available to control congenital toxoplasmosis in sheep and goats (Table 11.1). The vaccine Strain 48 of *T. gondii* was first isolated from an aborted lamb in New Zealand. Afterward, this strain of *T. gondii* was maintained in the laboratory by a continuous passage in mice, and ultimately following approximately 3000 passages, it lost its ability to differentiate into bradyzoites or oocysts and is not capable of establishing a persistent infection in the host animal (Hasan and Nishikawa 2022). This vaccine is available for use only in the United Kingdom, New Zealand, France, and Ireland (Buxton 1993; Zhang et al. 2022).

Toxovax is based on the incomplete Strain 48 of *T. gondii.* The S48 strain tachyzoites undergo limited multiplication within the host animal and thus induce appropriate cell-mediated immune responses involving CD4+, CD8+ T-cells, and IFNγ, thereby not permitting the protozoan to persist. Toxovax-primed immune system stops the spread of parasite within lymphatic systems and therefore limiting the spread to the circulation and ultimately prevents the parasite from reaching the placenta and producing disease in the pregnant sheep. Toxovax administered prior to mating confers immunity for about 18 months in susceptible ewes against *Toxoplasma*-induced abortion (Buxton et al. 1993). The vaccine is given by intramuscular route, and there is a meat and offal withdrawal period of 6 weeks after vaccination (https://www.noahcompendium.co.uk/?id=-476561). Besides preventing congenital toxoplasmosis, the vaccine is reported to be effective in reducing the risk of meat from the sheep harboring *T. gondii* tissue cysts (Innes et al. 1995, 2009; Sharma et al. 2015; Hasan and Nishikawa 2022).

In conclusion, vaccination is probably the most viable and affordable method to control parasitic infections in livestock including sheep and goats. Anthelmintic drugs act rapidly after administration, but their effect lasts for a short duration as animals become susceptible to further infection after stopping of chemotherapy. Unlike anthelmintic drugs, vaccines are required to be given only infrequently as immunity induced by vaccination is long-lived and there is a lack of occurrence of chemical residues in sheep and goats. The first most effective vaccine against *Dictyocaulus filaria* of sheep was an irradiated larval vaccine developed in India. Currently, helminth vaccines against *Haemonchus contortus*, *Echinococcus granulosus*, and *Taenia ovis* in sheep are commercially available. The vaccine against *Taenia ovis* has received registration for commercial use in New Zealand, but it is yet to be successfully marketed. In addition to helminth vaccines, the live protozoal vaccine against *Toxoplasma gondii* is also commercially available. The use of gastrointestinal helminth and protozoal vaccines could be part of coordinated worm control programs, along with other control methods such as chemotherapy, grazing management, and biological control. Further understanding on the immune responses induced by helminth and protozoal vaccines will pave way to improved antigen-adjuvant combinations and better vaccine delivery. Advancement in areas of molecular biology, immunology, proteomics, and recombinant technology has shown remarkable progress and success in developing vaccines for controlling parasites in farm animals.

References

Adduci I, Sajovitz F, Hinney B, Lichtmannsperger K, Joachim A, Wittek T, Yan S (2022) Haemonchosis in sheep and goats, control strategies and development of vaccines against *Haemonchus contortus*. Animals 12:2339. https://doi.org/10.3390/ani12182339

Alaro T, Dulo F, Wodajo W, Mathewos L (2023) Anthelmintic resistance of gastrointestinal nematodes of communally-grazing goats in Humbo District, Southern Ethiopia. Vet Med (Auckl) 14:185–194

Araújo JV, Braga FR, Mendoza-de-Gives P, Paz-Silva A, Vilela VLR (2021) Recent advances in the control of helminths of domestic animals by helminthophagous fungi. Parasitologia 1(3):168–176. https://doi.org/10.3390/parasitologia1030018

Arsenopoulos KV, Fthenakis GC, Katsarou EI, Papadopoulos E (2021) Haemonchosis: a challenging parasitic infection of sheep and goats. Animals 11:363. https://doi.org/10.3390/ani11020363

Babják M, Königová A, Komáromyová M, Kuzmina T, Nosal P, Várady M (2023) Multidrug resistance in *Haemonchus contortus* in sheep – can it be overcome? J Vet Res 67(4):575–581

Bassetto CC, Amarante AF (2015) Vaccination of sheep and cattle against haemonchosis. J Helminthol 89(5):517–525

Bihaqi SJ, Allaie IM, Banday MAA, Sankar M, Wani ZA, Prasad A (2020) Multiple anthelmintic resistance in gastrointestinal nematodes of caprines on Mountain Research Centre for Sheep and Goat at Kashmir valley, India. Parasite Epidemiol Control 11:e00163. https://doi.org/10.1016/j.parepi.2020.e00163

Buxton D (1993) Toxoplasmosis: the first commercial vaccine. Parasitol Today 9(9):335–337

Buxton D, Thomson KM, Maley S, Wright S, Bos HJ (1993) Experimental challenge of sheep 18 months after vaccination with a live (S48) *Toxoplasma gondii* vaccine. Vet Rec 133(13):310–312

Cachat E, Newlands GF, Ekoja SE, McAllister H, Smith WD (2010) Attempts to immunize sheep against *Haemonchus contortus* using a cocktail of recombinant proteases derived from the protective antigen, H-gal-GP. Parasite Immunol 32(6):414–419

Charlier J, Bartley DJ, Sotiraki S, Martinez-Valladares M, Claerebout E, von Samson-Himmelstjerna G, Thamsborg SM, Hoste H, Morgan ER, Rinaldi L (2022) Anthelmintic resistance in ruminants: challenges and solutions. Adv Parasitol 115:171–227

Claerebout E, Geldhof P (2020) Helminth vaccines in ruminants: from development to application. Vet Clin North Am Food Anim Pract 36(1):159–171

de Matos AFIM, Nobre COR, Monteiro JP, Bevilaqua CML, Smith WD, Teixeira M (2017) Attempt to control *Haemonchus contortus* in dairy goats with Barbervax®, a vaccine derived from the nematode gut membrane glycoproteins. Small Rumin Res 151:1–4

De Wolf BD, Peregrine AS, Jones-Bitton A, Jansen JT, Menzies PI (2014) *Taenia ovis* infection and its control: a Canadian perspective. N Z Vet J 62(1):1–7

Dhar DN, Sharma RL (1981) Immunization with irradiated larvae against *Dictyocaulus filaria* in young lambs. Vet Parasitol 9(2):125–131

Dhar DN, Sharma RL, Bansal GC (1981) The effect of inoculum size and betamethasone treatment on the course of infection with *Dictyocaulus filaria* in lambs. Vet Parasitol 8(3):219–228

Haenlein GFW, Park YW (2020) Fighting the deadly helminthiasis without drug resistance. Dairy 1(3):177–186. https://doi.org/10.3390/dairy1030012

Harrison GB, Heath DD, Dempster RP, Gauci C, Newton SE, Cameron WG, Robinson CM, Lawrence SB, Lightowlers MW, Rickard MD (1996) Identification and cDNA cloning of two novel low molecular weight host-protective antigens from *Taenia ovis* oncospheres. Int J Parasitol 26(2):195–204

Hasan T, Nishikawa Y (2022) Advances in vaccine development and the immune response against toxoplasmosis in sheep and goats. Front Vet Sci 9:951584. https://doi.org/10.3389/fvets.2022.951584

Innes EA, Panton WR, Sanderson A, Thomson KM, Wastling JM, Maley S, Buxton D (1995) Induction of CD4+ and CD8+ T cell responses in efferent lymph responding to *Toxoplasma gondii* infection: analysis of phenotype and function. Parasite Immunol 17(3):151–160

Innes EA, Bartley PM, Maley S, Katzer F, Buxton D (2009) Veterinary vaccines against *Toxoplasma gondii*. Mem Inst Oswaldo Cruz 104(2):246–251

Innes EA, Bartley PM, Rocchi M, Benavidas-Silvan J, Burrells A, Hotchkiss E, Chianini F, Canton G, Katzer F (2011) Developing vaccines to control protozoan parasites in ruminants: dead or alive? Vet Parasitol 180(1–2):155–163

Jabbar A, Crawford S, Gauci CG, Walduck AK, Anderson GA, Lightowlers MW (2010) Oncospheral penetration glands and secretory blebs are the sources of *Taenia ovis* vaccine antigens. Infect Immun 78(10):4363–4373

Johnson KS, Harrison GB, Lightowlers MW, O'Hoy KL, Cougle WG, Dempster RP, Lawrence SB, Vinton JG, Heath DD, Rickard MD (1989) Vaccination against ovine cysticercosis using a defined recombinant antigen. Nature 338(6216):585–587

Kalkal H, Vohra S, Gupta S (2020) Anthelmintic resistance in livestock parasites: Indian scenario. Int J Curr Microbiol App Sci 9(2):2839–2851

Kaplan RM, Vidyashankar AN (2012) An inconvenient truth: global warming and anthelmintic resistance. Vet Parasitol 186(1–2):70–78

Kumar RR, Yadav CL, Vatsya S (2017) Emergence of anthelmintic resistance in *Haemonchus contortus* on organized sheep and goat farms of sub-Himalayan region of northern India. Int J Curr Microbiol App Sci 6(10):790–796

Kumar A, Bhinsara D, Parthasarathi BC, Sankar M (2018) Paradigm of commercial irradiated vaccines for helminth control in livestock and companion animals – a review. Int J Sci Environ Technol 7(3):910–919

Larsen M (1999) Biological control of helminths. Int J Parasitol 29(1):139–146

Lawrence SB, Heath DD, Harrison GB, Robinson CM, Dempster RP, Gatehouse TK, Lightowlers MW, Rickard MD (1996) Pilot field trial of a recombinant *Taenia ovis* vaccine in lambs exposed to natural infection. N Z Vet J 44(4):155–157

Lightowlers MW, Flisser A, Gauci CG, Heath DD, Jensen O, Rolfe R (2000) Vaccination against cysticercosis and hydatid disease. Parasitol Today 16(5):191–196

Maqbool I, Malla BA, Mohmad A, Dutta N, Manzoor N, Kushwaha B, Saini VK, Para IA (2018) Vaccination against *Haemonchus contortus* in small ruminants: a review. Int J Curr Microbiol App Sci 7(11):2878–2895

Matthews JB, Geldhof P, Tzelos T, Claerebout E (2016) Progress in the development of subunit vaccines for gastrointestinal nematodes of ruminants. Parasite Immunol 38(12):744–753

McAllister MM (2014) Successful vaccines for naturally occurring protozoal diseases of animals should guide human vaccine research. A review of protozoal vaccines and their designs. Parasitology 141(5):624–640

McLeod RS (2004) The economic impact of worm infections in small ruminants in Southeast Asia, India and Australia. In: Sani RA, Gray GD, Baker RL (eds) Worm control for small ruminants in tropical Asia, ACIAR monograph, vol 113. ACIAR, Canberra, pp 23–33

Motran CC, Silvane L, Chiapello LS, Theumer MG, Ambrosio LF, Volpini X, Celias DP, Cervi L (2018) Helminth infections: recognition and modulation of the immune response by innate immune cells. Front Immunol 9:664. https://doi.org/10.3389/fimmu.2018.00664

Newton SE, Meeusen EN (2003) Progress and new technologies for developing vaccines against gastrointestinal nematode parasites of sheep. Parasite Immunol 25(5):283–296

Pourseif MM, Moghaddam G, Saeedi N, Barzegari A, Dehghani J, Omidi Y (2018) Current status and future prospective of vaccine development against *Echinococcus granulosus*. Biologicals 51:1–11

Ramaswamy K (2015) Role of parasite vaccines in sustained animal health and production. J Vet Parasitol 29(2):73–83

Rickard MD, Harrison GB, Heath DD, Lightowlers MW (1995) *Taenia ovis* recombinant vaccine—"quo vadit". Parasitology 110(Suppl):S5–S9

Roberts B, Antonopoulos A, Haslam SM, Dicker AJ, McNeilly TN, Johnston SL, Dell A, Knox DP, Britton C (2013) Novel expression of *Haemonchus contortus* vaccine candidate aminopeptidase H11 using the free-living nematode *Caenorhabditis elegans*. Vet Res 44(1):111

Rosa BA, Zarlenga DS, Fournet VM, Beshah E, Hill DE, Zarlenga A, Yee A, Liang X, Shandling AD, Oberai A, Urban JF Jr, Mitreva M (2023) Identification of broadly-conserved parasitic nematode proteins that activate immunity. Front Parasitol 2:1223942. https://doi.org/10.3389/fpara.2023.1223942

Sankhyan V, Thakur Y (2019) Migratory sheep husbandry in Himalayan state of Himachal Pradesh, India: status and headway options for improvement and sustainability. Int J Livest Res 9(8):71–77

Scarff CA, Thompson RF, Newlands GFJ, Jamson AH, Kennaway C, da Silva VJ, Rabelo EM, Song CF, Trinick J, Smith WD, Muench SP (2020) Structure of the protective nematode protease complex H-gal-GP and its conservation across roundworm parasites. PLoS Pathog 16(4):e1008465. https://doi.org/10.1371/journal.ppat.1008465

Sharma N, Singh V, Shyma KP (2015) Role of parasitic vaccines in integrated control of parasitic diseases in livestock. Vet World 8(5):590–598

Smith WD, Smith SK, Murray JM (1994) Protection studies with integral membrane fractions of *Haemonchus contortus*. Parasite Immunol 16(5):231–241

Smith SK, Pettit D, Newlands GF, Redmond DL, Skuce PJ, Knox DP, Smith WD (1999) Further immunization and biochemical studies with a protective antigen complex from the microvillar membrane of the intestine of *Haemonchus contortus*. Parasite Immunol 21(4):187–199

Smith WD, Smith SK, Pettit D, Newlands GF, Skuce PJ (2000) Relative protective properties of three membrane glycoprotein fractions from *Haemonchus contortus*. Parasite Immunol 22(2):63–71

Teixeira M, Matos AFIM, Albuquerque FHMA, Bassetto CC, Smith WD, Monteiro JP (2019) Strategic vaccination of hair sheep against *Haemonchus contortus*. Parasitol Res 118(8):2383–2388

Vargas L, Prieto L, Baquero M, Corredor W, Alcantara-Neves N, Jaramillo-Hernández D (2022) Vaccines for gastrointestinal parasites, a pillar of preventive medicine in veterinary practice: systematic review. RIAA 13(1):221–251

Vercruysse J, Charlier J, Van Dijk J, Morgan ER, Geary T, von Samson-Himmelstjerna G, Claerebout E (2018) Control of helminth ruminant infections by 2030. Parasitology 145(13):1655–1664

Verma R, Lata K, Das G (2018) An overview of anthelmintic resistance in gastrointestinal nematodes of livestock and its management: India perspectives. Int J Chem Stud 6(2):1755–1762

Waller PJ (2006) From discovery to development: current industry perspectives for the development of novel methods of helminth control in livestock. Vet Parasitol 139(1–3):1–14

Wang BB, Zhang N, Gong PT, Li JH, Yang J, Zhang HB, Zhang XC, Cai KZ (2019) Morphological variability, molecular phylogeny, and biological characteristics of the nematophagous fungus *Duddingtonia flagrans*. J Basic Microbiol 59(6):645–657

Wang D, Ma N, Rao W, Zhang Y (2023) Recent advances in life history transition with nematode-trapping fungus *Arthrobotrys oligospora* and its application in sustainable agriculture. Pathogens 12(3):367. https://doi.org/10.3390/pathogens12030367

Wilkins MF, O'Connell E, Te Punga WA (1988) Toxoplasmosis in sheep III. Further evaluation of the ability of a live *Toxoplasma gondii* vaccine to prevent lamb losses and reduce congenital infection following experimental oral challenge. N Z Vet J 36(2):86–89

Woollard DJ, Gauci CG, Heath DD, Lightowlers MW (1998) Epitope specificities and antibody responses to the EG95 hydatid vaccine. Parasite Immunol 20(11):535–540

Zhang Y, Li D, Lu S, Zheng B (2022) Toxoplasmosis vaccines: what we have and where to go? NPJ Vaccines 7(1):131. https://doi.org/10.1038/s41541-022-00563-0

The manufacturer's authorised representative in the EU is Springer Nature Customer Service Centre GmbH, Europaplatz 3, 69115 Heidelberg, Germany. If you have any concerns regarding our products, please contact ProductSafety@springernature.com

Printed and bound by CPI Group (UK) Ltd, Croydon, CR0 4YY
01/10/2026
02225936-0006